P9-ARP-617

FORBIDDEN SCIENCE

FORBIDDEN SCIENCE

Journals 1957–1969

Jacques Vallee

North Atlantic Books
Berkeley, California

Forbidden Science: Journals 1957–1969
Second edition, November 1992

ISBN 1–55643–125–2

Published by
North Atlantic Books
2800 Woolsey Street
Berkeley, California 94705

Cover photograph: the Horsehead Nebula,
© ROE/AATB, photography by David Malin
Cover and book design by Paula Morrison
Printed in the United States of America

Forbidden Science: Journals 1957–1969 is sponsored by the Society for the Study of Native Arts and Sciences, a nonprofit educational corporation whose goals are to develop an educational and crosscultural perspective linking various scientific, social, and artistic fields; to nurture a holistic view of arts, sciences, humanities, and healing; and to publish and distribute literature on the relationship of mind, body, and nature.

Library of Congress Cataloguing-in-Publication Data

Vallee, Jacques.
Forbidden Science: journals, 1957-1969 / Jacques Vallee.
p. cm.
Includes index.
ISBN 1-55643-125-2 (cloth) : $14.95
1. Unidentified flying objects—Sightings and encounters.
2. Vallee, Jacques—Diaries. 3. Computer scientists—Diaries.
I. Title.
TL789.3.V35 1992
001.9'42—dc20 92-8090
CIP

CONTENTS

FOREWORD

It is unusual for scientists to keep diaries and even more unusual for them to make them public. While we know much about the intimate lives and personal motivations of musicians, movie stars and literary figures, the day-to-day life of scientists remains carefully veiled, as if science somehow arose spontaneously by a process which superseded the mere activities of mortals.

Like most of my colleagues, I have followed this rule of silence for the last thirty years, never expecting that these Journals would be published before my death. But I have finally decided that I had no right to keep them private any more. Although they contain many passages that are very personal and some that are painful, they also provide a primary source about a crucial fact in the recent historical record: the appearance of new classes of phenomena that highlighted the reality of the paranormal. These phenomena were deliberately denied or distorted by those in authority within the government and the military. Science never had fair and complete access to the most important files. This fact has been alleged before, but never proven. The present book proves it.

Publication was not considered when the pages of these Journals slowly accumulated in the form of copybooks, loose pages, letters and marginal notes. I simply regarded it as a useful intellectual and spiritual discipline to review for myself the events of each period, if not those of each day. At first this exercise helped me cope with the uncertainties and the rapid changes in my life as a student in France. Later, when I moved to the United States, the Journal became a confidant and, more importantly, an adviser, a crystal ball, a tool to interrogate the future and to explore its potential.

It turns out that the thirteen years covered here, from 1957 to 1969, saw some of the most exciting events in technological history: the first space adventures, the rise of the computer, the electronic revolution, the invention of advanced software, the flight to the moon, the first detailed images of other planets. As a young scientist I was a minor contributor to some of these events, an avidly interested observer of others. These developments which changed our world are well-documented in countless books. Behind the grand parade of the visible breakthroughs in science, however, more private mysteries were also taking place. The paranormal, with its claims and counter-claims about telepathy, dowsing, astrology, healing and other effects, was a matter of sharp debate and secret passion among believers and skeptics. And there were even more exciting events taking place: all over the world people had begun to observe what they described as controlled devices in the sky. They were shaped like saucers or spheres. They seemed to violate every known principle in our physics.

Did these objects constitute the first signal of imminent contact with alien civilizations from outer space at a time when we were designing our own space probes? Governments took notice, organizing task forces, encouraging secret briefings and study groups, funding classified research . . . and all the time denying before the public that any of the phenomena might be real.

What the media and the scientific world were told by those responsible for public welfare had little to do with what was happening. Anyone reviewing that period and looking solely at the official story will have no chance of coming to grips with the truth about the unfolding drama. In fact, ***the major revelation of these Diaries may be the demonstration of how the scientific community was misled by the government, how the best data were kept hidden, and how the public record was shamelessly manipulated.***

Witnesses of the strange occurrences numbered in the millions. But the study of their observations had been forcefully driven underground. It had turned into a fascinating discipline in a hypocritical modern world that claimed rational thought and open inquiry as its highest standard: *it had become a Forbidden Science.*

No reminiscences of that era can be credible unless they are supported by the daily record of conversations, meetings and research results made by a participant in the actual events. I kept such a record and I was such a participant, first as a direct witness to the phenomenon in 1955, then as a French Government astronomer, and later as a computer scientist who played a significant role in detecting and publishing some of the major patterns behind the mystery and in arguing for its reality. In that phase of my work I was a close associate of Dr. J. Allen

Hynek, the man who was scientific consultant for the U.S. Air Force on the UFO problem for nearly a quarter century, specifically from 1947 to 1969.

Several factors make it important to bring these notes, however personal and fragmentary, to the attention of the public. Only one book was published by a professional historian who took an interest in the field, but it is marred by distortions and errors of omission. And there is a growing misunderstanding of the actual role played by Dr. Hynek in the study of unidentified flying objects.

Allen Hynek liked to remind us that beyond today's science there would be a twenty-first century science that would have to take into account phenomena that seemed paranormal to us simply because of our parochial mental attitudes and the limitations of what he aptly called our cultural provincialism. I hope to bring him back to life here, along with Dr. James McDonald and other figures of that era.

The record stops twenty years ago, as I arrived in California where I now live with my family. I have augmented it with an Epilogue that brings the reader up to the present. Indeed, many important events that have taken place in the intervening period throw new light on the theories I formed before 1969. Some of these theories have turned out to be quite accurate; some were wrong, and the true facts were only revealed later. Other facts are still hidden. When they eventually come to the surface, as they must, it is my hope that this statement of the early years of our research into *Forbidden Science* may serve to highlight their true significance.

I fully recognize that this is only one man's perspective on a series of very complex events. Because this book is a compilation of diaries, it contains opinions that are no longer mine and judgments I now regret, along with much evidence of mistakes I made along the way. I owe many thanks to Janine, to Richard Grossinger and especially to Lindy Hough at North Atlantic Books for their guidance in editing, pruning and streamlining the text. However it was not appropriate, of course, to change the record. At this late date I can only beg the forgiveness of those who may feel that my pen, often "hurriedly dipped in the inkwell of frustration," was overly rash.

Jacques Vallee
San Francisco, January 1992.

Part One

SUB-SPACE

1

Pontoise. Christmas Eve 1957.

Never again will I wait for Philippe near my house on Saint-Jean street. Such is the sudden realization that fills my mind, and these words seem to match the color of my wall covered with red ivy, the color of my whole childhood.

Philippe is a high school classmate, an old friend. When I lived here in my parents' house, my house, he used to come and pick me up every morning precisely at ten to eight, on the way to school. I would already be in the street, walking ever so slowly, to give him time to appear at the turn beyond the grocery store. This went on for years, when we were eight, when we were fifteen. The perspective of things changed gradually without our noticing it.

Now life is separating me from all the things I have known. I suddenly realize that this little town where I grew up is no longer my town; that I do not belong in the streets along which I walked in years past. Philippe is going away to study for a bachelor's degree in physics, my other friends are scattered far and wide. As for me, I am eighteen years old now. I am already forgetting the speeches of Cicero and the art of scaring away the neighbor's cat with my slingshot. Only last week, in Paris, I parted with my first mistress, a tiny girl from Britanny who cried at the movies.

I am trying to enter the life of a scientist, of a man who peels apart new concepts like the skins of an onion to remove each layer. Yet on the other side, on the side of my ivy-covered wall, I have a hard time giving up the slingshot kid.

This anguish of a Christmas night filled with the books of old and the tree of tradition, this anguish is born of sorrow. There was a need for something that would mark the transition. Indeed everything will change again: my father will not live long. A certain pain in my brother's voice, a pain he could not hide over the phone, was the signal.

I will not be very demonstrative when the end comes. But this is Christmas and the bells, the obsolete bells of nearby Saint Maclou, ring at midnight, urging me to write.

Consider the whole existence of a man, with all of the ramifications that implies in the existence of others, in their minds, in their consciousness: How can all that be annihilated by more or less simple chemical degradation of the cells? I understand why people still need to erect a God to store within it this kind of dilemma. Never again will I wait for Philippe. And soon I will no longer see my father, my old father, walking from one room to another in my old house. And my old house on Saint-Jean street must disappear forever as well, with all the stars that are above and the trees around it. How am I expected to find a grave big enough for all that?

Paris. 10 January 1958.

I spent New Year's Eve at the Mexican café, our Headquarters on *Place de la Contrescarpe*. It is a tiny square on the eastern edge of the Latin Quarter, surrounded by quaint shops and picturesque buildings, home to tramps and winos. The whole gang was there, including Granville and the Baron, my friend Claudine and others, unavoidable others. Now classes are starting again at the Sorbonne.

My brother is a medical specialist. He has discussed our father's illness with his colleagues in Pontoise: there is no hope.

Paris. 20 January 1958.

A frozen impression, a strange release: My father is dead. Oddly, I don't have the feeling of having "lost" anything, of having less substance. On the contrary, I have this absurd sense: I have come closer to a certain reality. But here is the sorrow, to have lost in potential what I gained in knowledge. The new emotions I have just gained are useless.

After I watched him die, and kissed him, I went out into the street. The first snow had fallen, in fine heavy layers, pure perfection against a great blue silent sky. I was astonished at the sudden beauty of the world.

In today's society there can be no harmony among people like my father, my brother, and me. We disagree viscerally on too many subjects, from morality to music, or the war in Algeria. He was a stern conservative and we yearn for change. That's normal, and also cruel. The Norm excludes any tears: it rules, that is all.

Will things go the same way for me and my children? Probably. Unless I find what I am looking for: well-defined substance, unbounded potential within myself. Is that possible?

Back in Paris, at the student house where I live, I am pretending that life carries me along normally. I have not taken anyone into my confidence. I want to keep all my strength for the future.

Paris. 15 February 1958.

The French Astronomical Society has just published an account I sent them of an unusual sighting of the first *Sputnik*. It's an observation I made last year in Pontoise, from the terrace behind the house. The event took place three months ago on Sunday, 24 November 1957 at 5:54 p.m. Watching the object, probably part of the booster rocket, I found it similar to Jupiter in apparent size and luminosity as it passed through Cassiopeia. It got lost in the Southeast in less than two minutes.

Having heard that the booster of *Sputnik* had broken up into several pieces I waited for any other object that might follow on a similar orbit. Indeed at 6:10 p.m. I saw a faint luminous trail with the naked eye. It was rising between the first two stars of the Big Dipper, in the direction of Polaris. I looked at it with a telescope given to me by my uncle Maurice, my father's brother. This instrument is an antique World War I artillery refractor with a magnification of 25 which enabled me to see a small orange point at the tip of the trail. I lost it after about 15 seconds, but the trail remained clearly visible in the sky and it drifted to the zenith at 6:30 p.m. I sent an account of the whole thing to the French Astronomical Society and to Paul Muller, head of the artificial satellite service at Paris Observatory in Meudon. Now it turns out that another amateur astronomer saw precisely the same thing from Joinville, and our two observations appear together in *L'Astronomie.*[1]

Paris. 20 March 1958.

The world is changing. This city has been shaken by sudden political upheaval. History is accelerating. I can feel it turning from its usual elusiveness to the consistency of a liquid or a jelly. The Fourth Republic is threatened from the Right, as a result of the lingering, impossible war in Algeria where a large, conservative and increasingly militant French population remains. My friends among the students expect things to turn nasty. Trucks covered with greenish-brown fabric are in evidence throughout the Latin Quarter; jeeps equipped with radio transmitters drive up and down the Boulevard Saint-Michel. I have just seen half a dozen trucks filled with gendarmes parked in front of the Panthéon. They seem to be expecting a full-scale riot.

French democracy may be about to pay for all the mistakes it has made over the years, especially this stupid war. Like my fellow students, I am outraged at all the lies spread by the bureaucracy, the censorship, the denial of the tortures committed in North Africa by the French Army: our country is engaged in the same kind of actions that we were taught would forever designate the German Nazi to the shame of the whole world. We wonder what this means for us. What kind of future are we studying for?

Paris. 27 March 1958.

I just wrote a letter to my mother, who is now alone in that big house my parents have rented in Pontoise for the last sixteen years:

> I got back safely last Sunday on that excellent train. I was back exactly at 9:30 p.m. Some thirty cops were stationed at the subway entrance. They had machine guns and everything that is necessary in order to kill people. They were systematically stopping anyone who looked like an Arab.
>
> My work goes along well. I am studying hard for the Analysis exams, and now it's only a question of spending more time with the books and the homework problems. My goal is to pass the written part in June.

I will make arrangements to spend more time with her at Easter. It makes me sad to imagine her alone in Pontoise. She has always looked at the world through the eyes of a great lady. She is quick to assess people's character, quick to rescue the lost child, to feed the poor beggar, to get angry at injustice. She comes from an industrious family of Protestants whose various branches extend all over Europe. Her parents went broke when the flood of 1910 wiped out their fur and pelt trading business. She raised her thirteen brothers and sisters by herself, and there was no opportunity for her to finish school. But her heart is as big as the whole world and her mind has the direct intuition that needs no schooling.

Recently I came across a picture of us taken when I was about eight years old, on the terrace in Pontoise. My father is dressed in his Sunday best, a three-piece suit and a tie. My mother holds my hand. I lean against the wall, without a care in the world.

Paris. 16 April 1958.

Normal work has become impossible, life is suspended. The Government has fallen. Socialist leader Guy Mollet warns of "a crisis of Régime" and calls for a new Popular Front, while powerful appeals to a neo-fascist "Comité de Salut Public" (Public Salvation Committee) are heard from the Right. All this is drowned in idiotic commentaries by our well-informed media: "the crisis will be long and painful," a political journalist has stated in all seriousness; "the President expects to solve it rapidly." The President in question is Monsieur Coty, a nice old man who has never done anything rapidly in his life. I have heard another politician, Le Troquer, adding ponderously that "depending on the circumstances, the crisis may be long or short." Only one thing seems clear to me: if the Assembly does not come to some decision soon, time will work in favor of an overthrow of the regime: we will either get General De Gaulle, or a new Popular Front.

Paris. May Day 1958.

The crisis has entered its fifteenth day. Fights have started. This afternoon I found myself on rue Mouffetard returning from a demonstration in protest of the execution of a young Algerian. The Latin Quarter was full of sunshine. Around us the market was bustling with activity, with its open-air displays of fruits and vegetables, the stalls selling meat and fresh fish, flowers, ham and sausages. Suddenly, frantic screams made us whirl around: A struggle was erupting. Foolishly, we were tempted to watch and we came closer. A dozen men were engaged in a brawl in the narrow street. One of them, a fellow in his forties, produced a heavy stick and started swinging, but others jumped on him and the stick rolled away. As he freed himself, a gun in his hand caught the sun. Stunned, his assailants took a hurried step back. We did the same, with that sick feeling: who would catch the first bullet? I was less than ten feet away. He turned and rushed ahead into the crowd.

Someone yelled: "The cops!" The participants scattered down the side streets.

Soon the entire area was surrounded, from the Gobelins to rue d'Ulm. Police buses blocked every corner, machine guns and radio transmitters in evidence. I saw a car and two motorcycles coming down the medieval streets of Contrescarpe. We ran away from them towards Place Monge. A helicopter flew low over the rooftops. Now there is a rumor that the man was a provocateur, that he worked for the police, who were seeking an excuse to come into the area in force.

A huge black bus full of cops, unable to wedge its way down the narrow street to the church of Saint-Médard, was forced to drive backwards all the way up rue Mouffetard under the catcalls and the jokes of the shopkeepers, the peddlers, the old women of the market.

The Latin Quarter, which has seen many a revolution down through history, remains effervescent tonight. The helicopter keeps flying in narrow circles.

Paris. 12 May 1958.

Things are getting worse. A very bland politician named Pflimlin is attempting to form a new Cabinet. The Far Right seems ready to take drastic action to overthrow him and seize control. In our section of Paris there is an intense war of the walls. Graffiti of both sides, childishly, cover every fence, every available space. When we walk back from our evening coffee at Contrescarpe we can't resist scribbling over the rightists' slogans. Thus "Vive Le Pen" becomes "Vive Le Penis!" But we worry about the future, even as we confidently sing "Fascism shall not pass!" People look at the empty sky, naively expecting it to fill up at any time with paratroopers from Algiers in full battle gear, red berets on their heads, machine guns at the ready.

Paris. 13 May 1958.

The Prime Minister seems to have gained the upper hand: "It is an insult to suggest that I would permit Algeria to be lost," he says. "Algeria shall remain French."

In the meantime, back in Algiers, an anti-Government demonstration initially scheduled for the middle of the afternoon has been delayed by two hours to allow it to gain strength by merging with a rally planned for the same evening. The general strike is beginning. French troops have been ordered to remain in their quarters.

10:10 p.m. The creation of a Public Salvation Committee has just been announced. Predictably, it is headed up by rightist General Massu. The French who live in Algiers are taking to the street to greet Soustelle, an ultra-conservative politician. The University is in such turmoil that normal studies are out of the question, many classes have been called off, others are constantly disrupted by demonstrations and political meetings.

Paris. 14 May 1958.

It is 9:40 a.m. Telephone communications between Algiers and Paris have been cut off by the Government. Maritime shipping traffic is being detoured to Tunisia. Several people have been arrested in Paris. In the Latin Quarter the excitement I witness is unprecedented. There is an air of insurrection in every gathering.

Yesterday I joined a demonstration in the courtyard of the Sorbonne. Fine speeches were made, announcing great imminent movements on the part of "The People" and "The Masses." But it was hard to find two individuals in the crowd with the same interpretation of current events. Pflimlin, who is still in charge of the Government, has banned all political gatherings. De Gaulle is rumored to be in Paris, in an office located on rue de Solférino.

Paris. 15 May 1958.

Like every French citizen I am staying close to the radio to follow events in Algiers minute by minute. The Soviet Union has launched its third *Sputnik*, but in the current political frenzy no one seems to care that a major new step in the conquest of space has just taken place. The satellite weighs a ton and a half, and the booster rocket is in orbit along with it.

General Salan has stunned the country by delivering a speech which ended with "Vive la France! Vive l'Algèrie Française! Vive De Gaulle!" In Paris the government called this statement an "optical illusion." But De Gaulle has answered the call by announcing that he was indeed ready, "as in 1945, to assume the widest responsibilities." What we are seeing is the unfolding of an obvious conspiracy to bring the General back to power, and to bury the Fourth Republic.

Paris. 27 May 1958.

It is 1:40 p.m. Planes carrying paratroopers and units of the Special Forces are said to have landed near Paris. The French fleet is in Algerian ports. De Gaulle is definitely in Paris. Pflimlin vanished during the night. Has President Coty met with De Gaulle? There is talk of an insurrection in the Southwest.

Pontoise. 29 June 1958.

I have learned that I flunked General Mathematics. This throws my life into further uncertainty. Yet passing this examination is a crucial requisite for me. Without it I can do nothing. All the recent political turmoil in Paris, the demonstrations, the strikes, have not helped my studies; neither has the life I have led, these last few weeks, with Claudine. She creates a feeling of impossible nightmare. As dawn arrives, a weak whitish daylight leaks into her room through the curtains. We wake up in the low bed. I lose all sense of time. Since the first day, there has been an invisible barrier between us.

It was on Monday evening that I found out that I had failed, after a huge scuffle to fight my way through the crowd of students and to get near the posted results. I spent Tuesday night with Claudine. This time I found her less tormented, more accessible. On the bedside table there was a love letter in fine handwriting, addressed to her: "My Darling Claudine . . ." I did not read it, but I was indiscreet enough to glance at the signature: it was from another woman who lives in the Midi.

Paris. 21 July 1958.

Finally, De Gaulle is here. What seemed unthinkable has happened quite naturally, in spite of all the Leftists who were clamoring that his return would surely trigger an insurrection, a terrible civil war. In fact, after a few days of disorder, during which madness did rule and newspaper headlines became huge, all the political parties have simply resumed their old intrigues as if nothing had happened.

I saw Claudine on rue Monge.

"Do we say hello?" I asked her.

In response she simply gave me her hand.

"You would come and have a drink with me, if you were a true friend," I added.

We went to the corner bistro and we had a cup of coffee. We were very close again, very tender. The next morning we took the train and went to Pontoise, where I now spend every week-end visiting Maman. My brother was there; his children cheered up Claudine.

On Friday we had dinner with my friend Granville, who studies for a degree in pedagogy. I was rather somber at first, but I soon found it funny to watch our strange bohemian group. We looked like the survivors of a wreck, a band of drifters united by their uncertain destiny. Claudine was terribly out of place in her red party dress. Granville had plastered some sort of white powder over his face. I was wearing a dirty old jacket. I had been painting my walls all day, fixing up as best I could the little room into which I will soon be moving, at the other end of Paris. My fingers were still spotted with paint. To make things worse we decided to eat at a fancy Chinese restaurant, where the waiters looked down on us in disapproval. Yet I felt this pantomime was a fitting way to bring to a close my two years of wandering in the old Latin Quarter, two years devoted more to the vibrant streets than to serious study on the hard benches of the Sorbonne.

When we got out of the restaurant we danced on the sidewalk like three idiots, not caring about tomorrow. Yet later, on the Métro, Claudine held my hand in a strange, serious, almost desperate way.

Paris. 7 August 1958.

My friend Marcel was right the other day when he asked me: "Why is it so damned important for you to study science?" He was right, but only in asking the question. It would be a drastic limitation to dedicate myself exclusively to the study of science, like a priest dedicating himself to God. I will indeed study science, but I will do it with the knowledge that an appreciation for art, fantasy and sensitivity is not a "negative trait" that I ought to suppress within myself.

During my first year at the Sorbonne I was frequently discussing these lofty topics with a girl who had befriended me. One day she brought a small package: "This is for you," she said. "It was among some books my grandmother left when she died. I think you should have it." It was *Histoire et Doctrines des Rose+Croix,* by Sedir (1932). I lost sight of the girl, but I have treasured the book ever since, and it is with me now, a source of inspiration and a tangible link to the deeper questions I long to explore.

I want to look behind the scenes of our human existence. Unfortunately I have found no one who is able to answer my questions about forbidden things: What is research? Does it consist simply in tiring our minds while looking for impossible solutions? Could one find the ultimate secret by simply giving up the search, satiated with the pointless, superficial agitation of life, and looking instead at the infinite void beyond it?

When we discuss love, sex and destiny Claudine cautions me: "You're only nineteen. At twenty-two you will run the risk of discovering that you have already known what most men only experience at thirty or even, for most of them, never."

Funny how she still uses the formal *vous* with me. Perhaps it is true that I have been here, inside this particular body, for nineteen years. But in reality I feel that I have always existed. My brother is a hard-boiled physician, an agnostic and a cynic. But for his attitude towards life to be justified, the ancestral terrors I hear blowing through my soul would have to stop, the universe would have to become limited, time finite. Everything would have to die and go away.

Paris. 8 August 1958.

Who will tell me what death is?

My father has ceased to think, to hear and to see. In the last few hours before he died he thought he heard music. He asked my mother if it was a piece by Bach playing on the radio. But do I feel any call from him? No: nothing but the whispers of eternity itself, which I cannot hush within me. Night beckons to me in a similar way, can I deny it? I can almost hear the night, falling in fine drops around me, when I am holding Claudine's sleeping body against mine; and the starry night calls out to me too, a living mystery filled with other worlds. What is this attraction of nothingness we feel, beyond the fabulous amount of matter radiating in the dark sky? So much substance, metals, energy, explosions, just to create the tiny point of a star in my eye! Nature is multiplying these orgies of time and distance beyond the understanding of my poor human spirit, and to prove what? The existence of Nothingness?

Claudine, I sought the answer in your own life, in a tenderness that remained beyond reach. This strange privilege you afforded me, giving me your body without letting me touch your soul. . . . What were you afraid of?

2

Paris. 25 August 1958.

Now I have my own room in Paris, close to Porte Champerret. It is one of the small rooms on the seventh floor of the building, under the roof, which in more elegant times used to be allocated to the chambermaids of the bourgeoisie. The elevator only reaches to the sixth floor, then I walk up one more level up the servants' stairs. This is a tiny place, barely ten feet long and seven feet wide, into which the slope of the roof cuts an angle. But it is mine. I am in bed at last, lying under a blue blanket. It is 10:20 p.m.

At first it was nothing but a dusty mess, to be truthful, my little *mansarde.* I was thrilled two months ago when Claudine told me it was available, because I have no money and I certainly cannot afford an apartment or even a studio. I have made some improvements: a few electrical connections, a movable lamp. I cleaned up the floor, I installed a small water tank above a wash basin (there is no running water, no sewer: I carry the waste water and empty it in the lavatory down the hall). I put up shelves for my books. I nailed a piece of plywood to the wall and I painted it black to serve as a chalkboard for math problems.

This part of the city was unfamiliar to me, but it is now coming alive through many tiny scenes, as I wait for the 84 bus every day, or as I take my breakfast at the *Café des Sports.*

On Wednesday I found a letter from Claudine, so unsure of herself. So direct: *"Pas ma fête a moi."* Not my day. Write to me, she was asking.

Is there another level of life and awareness? I have long been aware that I could pass almost at will from the plane of normal consciousness to . . . another plane. There are dozens of examples: all those circumstances when something like an electronic relay suddenly seems to close deep inside me, when time starts flowing at a different speed, when new angles of reality are revealed. My strengths become more clear then, my body goes on automatic. The spirit flies off.

Whoever possesses this "other kind" of thought recognizes it at once. It comes with the feeling that we do not really "exist" any more in this world than a single note in a symphony exists, or a single spark in the fireplace. We are both creators and tributaries of the universe we perceive.

A chance meeting I recently had with one of my neighbors, a strange mystical man from the Middle East with an advanced degree in engineering and a passion for ancient texts, makes me experience once again this unusual ability of my mind. He noticed the urgency I was putting into my work. He told me: "You seek to create in order to fulfill something within yourself. That's absurd, my friend. What is zero plus zero? Instead you should create through the mere desire to create: inspiration pure and simple. Never look at your own work. If you want to be a master some day you must find pleasure in creating without having a precise objective, without pursuing a rational goal."

Paris. 27 August 1958.

I am now reading a book called *Mystérieux Objets Célestes*,[2] which is challenging the very depths of my mind. It was while browsing at the *Bazar de l'Hôtel de Ville* department store that the title caught my eye. I grabbed it immediately. At last, an intelligent book about flying saucers! Yet I suppose that for those who are rooted in the ordinary world, it does not matter if a few researchers have found that the immense contour of other shapes, other civilizations, could be discerned beyond our world.

Will these strange events begin again soon? Deep within myself *I passionately want them to wait for me*, and to find me established in my future life as a researcher. This is an ironic thought, knowing as I do that I will probably die without seeing any solution to this immense problem, or without being able to contribute to it.

On a more finite level I have a new girlfriend named Juliette. Something tells me that some serious developments could take place between us. Claudine has awakened my instincts in this domain where I was blind, deaf and mute. Yet there has been nothing said between Juliette and me, not even a hint of a flirting gesture. Only the atmosphere getting heavier.

Paris. 1 September 1958.

My interest in "flying saucers" goes back to the Fall of 1954 when there was a deluge of sightings in France, and indeed throughout Europe, from England to Italy. Every day the front page of all the newspapers, from *L'Aurore* to *France-Soir*, carried big headlines and surprising claims which the radio amplified with commentaries and on-the-air interviews. My father, a respected magistrate, a former investigative Judge who had been promoted to Paris as a Justice of the Court of Appeals, would scoff at such reports: in his profession, he pointed out, he had become leery of the weakness of human testimony. Especially that of experts.

As a kid I remember hearing one of the earliest French witnesses, a railroad worker named Marius Dewilde, telling his story to radio broadcaster Jean Nohain in a live interview on the evening news: "I had gone out to piss . . ." he bluntly told the whole nation. He had seen two little robots next to a dark machine resting on the nearby railroad tracks. The air police found traces of a large mass. A strange ray issued from the object and paralyzed Dewilde. I believed his story at the time. I still do. During the three months the wave lasted I carefully gathered such clippings and glued them into a fat copybook.

It was during the following year, a Sunday in May 1955, that I observed a flying saucer over Pontoise.

My mother saw it first. She had been working in the yard, pulling weeds and caring for her flowers. She was getting ready to put her tools away to prepare the afternoon coffee, a sacred tradition in our family. She had to scream to get our attention, because my father and I were up in the attic, where he had his woodworking room. He was busy and did not consider such an event significant enough for him to come down. I rushed to a window that had a Southern exposure but could see nothing. I ran down three flights of stairs into the yard to join my mother, and then I did see it.

What I observed was a gray, metallic disk with a clear bubble on top. It was about the apparent size of the moon and it hovered silently in the sky above the church of Saint-Maclou. I have no recollection of seeing it go away. My mother says it flew off, leaving a few puffs of white substance behind. Remembering the war years, she first thought they were parachutes.

I was left with the single strong impression that *we must respond;* that human dignity demanded an answer, even if it was only a symbolic acknowledgment of our lack of understanding. I realized then and there that I would forever be ashamed of the human race if we simply ignored "their" presence.

The next day I met with my closest friend Philippe at the College, where we were "cramming" for the Baccalaureate examinations. He mentioned seeing the same strange object from his house, half a mile North of my position, on higher ground. He had watched it through binoculars, and confirmed my description.

My father was sternly opposed to our making any kind of report. The family of a distinguished judge does not get his name into the papers with some flying saucer story. What we had seen must be some kind of new aircraft, he insisted, something explainable. I convinced myself that he must be right.

Now Aimé Michel has reopened the whole question: studying all the sightings of 1954, he has found that they fell along straight alignments that criss-cross the French territory. He calls this pattern "orthoteny," from Greek words that mean "drawn along a straight line."

Paris. 9 September 1958.

I have written to Aimé Michel. My letter begins:

> I have just put down your book, and this is far from a gratuitous act.... On every fundamental point you bring reason where the best people who came before you gave us nothing but a multiplicity of excuses.

It is only with a few of his conclusions that I argue, when he despairs of our position with respect to the beings who control the objects. I find two arguments against this despair:

> 1. Faced with "orthoteny" (the fact that saucer sightings seem to occur along straight lines), you compare us to an eight-year-old boy standing before Einstein's blackboard. Yet the boy, when he grows up to be thirty and is educated in math and physics, may be an even greater genius.... How can we believe that beings with the degree of evolution we can reasonably ascribe to "them" would not have methods of education superior to ours?

Of course, in order to educate us, they would have to find us worthy of a dialogue with them. When we probe their behavior, what do we find? The gap between their knowledge and ours does not appear to be so enormous. And they seem to reason along a set of concepts analogous to ours.

> 2. If we believe the flying saucer witnesses who speak of seeing small hairy beings, we should also believe them when they claim to have seen these same beings along with others who were morphologically human. This implies a similarity of level between us and them. There are indeed differences, but mere "differences" can be bridged. I am only a math student, and my nineteen years do not give me that right to prophecy that some scientists are so quick to claim for themselves. But it seems to me that if we extrapolate our civilization by fifty or a hundred years, we could well find round flying machines in our own future, as well as excursions beyond the solar system.

I close my long letter by thanking Aimé Michel for writing the book: "It gives us a reason to face the problem. It enables us to begin valuable research quietly. Serious work can start at last, because of you."

Paris. 13 September 1958.

How should one speak of a night of love? What is the use of words beyond meaning? I only want silence, warm lingering rest. My room has lost its arid, awkward face. Upon waking up, next to Juliette's long black hair curling up over the blanket, I cast an ironic eye on the word *"Ascèse,"* asceticism, which I had painted on a curtain in purer, lonelier days. This bed lost all shape last night, this narrow single bed torn away by passion. Why should I describe our trust on paper, when I can still taste it on my lips?

Paris. 16 September 1958.

There are diverse ways towards life. I need to find one, and I need new ground on which to build, to open new roads: I must give free rein to a new intelligence within myself. I am not speaking so much of building my own life as of achieving the final destruction of the lives of others within myself. I long for the end of adolescence, that worthless tumult.

How lonely I have been all these years!

Paris. 21 September 1958.

Aimé Michel has answered me. He thinks that I underestimate the problem of communication between us and X:

> My book gives a misleading feeling of simplicity because of the restrictions I imposed on myself. There are some sightings, as credible as those I have quoted, where the witnesses saw the object disappear instantaneously without any spatial displacement. There are others where a solid geometric object changed shape in a fraction of a second. Can you imagine a pyramid turning into a cube? Remember what Poincaré said about the fundamental importance of solid bodies in our logic.

He points out that there are some domains in which no one will ever do better than man, not even God if He exists: Mozart's oboe and clarinet concerto, for instance. "There are human absolutes," he says.

> By the way, allow someone who was your age when you were born to give you some advice: you have a remarkably gifted mind. Do not let yourself get abused by the idea of "getting to the bottom of things," which is only a mirage. Cultivate your mind like a flower but be careful: the pavot is a flower.

Pontoise. 12 October 1958.

The qualifying examinations come two days from now. I have another shot at General Mathematics. When I read all this again later, I am not sure I will remember my obscure battle against the wind and the mud, the stupid fight in which I am now engaged as I try to get out of the quicksand of these studies.

Today I have given myself solemn instructions for the creation of a new being. In two days I will go to this exam, this last fight. Do not be concerned with it. Born from me, leave me quickly. Bury me deep within the memory of Granville, Claudine and my other friends. I will be at ease there. Be free and go, as a little Sphinx who already bites and flies. I do not understand your enigma, but I believe in an escape towards the new dimension you represent.

Paris. 17 October 1958.

I took the General Mathematics examinations on Tuesday and Wednesday. I spent both nights in chaste, peaceful sleep near my "sister" Claudine. Everyone says that the examination was tougher than at the June session, when I failed.

Pontoise. 27 October 1958.

Success! I have passed the test. Egotistically, I savor this victory. I feel that I now belong to a new world, and I am proud of it. It is the same happiness I experience when I am patiently scanning the craters of the moon, or watching whirling counters in the physics lab and when I think of all the people whose lives are confined to the weekly movie, the soccer game and the nearest bistro. My inner happiness doesn't come from being different from them, and I certainly do not feel superior to them. But I am proud to have gained a wider vantage point on the world.

Paris. 12 November 1958.

My next goal: a bachelor's degree in science. How can I describe our crowded lecture rooms at the Sorbonne? Four hundred seats and eight hundred students, people sitting on the ledge of the windows, on the stairs? How can I describe this wretched French University system, against the backdrop of our continuing colonial wars which consume most of the available funds the government should be putting into education and the modernization of this old country?

Our generation will have to re-invent everything. Centuries of civilization and philosophy seem of little help here. Contemporary artists from Varèse to Pierre Schaeffer and from Dali to Miro have already destroyed the old standards and the old morality, bringing the blast of their dynamite all the way into the exploded language, freeing up design and painting, yet science still follows the ancient models. It too will have to be shaken up. Then everything will have to be rebuilt within a society that doesn't provide us with any useful models.

Paris. 13 November 1958.

A kind of quiet harmony is spreading around me. Perhaps it comes with the fog over Paris, which drowns the trees and the car headlights; or with the emotion in a friend's handshake in a Paris café, or the fall of dead leaves swinging all the way down to the wet pavement like a jazz melody. Perhaps it comes with the words of scientific reality exchanged among young men in impeccable white coats in the corridors of the Radium Institute, where I now attend some of the lectures.

There are many lessons to be drawn from ancient Magic. For we are still in the Dark Ages: Consider our churches, our Lords! Look at the serfs here, the Baroness passing by in her beautiful coach, our men-at-arms swinging their sticks! See our fortresses, our quaint coins, our narrow minds! See the little compartments of our science humbly growing in the midst of public indifference. The only new fact is the uncontrolled use of this science by the government and the military. The wise men of the Middle Ages, at least, knew how to hide their discoveries behind obscure Latin paragraphs. If necessary they took them into their graves.

I spent a long time talking to Claudine the other evening, in an ugly bistro on rue Saint Jacques. The place looked like the inside of a submarine. But we were warmly squeezed against each other, like good close friends.

Paris. 22 November 1958.

The little *café* is very poorly lit. We have made it our headquarters because we are used to the fare. We bring along our mistresses, the girls bring their lovers on the back seat of little Italian motorcycles called "scooters." We are these peculiar, privileged creatures, *étudiants.* We have friends who arrive from Japan or China. They speak slowly, with the peculiar tone of voice that becomes those who have travelled far in spite of themselves, and have seen much. They play absent-mindedly with the matchboxes left on the table, they drain their cup of coffee, and go back to the Sorbonne to apply for another travel grant. There is nothing here of the intense discussions I used to have with Granville and with Marcel, from which arose something mystical. Instead we confront serious, rational ideas. Coffee and conversation are thick with the dust of learning.

Occasionally I drag Claudine here, literally, by the hood of her white and blue coat. She is older than I am. She laughs at being treated like a kid. I write lyrical things, strange poems. People tell me I'm young, with the tone of an insult. Since I am always hungry (at the student restaurant they serve us pure shit), I stay here to fool my stomach with coffee and to work on topology problems. I draw funny shapes on ashtrays. My heart isn't in all this. I am growing tired of all the silliness of Paris.

Paris. 5 December 1958.

Instead of promoting mass communications one should isolate each man, isolate him inside himself, in order to build up his spiritual life. How useless, stale and empty is the intellectual life of this famous Rive Gauche! How flat are the sex stories, how uniform are all these "original" people, gossiping about the obsolete Absolute!

Pontoise. 13 December 1958.

Since the death of my father almost a year ago, Maman has been living in this large house on Saint-Jean street. Her neighbors are provincial *bourgeoises* who share nothing of her enthusiasm for space exploration. The other day she heard on the radio that a team of English astronomers had bounced a signal off the moon. "Hello!" They said. "Hello!" answered the lunar surface a few seconds later. When Maman told her neighbor she had heard the exchange the lady looked at her skeptically: "My goodness," she said, "you must be spending all your time at that window!"

Pontoise. 14 December 1958.

Slowly, I am beginning to understand the feelings of people, I appreciate better their complexity. Could I have been touched? No, who could be touching me? Juliette has disappointed me, and Claudine is "just a friend." But I am beginning to understand those who love and to realize the complexity of the relationship between spiritual and physical pleasures. I seek the terrestrial foods, without flaw or complication or pretense.

Paris. 16 December 1958.

Now I am fed up with our little group which always meets at the same *café* near Port-Royal. Fed up with the people themselves, their humorless lives, their habit of talking forever about the same meaningless details of their petty lives. I want to move away.

Every chance I get I rush to a little open-air bookshop on the Boulevard des Italiens which sells used science fiction. I devour everything, from Van Vogt to Heinlein and from Jimmy Guieu to Asimov.

Paris. 22 December 1958.

Everything seems to confirm a single observation: we are living fake lives, absurd lives in today's cities. Nothing actually exists of these so-called "acts" and "opinions" of ours. Truly important decisions are made beyond our observation, beyond the control of ordinary citizens. Everything we see is fake, a stage drowned in movie fog. We come and go like puppets in search of their own strings.

I long to send this message to a wiser man somewhere in time, far away: "You should know that down here we are managed, surveyed, and classified like insects by police and publicity men, or simply by the mechanical stupidity of our own bureaucracy."

Slowly, revolt after revolt, torture after torture, this earth will eventually emerge into its true history. In the meantime I am eager to learn what is outside all these events, I want to see the mechanism beyond time itself.

Paris. 6 January 1959.

Juliette wrote to me today: "Do not wait for me tonight, or tomorrow, or ever. It is too hard for me to start again, to rebuild something."

I felt deeply hurt. Everything seemed to be collapsing. But the storm has now swept the sky clean.

3

Paris. 10 February 1959.

A proposal: To go straight ahead, wisely and quietly, without jealousy or hate . . . To walk through one's life in long equal steps. To put everything we are, especially our love, into our gestures.

At night I try unsuccessfully to travel in spirit through the whole night of Paris: I am quickly brought back to reality when a ten-ton truck rumbles down the canyon at the base of this huge building; in the next block a tall chimney throws up torrents of black smoke; hideous yellow dogs, taking hideous old yellow ladies on a routine walk at the end of a leash, piss all over my scooter parked on the sidewalk . . . I have to pull my thoughts away from these sordid scenes.

I drape the covers over the shoulders of the girl sleeping next to me. We commune in warmth and tenderness. Some day we will leave this city for a place where we will no longer be cut off from nature.

Janine is a schoolteacher from Normandy studying for a Master's degree in child psychology under Professor Piaget. She has moved into the room next to mine, a pretty brunette with green laughing eyes. For some reason she thought I worked as a photographer. We happen to own the same records, easily heard through the thin wall at night. We made love for the first time a week ago, and we have been together since.

Paris. 28 February 1959.

Without forethought I have started to write a novel I call *The Praxiteles Network*. It has to do with the adventures of a group of kids amidst the ruins left by war. The idea came over dinner with Granville, who told me of "something silly he was writing for a publisher, hoping to make a few francs." I decided to do the same for fun, without any plan. I am letting the story develop. A year ago I found it hard and painful to write. I am surprised to see how much easier the process is becoming.

The other day I found a note from Janine under my door:

> I went away from your room utterly distressed, probably because what you say resonates deeply within me. You will get through because you see things, not in terms of yourself but in a detached, impersonal way. I have not reached that point yet. I feel I will only be able to achieve this after resolving some conflicts that I do not master, because I don't know where they come from.

Paris. 2 March 1959.

I am in love with her. I was speaking of a high point, of new horizons, yet I wasn't even able to see the landscape. Now I feel like a pilot in flight who suddenly breaks through the clouds and watches the sunshine illuminating some wonderful island below....

We are crazy. We are two crazy lovers. Janine has set a new machine into motion within me. She is holding my life on the highest wing of the storm.

I have finished and set aside the manuscript of *The Praxiteles Net-*

work. I have begun with even greater enthusiasm a science-fiction novel called *Sub-Space.*[3] I condense my current life within it: in the middle of a big stupid city there are people who love and search; they are forced to go beyond the limits of the world both outside and within themselves. It's a fun process, because the story writes itself in fury and disorder, carrying me at a gallop pace. When I see a new protagonist, he often moves without warning out of the context I had prepared for him. I am just as unable to say what my character Alexis Nivgorod will do in fifteen pages as what I will do in thirty years.

Janine poses a deep question to me. She is carrying within herself a powerful secret anguish. Our love goes faster than light.

At night the glow of the record player scatters iridescent droplets all around my room, and over the night shirt of simple white fabric she has dropped on the dark red carpet.

Paris. 13 April 1959.

The weather has turned hot and heavy. I have to stay in Paris to study in the midst of mirages. The random notes I write down in these pages are only useful because they provide me with a standard, a reference point among all the illusions.

It is so hot today that the asphalt melts, sticking to the feet and to the mind like caramel candy. The ugly buildings, with their rococo style, seem to crush our lives under the weight of their dripping ornaments. Dust is flying, soft and sour, over Place du Chatelet. Tourists stare at the column through their binoculars. Who needs science fiction? No telepathic Martian with green tentacles will ever be more weird than they are. This city is only livable when you walk along with blinders on, going about your own business. You pay a high price for trying to get out of the maze, to think different thoughts, to discover an alternative to common customs.

Within French society are lines of equi-stupidity which cannot be crossed without much pain and a huge energy quantum. This is an atrocious, absurd, unjust system.

In the meantime I am within it, whether I like it or not. I have no choice but to get a bachelor's degree, sitting on the benches of the Sorbonne next to a few battalions of armored girls from the Catholic Center and a bunch of stubborn, narrow-minded fellows whose sole ambition is to graduate quickly to earn more money and buy bigger cars. Fortunately I have the greatest teachers I could hope for: the whole *Bourbaki* school of mathematics[4] has come back to Paris now that the old guard of French academics has died off. Thus I study under Godement in Analysis, Chevalley in Set Theory; a team of internationally known mathematicians has decided to train the mass of the younger students like me rather than concentrating on a small group of higher graduates. They are an exciting Faculty, but I am tired: I haven't had a vacation in three years.

Pontoise. 25 April 1959.

I just saw a movie about volcanoes directed by Tazieff. It contains a strong, appealing idea: No point on earth is immune to a sudden eruption. In space the most ambitious realizations of Man only rest on a thin layer of the planet; in time they can be encompassed within the first page of the topmost book in a pile of volumes as tall as the Eiffel Tower. I do like this idea. It satisfies me to think that the Arc de Triomphe, for instance, that "bearer of eternal symbols" of military glory and horrible death is actually resting on the original boiling magma of the planet which makes a mockery of this exalted hoax. Thought and sex are the only human activities which are not totally ridiculous. As soon as man makes a gesture which is not intended for love or for discovery he is nothing but a dirty little beast, a swindle, a pest unto the universe.

Pontoise. 1 May 1959.

Fantasy alone is what should drive me forward. It is a tumultuous torrent, but my boat is sturdy enough not to capsize within it.

I had forgotten what Spring could be in blessed Ile-de-France, with these towering branches in bloom, these multiple levels of sumptuous colors in the leaves around us, and the majesty of the huge pine tree in the neighbor's yard.

There is a song by Jacques Douai:

> Que sont mes amis devenus
> Que j'avais de si près tenus,
> Et tant aimés?
>
> *(What has become of my friends,*
> *Those I loved so dearly,*
> *And held so close?)*

My friends have melted away in the breeze, in the mud of the fields, the sand of holiday beaches.

Pontoise. 13 May 1959.

At twenty years of age my contemporaries are attracted by powerful myths: the myth of intellectual comfort, of material riches, of a "career path" to success and respectability among the bourgeois, leading to quiet retirement. At forty their minds are sclerosed. Rare are those who keep a strong spirit till the end. I think of my father as an exception. His spirit came through in the way he would open a book by Barbey d'Aurevilly, the way he taught me the difference between oak leaves and aspen leaves as we walked through the woods.

It was he who showed me how to make a slingshot, bows and arrows out of branches of hazelnut tree, whistles out of reeds. He had grown up between Caen and Cèrences. He knew all the tricks of the clever Norman farmers.

Once he took me above the Pontoise railroad station, along the road to Rouen, to show me some large flat stones buried in the fields. They were the remains of an older highway: the road built by Julius Caesar when he crossed the Oise on his way to Great Britain, nearly two thousand years ago. From these trips with him I learned about the depth of time and the widely scattered wonders covered with moss and weeds, just waiting to be discovered and pulled out.

The changes of the post-war period took my father by surprise. He was shocked by the selfishness of politicians, the brash explosion of the media. He regarded the new movies as a social evil. French society was shaking free of its older models: the songs became irreverent, the moral references imposed by the established order collapsed. A free-thinker, a fervently independent mind, deeply devoted to the ideals of the Republic (a framed *Declaration of the Rights of Man* was hanging on the wall behind his desk), he was not part of the traditional Old France of religion and wealth. Yet when I confronted him about the stupidity of the Algerian war, as my brother before me had confronted him about France's attempts to keep its colonies in Indochina, he sought refuge in simple-minded slogans ("My country is never wrong") which infuriated me.

It is very important to refuse to take any of the predetermined paths society offers us. There are no predetermined paths in nature, only relativity of directions and goals. True initiation deals with the Whole, and with love.

What is Destiny? Are some individuals just carried along by events, while others spin around in narrow circles, unable to find a solution to the simplest problems? Events do happen at the right time if one knows how to place himself at the spot where their greatest probability of happening lies. Destiny might simply be a measure of this ability.

Paris. 5 September 1959.

These pages are nothing but a schoolboy's notebook, in the strange classroom we call life.

Over the bridge at Bezons, the huge deserted bridge, twilight made the sky gray and mauve as I was riding through tonight on my way home. The wide road swept up and I could see nothing beyond it, only the tall street lights flooding the wide pool of asphalt, thin lines of sodium over the moribund Summer. Suddenly this huge bridge appeared to me as a fine and rich thing, a novel image of beauty. I felt the approach of Winter. Under the rain or under the nourishing fog the most ordinary objects suddenly seem able to meditate and resonate beyond our wildest thoughts.

Life in my little room is reduced to bare threads. I have passed a new batch of examinations. I will gladly walk away from the Sorbonne, clutching my science diploma. Janine only comes to see me in the evening, when she mends my unstable existence. She has completed her Master's degree in psychology, and the Administration has reassigned her to a school in Amiens. Emotionally, we are both ready to leave Paris behind. Already, she has moved her things out of her own room.

My thoughts are already shifting to the flat landscapes of the North of France, to the city of Lille where we have already made a few quick explorations together. Lille is not too far from Amiens. Janine will be able to come and visit me often. I relish the idea of moving to a city where I do not know anyone, where nobody expects me to be. I like the sad, gray, quiet suburbs of Lille and its small, one-dome observatory where I will be studying towards a Master's degree next year. In Paris they teach astronomy without ever showing you the sky. The lecture halls are crowded with hundreds of students who have no interest in the subject but need the credits.

I was created in the form of a man. This is supposed to be obvious: "I am a man." Yet there is an infinite distance between "me" and "the man I am." We can only shape our life through the control of everyday acts, with that fine knowledge of the structure of destiny which is provided by the constant proximity and vibrant awareness of death.

Science, too, is supposed to be obvious. But it is nature which is important, not science. Physics is nothing but a user's manual, a cookbook based on a narrow conventional language. Physics is a confession of weakness. I can only believe in simple, beautiful things. Why do they show us science as such a complicated structure? There is nothing very complicated in the world, only states of mind which get in the way and complicate simple things.

Paris. 24 September 1959.

Today is my twentieth birthday.

I am only myself when I am with you.

Paris. 8 October 1959.

Alone in my room, I wait for Janine. Winter is coming, my winter. This city is moribund, except for two movies by Ingmar Bergman that have just been released. It is odd, how his songs of death are supplying the only spark of life in these dull gray buildings, these idle masses of stone.

4

Lille. 24 October 1959.

Childhood: the spring itself is forgotten when we can see the river rolling along. Yet it is the same water, in color and in taste. The memories of man, as far back as he goes, are lost in question marks. And the very source of being, one's origin, remains as exciting a mystery as the state that follows death.

The records of our ancient dreams are the most fascinating of all bedside books. They paste an ironic smile over today's freshly hatched plans.

I was born in an interesting year, 1939, a point of low birth rate for the French nation because of the combined effects of brewing international tension and the lack of marriageable males, a long-term consequence of the massive killings of the First World War. It was in 1939 that the first digital computer was demonstrated,[5] and that Roosevelt learned from Einstein's secret letter that an atom bomb could be designed and built.

Even the timing of my birth, on 24 September, was poorly chosen. The Second World War had been declared just three weeks previously. German planes were preparing future campaigns by executing bombing raids against strategic objectives. Pontoise was high on the list. It had both a highway bridge and a railroad bridge, and held the key to Normandy. The Luftwaffe was pounding away at the little town. The midwife, who lived on the other side of the river, was unable to come. It was the doctor who delivered me amidst the sound of the first air strike. My father was fifty-five, my mother was thirty-nine, and I had a brother who was finishing High School.

A few short months, and the invasion of Hitler's Panzers came from the East and the North. A great panic swept the French population into a mad exodus. My parents left Pontoise on 10 June 1940 to seek refuge among our cousins in the safer province of Normandy, where they spent several months. I naturally have no memory of it. One day a former neighbor who was passing through Normandy told my parents what had happened to the fourth-floor apartment where they lived, and where I had been born: the Germans had entered Pontoise on June 11th. The day had been marked by numerous incidents. It was alleged that a sniper had hidden himself in the attic of our building. Wehrmacht soldiers rushed in and threw incendiary devices into every apartment, then they just watched the whole structure turn into ashes. The fire destroyed my parents' small treasure: a few pieces of furniture, many books; but we had escaped with our lives. Amidst the great tragedy of Europe that was known as being lucky.

When they returned to Pontoise they rented a small house on the hill that overlooked the river, a rocky escarpment where medieval walls used to defy the invasions. A little knowledge of history would have discouraged such a move. The medieval fortified site of Pontoise was the birthplace of alchemist Nicolas Flamel. It was once so prominent and wealthy under the banners of Saint Louis, the White Queen and Philippe-Auguste that its cathedral was larger than Notre-Dame in Paris, but it was consistently attacked throughout the centuries because of its key position on the river. Every invader since Julius Caesar had crossed the Oise at that spot. The British and the Americans now needed to destroy the bridge to cut off Hitler's forces from their reinforcements. They bombed the river, reducing many houses in the vicinity, including the one we rented, to mere dust.

I remember my mother picking me up in her arms and walking through the rubble. I stared at a door frame still standing amidst the destruction: it was all that was left of our previous home. Fortunately, just the day before, we had been evicted by Anna, our greedy landlady who was hoping to rent her house more profitably to some German officers. In fact that scene, too, has remained as one of my earliest memories, my mother holding me with one arm while striking Anna with her kitchen rag in utter frustration.

This time my parents wisely moved farther away from the river, to the house on Saint-Jean street.

One evening the *Résistance* blew up the switches of the railroad station. We went up to the attic to watch the drama unfolding. My father, who had been at Verdun in the previous war, had a very sure sense of danger. He also knew that our cellar afforded no realistic shelter and that we were better off watching the battle. For a five-year-old child the spectacle of war was a fantastic game, a splendid education in the unreliability of the world. I remember the rails being thrown up in the air like matchsticks. In later weeks and months waves of airplanes came from the West, flying towards the Ruhr in triangular formation. I watched aerial dogfights in which wings were torn off. German batteries would fire pitilessly at the bodies of helpless Allied pilots swinging down from the bright blue sky at the end of their white parachutes. Every day in the garden we gathered tinsel and radar-fooling "chaff." Soon came the mighty rumble of Patton's tanks, behind which marched tall laughing Americans with chewing gum in their mouths and nets over their helmets. Interminable truck convoys took over the main roads.

Slowly my parents resumed their existence, watching every franc and every sou. In my father's papers, under the heading "rebuilding," I have found an official document dated from 1953, eight years after the end of the war. It read:

> The undersigned expert went to the new domicile of the person who has suffered the loss. He observed that no professional effects were in existence, except for a few books. The furniture currently used by this individual and his wife is old furniture coming from his mother's home.

My father never owned a house, a car or even a telephone.

Lille. 27 October 1959.

I am early for a physics class. Sitting in my car near the University I can see the wet cobblestones, the kids in their hooded coats fighting against the wind that shakes my old Renault 4CV with violent blows. Behind me the storm sweeps the Square Philippe Lebon in the gray morning. It picks up dead leaves and sends them flying clockwise around the statue. A shiny black car slowly drives by. The driver lowers his window and calls out to his girlfriend on the sidewalk. She stops, turns around; I see her eyes moist and wrinkled with stormy rain. She laughs as she joins him. I cannot hear what she says. They are lost down the street, absorbed into the mystery that lies beyond the corner where an orange light continues to blink stubbornly.

The rain batters the roof of my shelter. The howling wind blows, as it did last night in the fireplace of the room I am renting in a drab suburb South of the city. Girl students walk close to the walls like ghosts in their white raincoats. They look like the Touaregs of Morocco, their faces hidden by scarves and high collars.

Here in Lille abstract things seem easier to grasp. Abstraction lies in wait within every object and every gesture; every wall seems to contain an idea, fine and straight as a javelin, luminous and precise as the spot of an oscilloscope.

Aimé Michel writes to me:

> Any progress of the mind consists in gradually stripping away the preconceived ideas, the systems you have inherited. You are right to stir all that up. But do not expect to find the idea that will reassure you, "the Truth" if you will. Above all, Truth means understanding why we don't understand. Wisdom is to be able to measure what is certain and what is uncertain in science, a feat most so-called "scientists" are incapable of accomplishing.

Lille. 21 March 1960.

This evening we opened the dome of the observatory at 8:30 p.m. and we took three spectra of Arcturus (each a twenty-minute exposure) with the refracting telescope, a fine old instrument twenty-one feet long. We worked until 3 a.m.

Lille. 15 April 1960.

My first job: I have been making a little money by doing lengthy hand calculations of the integrated energy of eleven open "galactic" clusters, among them the Pleiades, Praesepe, Hyades and Coma for a research project headed up by Professor Kourganoff. I also compute the integrated color. Each calculation represents a long set of operations with a Marchant tabulating machine.

Although Lille University is one of the first campuses in France where programming of electronic computers is taught, we do not have any means of computation at the Observatory. Everything is still done by hand. Even the decision to teach about the new technology has caused something of a scandal: there was no one on the Faculty with any experience in programming, so we have to take a course which is taught (and taught quite well) by a local IBM engineer. To save face the University has asked a prestigious academician, Professor Kampé de Fériet, to come from Paris twice a week to lecture on information theory. This gives the course on computers some semblance of respectability in this traditional, conservative institution....

Lille. 26 April 1960.

One o'clock in the morning. From my little whirring car I step out into the silence, cross the deserted street to my room. I will sleep like an animal. Tomorrow I must work on the math problem with a friend.

Last Sunday I went to Paris and by chance I met Granville. Has he changed? Not at all. He is so identical to the image of him I have kept that he is able to say:

"You see, I have been waiting for you, on this park bench, for the last three years...."

Lille. 8 June 1960.

The crowd moves ahead like a mechanical storm, tough and swirling, heavy with obscure prophecies. We follow it, walking down the wide streets. We pass houses where landladies sit, knitting in the moist darkness of their hallways like spiders in their holes. The new sun is crushing the city. Men in shirt sleeves are taking the bus.

Lille. 10 August 1960.

Some months ago I sent an application to the Rosicrucian Order,[6] whose French branch has its headquarters in Villeneuve-Saint-Georges. Now I find their documents to be an interesting spiritual complement to my scientific training. Every month I receive a set of course material through the mail. It includes both theoretical reading and instructions for simple rituals, promising insight into higher realities.

Lille. 19 October 1960.

Janine and I were married today at the Lille City Hall. We had made arrangements with a friend, a research associate at the Observatory, to be our witness, but the law required a second one. On the sidewalk before the University entrance I found one of my classmates:

"How would you like to be a witness at a wedding this morning?" I asked.

"Whose wedding is it?"

"Mine!"

He laughed and looked at me with the indulgence reserved for lovers and other simple-minded folk.

"When is it?"

"In half an hour!"

We were married at 10:30. It was a very relaxed and informal affair. It isn't the external trappings that count, or how much one spends, but what we feel inside. We took our witnesses to the local student bistro for a couple of beers before returning to class.

Lille. 12 February 1961.

I've written a letter to Georges Gallet, the editor of the science-fiction collection for Hachette, the large Paris publisher to whom I have submitted my manuscript of *Sub-Space:*

> My protagonists are simple researchers coming out of the crowd under pressure from fantastic events. They are "awakened" to the absurdity of the world in which they have lived until then. They have to conquer it, not only outside but within themselves. The monsters from sub-space illustrate this transformation. . . . At the end of the book the world appears no less absurd than it was at the beginning, but a dozen scientists have understood its genuine depth.

I end my letter with a timid request for a personal meeting.

Lille. 20 February 1961.

We now rent a single large room on the fifth floor of a hotel near the railway station, in the center of Lille. In one corner is a small kitchen with a primitive stove. There is no elevator. The bathroom we share with other tenants is five floors below, behind the bar. Our favorite dinner is a plate of Frankfurt sausages with French fries and mustard, eaten in some cheap bistro, listening nostalgically to Edith Piaf who sings Milord: *"Vos peines sur mon coeur, et vos pieds sur une chaise . . ."*

Many thoughts are rushing through my brain. The tension between the high potential and the petty reality of yet another series of examinations to prepare causes me both exaltation and pain. Early this morning Janine has gone to her work in Amiens, leaving my room full of sunshine. I am immersed in my notes from the astrophysics course, taught with great wit and wisdom by Vladimir Kourganoff.

Lille. 22 February 1961.

Janine and I have good long talks in the evening, until midnight. We discuss the course of our earlier studies and I see a similar force within both of us, a vision of life, an anticipation of the future, a certain way of committing ourselves to it. Something inside us seems to know where the path leads. It is as if we were marching towards another world, and as if we knew that other individuals throughout the earth are going in the same direction to meet us there. Perhaps we are going towards Paul Eluard's *other universe* which, as he says so eloquently, "lies within this one."

Each new discovery brings with it renewed silence. For those who have pierced the barrier, words have never represented more than the emerging part of thought. Beyond words is the second meaning, the third meaning, the true ones.

Since September I have been working on a new science-fiction novel entitled *Dark Satellite.*[7] I am writing very fast, swept along by passion and Janine's kisses. Every evening I am anxious to read the new pages to her.

Lille. 23 February 1961.

Georges Gallet of Hachette has answered me:

> I will be happy to see you if you come to Paris, any day next week at your convenience, as long as you let me know forty-eight hours in advance. In the meantime do not worry about your *Sub-Space.* You will probably be pleased to learn that it is among the manuscripts we have selected for the Jules Verne Prize.

Lille. 20 April 1961.

Too many ideas at the same time. I lack a roof, a job, in short life itself. How simple everything would be if I could just accept to be "a student." But I am not content with University life. Quite different is the world I see when I visit Hachette on Boulevard Saint-Germain in Paris, where *Sub-Space* is now under serious consideration for publication. I meet other writers there, free spirits, thinking minds like Jacques Bergier, who has recently published, with Louis Pauwels, *The Morning of the Magicians.* Reigning over this world of creative confusion Georges Gallet is warm and generous, jovial. He becomes enthusiastic as he talks about the early days of sci-fi, when he discovered that novel form of American popular literature as an interpreter with Allied troops during the war. He has met most of the classic U.S. authors and is a friend of Forrest Ackerman, the legendary collector of weird tales, who lives in Los Angeles.

Lille, 23 April 1961.

Psychology has to start from the relationship between man and the tools that link the psyche to the environment. But our tools have changed, changing the environment in turn and offering new paths of development for man's intelligence to follow. Properly speaking, man has not "gone into space" yet, contrary to what journalists are fond of saying. Gagarin and Sheppard have simply flown higher than the others, but they did reach a new realm outside the atmosphere of Everyday Life. Tomorrow we will truly take mankind away from the earth, squirting humanity into the blackness of real space, deep space, splashing the universe with our childish laughter and our profound terrors. But that adventure may be a devastating test for us: The old type of man will not survive it.

The human being who will be able to function in space is already living among us. He has always been within ourselves, in every cry of despair of the soldier cursing the sky, every heartbeat of the young girl looking at the rising moon. Our scientists, romantics of the Sputnik era, may speak eloquently about the future, but they fail to see the mysteries buried in the present: infinity brushing against us in the anonymity of crowds, galactic trapdoors at the street corner. . . .

Those who are dead to the potential of the earth will also be dead when they fly into space. Those who have not eagerly expected the sudden glow of a star amidst fast-changing clouds will never be able to grasp its reality, even if some day they happen to hurl their rocket ship into the burning blast of a supernova.

What seems like the most opaque fabric from a distance appears under the microscope as a network of loosely textured threads. The human eye can see through the toughest of metals, if it wants to.

Lille. 26 May 1961.

One of Janine's psychology teachers at the Sorbonne, Professor René Zazzo, wrote in 1946 in *The Future of Intelligence:*[8]

> The metaphysics of the irrational is an excuse for every shameful act, every surrender. I know that it seems to answer a need for spiritual liberation, and I also know that this need has led the poets of surrealism and the philosophers of supra-rationalism to some magnificent attempts to reach beyond the world of everyday habit . . . but beyond reality what we find is still reality, only richer. Beyond reason what we find is still reason, only wider and deeper.

He adds:

> Nothing is obscure except through our ignorance; nothing is fantastic except through our terror. It is by refining our reason that we will discover the laws and the rhythm of those things that are still hidden, not by some indescribable experience, some vague ecstatic intuition, some Dionysiac drunkenness. . . .

It would be absurd to imagine that we will simply be able to transplant our old mental habits to Mars or Venus. There will be a major change this time, perhaps even a conscious one. In this sense, Zazzo's faith in reason is well-founded.

We are just entering into the world. We are going into a universe from which technical difficulties had kept us. Until now we could be satisfied with a contemplative, idealistic or romantic attitude. Planets, solar systems, comets and galaxies were topics to be discussed under the heading of speculative philosophy or position astronomy. This will no longer be the case tomorrow. They will become our familiar landscapes, our everyday risks. In that new world we will have to learn again how to love, to cry, to laugh. The earth will no longer provide appropriate standards.

The European countries are making a number of grave mistakes right now in this domain. They are incapable of resolving their ancient cultural contradiction between "literary" and "technical" modes of thought. They are headed towards asphyxiation of their creative faculties. Any nation that approaches space research with the mindset of a graduate of *Ecole Polytechnique*[9] is a nation that will not survive the very first discoveries from that research without some major changes.

Sources of imagination are drying up in Europe. This is an amazing fact. Literary people are cut off from reality because they stubbornly remain blind to the technical underpinnings of the modern world. They cannot feed their younger public with new hopes or new images. As for the scientists, they deny or they ignore the cultural value of their own work. The way they look at science fiction is a case in point. Many view it with contempt, as a lower form of literature. Their science has lost all contact with the public. The chasm that a man like Flammarion[10] had once succeeded in bridging between the man in the street and the scientists has been opened again, and widened: modern scientists are becoming needlessly, hopelessly specialized. Zazzo writes that "nothing is fantastic, except through our terrors." But our terror is a force that propels us forward. This new terror may be the most dynamic, generous attribute we have, and the only thing about us which is still pure.

Lille. 31 May 1961.

We will soon be going back to Paris. I have completed my examinations for the Master's degree here. The time has come to finish the operation.

Kourganoff has told me that the artificial satellite department at Paris Observatory was looking for people: I am applying for a position under Paul Muller, the same man to whom I had sent my *Sputnik* observation of November 1957. But in the long run Kourganoff tells me that I will be wasting my time if I stay there: "Artificial satellites do not have any real significance for science," he insists. Already last year he tried to discourage me from taking courses in programming: "Electronic computers are just toys for engineers."

In saying this he is merely reflecting the opinions of the academic mainstream. Four years ago the world was amazed to learn that the Russians had launched the first artificial satellite. My father did not believe in it any more than he believed in flying saucers: "Communist propaganda," he stated immediately. It was only when the conservative papers confirmed the news that he was forced to admit the evidence. But he never agreed that an astronaut would some day orbit the earth: "Man cannot get out of his sphere," he would say.

Even professional scientists were caught completely unprepared. No less an authority than the Astronomer Royal of Great Britain had said, a mere four months before *Sputnik*, that "Space Travel is Utter Bilge."

Lille. 9 June 1961.

In *Dark Satellite*, which takes place on Venus in the twenty-first century, I had to invent a social system that would replace both Marxism and Capitalism, since it is hard to imagine that either of these could support the expansion of human activity beyond the earth. I have dreamed up a new political organization called *Peripherism*. Under Peripherism the world would be structured, not as a consortium of big nations, but as a network of smaller regions. Most of Europe, for instance, would break up into areas similar to the old Provinces, with autonomy for well-defined cultural, linguistic and economic units which would receive key services from the larger world. I imagine autonomous regions linked together through a giant computer that could manage Martian bases and lunar cities, as well as terrestrial units. The key idea in Peripherism is that by breaking down the major countries into their local components it would be much easier to defuse cultural antagonisms, to force people to be responsible for their own destiny, and to create an enormous federation serving everyone's interests.

I do not know what the destiny of *Dark Satellite* will be, but Georges Gallet has told me that *Sub-Space* had won the Jules Verne Prize. I have just signed the contract with Hachette. There will be a first printing of 15,000 copies, and a prize of 100,000 francs.[11]

Paris. 14 June 1961.

The medal of the Jules Verne Prize was given to me today at the Eiffel Tower by the pretty hands of actress Mylène Demongeot, who kissed me on one cheek while Janine smiled. More formally, someone from Hachette gave me my check, photographers took pictures and the *Tout-Paris* gorged itself on caviar and *petits-fours.*

Georges Gallet, kind and good-humored as usual, was greatly amused by my shyness. He took me aside: "Do you have any idea how much all this costs, renting the Eiffel Tower for a day, and all this food?" I replied that I did not. "Eleven million francs," was the answer.[12] Suddenly the check of one hundred thousand I had folded into my wallet had taken ridiculously small proportions. It was clear that all these people had not come to see me, or to read my book. This was purely a promotional event for Hachette. If I suddenly dropped dead they would all go on eating caviar and *petits-fours* as if nothing had happened. A wonderful, humbling lesson in the true meaning of literary glory!

The high points of the afternoon for me were the chance to meet Forrest, the "father" of Barbarella, who has designed a magnificent cover for my book, and Daniel Drode, last year's laureate, who had an even stranger experience than mine:

"I was a soldier in Algeria," he told us, "stationed at the edge of the Sahara. I was staring at acres of hot sand beyond the nose of my machine gun. The field telephone rang. It was my Colonel. *Drode,* he said, *you have just won the Jules Verne Prize. Report to Paris immediately.* He gave me an exceptional two-day leave."

Poor Drode barely had time to fly to Algiers, and from Algiers to Paris, where he arrived without any sleep, changed into civilian clothes, rushed to the Eiffel Tower, was kissed by a lascivious movie star, watched the *Tout-Paris* gorging itself on eleven million francs' worth of delicacies, and flew right back to the front lines.

"I thought the sun was giving me hallucinations," he added. "Periodically I had to take my eyes off the sights of my gun, pull out the bronze medal from the pocket of my uniform and stare at Jules Verne's face to be certain all that had actually happened!"

5

Paris. 12 August 1961.

I believe in terror. I believe in glimpses of beings from Beyond seen in dreams, reflections on layers of time. I have become interested in the saucer phenomenon again.

The flying saucers! For years I have thought about what we saw in Pontoise, that bright afternoon in 1955. I have tried to fathom what kind of research may be going on in "high places" and what data must be kept in the official files. I wrote to Aimé Michel when he published the book in which he announced the claim that apparently unrelated sightings of a single day occurred along straight lines. Our correspondence has continued, without reaching any conclusion about the nature of the problem. Now the subject is taking on a new aspect.

This is not at all a propitious time for me to become actively interested in this particular problem. Since June I have become a government employee serving on the staff of the artificial satellite service of Paris Observatory. Actually we are part of the recently created Space Committee, which reports to the Prime Minister. I have a beautiful card in my wallet, where my picture is struck diagonally with the official tricolor. Naively, I started work here with great enthusiasm, assuming that we would be engaged in genuine research, in the highest quest for truth. That is not what I found.

The artificial satellite service is located at Meudon, on a high plateau near a fine forest from which one can see the whole of Paris. The staff is composed of three scientists under Paul Muller, with three secretaries and a computer programmer. Our mission is to track as many orbiting space objects as possible to keep every ephemeris up to date. The orbital elements we calculate are later used to improve theories of the shape and weight distribution of the earth, and to advance the calculation of satellite trajectories. Our equipment is primitive. During the day we compute the visibility windows of each satellite to build up an observing schedule; we answer the mail from the public; we plot the passages of the objects over a map of Europe; we run the programs that reduce the observations of the previous nights, using an old IBM 650 located in the former stables of the King's mistress in Meudon castle. We use theodolites, small but highly maneuverable telescopes that can be pointed quickly and with high precision. Every night we set them up in the dewy grass under the open sky. We patiently wait for our satellites to come and we aim our instruments at them as they cross the sky: "It's like hunting rabbits," says Monsieur Muller. We use a red traffic light down in the valley in Clamart as our reference point, azimuth 265 degrees 19 minutes. Electrical cables power the lights on the circles of the theodolite, the tape recorder, the precision chronographer.

After each series of observations I go back inside a nearby dome where we have set up our field office. Under the creaking floor I can hear the rats scurrying around. Rubbing the sleep from my eyes I reduce the data to punch up a Telex tape which will be transmitted to the U.S. Navy in Paris and, from there, to the Smithsonian Astrophysical Observatory in Cambridge, Massachusetts. In the nearby woods strange animals wail and scream. There is an owl that shrieks with an especially disturbing, strident cry, like a slaughtered baby.

Occasionally we observe objects that remain unidentified. Thus on 11 July at 10:35 p.m. I saw a satellite brighter than second magnitude. I had time to log a few data points. On another occasion several of us recorded no less than eleven points. The next morning Muller, who behaves like a petty Army officer, simply confiscated the tape and destroyed it, although a similar object had just been tracked by other astronomers at Besançon and by Pierre Neirinck, a satellite expert based in Saint-Malo.

"Why don't we send the data to the Americans?" I asked him.

Muller just shrugged.

"The Americans would laugh at us."

He seems terrified at the idea that the morning papers might come out with the headline: "Paris Observatory tracking something it cannot identify." Muller is a tough man who believes in discipline and a simple world where everything is neatly labeled. His previous career in astronomy was based on measuring the angular separation of thousands of double stars, the most painstaking work imaginable. He approaches artificial satellites in the same spirit.

We receive many letters from the public because artificial satellites are a hot topic. Many contain mad theories, deranged proposals complete with convoluted color diagrams and prophecies of the end of the world, of imminent doom. Many others simply submit sincere and accurate observations of various satellites which we can readily identify. I spend about an hour every day answering such letters, since we are a publicly financed service of the French government, with a duty to respond to public inquiries.

One morning Muller read to us passages from a letter he had just opened. It came from none other than Aimé Michel (who doesn't know I work here): "A few years ago I have been unfortunate enough to publish a book about flying saucers," the letter said. "I have been the recipient of hundreds of reports describing observations that are of potential interest to science. I am in poor health and I have reason to believe that I suffer from a brain tumor. I would like to turn over all my records to an institution such as your observatory, where they can be preserved. Even if you do not agree that research on these phenomena is warranted, at least the records should be protected from potential destruction after my death."

"You see," said Muller with contempt, "that's another letter for the crackpot file. Although properly speaking Aimé Michel is not really a crackpot, he is a crook." *("Ce n'est pas un fou, c'est un escroc.")*

What could I reply? I felt incensed at the narrow-minded stupidity, at the injustice of this remark. Aimé Michel sought no money, no publicity in return for his offer to turn over all his files. It was a proposal from a truly desperate man.

I went home to this one-bedroom apartment we are renting on the edge of *Les Halles.* I sat at my desk and I wrote a letter to Aimé Michel, suggesting a meeting.

Paris. 20 August 1961.

I am just coming back from my first meeting with Aimé Michel. He lives in an apartment in Vanves, just South of Paris, on the second floor of a building that overlooks the park. I barely caught sight of his wife, who opened the door and ran off shyly into the darkness of the hallway without speaking to me. He greeted me and took me into his office, a warm little room with a desk overloaded with papers, piles of books, articles in various languages, many letters. Notes are pinned to the fabric which covers the walls.

In control of that mass of information is an amazing gnome of a man, short and deformed, who barely reaches to my stomach. Yet he radiates a kind of beauty that is unforgettable, a beauty that comes from the mind and from the nobility of his piercing eyes. He shakes his bald head and lights up with a wonderful smile as he tells me: "You know the worst thing about being crippled as I am? I will never be able to kick some of these arrogant scientists in the *derrière* as they deserve!"

Yes, he tells me, he wrote this letter to Muller because he is fed up with flying saucers.. He has amassed so many documents that he will never be able to process them all by himself. He thought he was close to a breakthrough a few years ago when he discovered his famous alignments among the sightings, but the underlying order, if there is one, eludes him to this day. All of that cries out to be checked by professional scientists. If he turns out to be wrong, so be it. At least he will be free from the anguish of bearing this awesome responsibility by himself.

He is utterly disheartened by the onslaught of bitter criticism he has received from the so-called "rationalists." They go so far as implying that he has actually invented the observations, drawing the lines first and then writing to newspapers in the towns that fell on each alignment to inquire about possible sightings! One of the most rabid of the group, cosmologist Evry Schatzmann, has even told him: "Your alignments cannot possibly exist, since flying saucers cannot exist!"

"Well, let them take all these letters," he says as he lifts piles of envelopes from his cluttered desk. "Let them check, they will find that all these people do exist. Then correlate what they have seen with the newspaper clippings, with all the published documents!" He pauses, looks me straight in the eye:

"They will find out that I am not lying."

He is surely not lying, I tell him. But truth will not impress those people. I describe what has happened to his letter to Muller, and I quickly add:

"Don't worry, those who do all the real work at the observatory are the young researchers. We program the computers, we track the satellites at night. You don't need official recognition from the old guard, what you need is some people who will roll up their sleeves and do the real work. What do you want me to do?"

"Are you serious?"

"I wouldn't have come to see you if I wasn't serious."

He whistles softly.

"That changes everything."

He remains silent for a few minutes. He looks out at the park, then brings his eyes back to me:

"There is one thing you could do. You see, my alignments don't really make any sense if they stop at the French border. If they are real, they must indicate a world-wide pattern. Every one of them must be a section of a great circle. It would be interesting to take some of the best-authenticated straight lines and to extend them around the world, to see where they go. Would you know how to do that?"

"I wouldn't be a very good astronomer if I couldn't compute a great circle."

"That would be a lot of work."

"I'll use an electronic computer."

I assure him I will find a way to discreetly use the IBM 650 at Meudon to perform this calculation. I can see he is skeptical that I will actually get to work on the problem. Many people have promised to help him before, and he has never seen them again.

Even the Americans, with their enormous technical means, don't seem to have done any real analytical research.

On my way home I have started to think about the data reduction in spherical coordinates. I have decided to begin with the Bayonne-Vichy line which links no less than six independent sightings made on the same day of September 1954, as the French wave was just beginning.

Paris. 28 August 1961.

A letter mailed today to Aimé Michel provides him with the track of the Bayonne-Vichy line computed as a great circle of the earth. I did the first calculations by hand, using only the two end points. But that will already enable him to plot the coordinates on a map and to verify his hypothesis, using foreign cases this time, well beyond the French borders.

Paris. 24 September 1961.

Aimé Michel has answered me: the great circle was remarkable, he wrote. Intrigued, I called him on the phone this afternoon from the subway station. (We don't have a phone at home. It would take money to get on the waiting list and three years to obtain the equipment.) He explained to me why he was so excited: the line goes through three major areas of high concentration, linking saucer waves in Brazil, New Guinea and New Zealand. In the latter case the line goes exactly over the harbor at Wellington, where a celebrated observation was made. Is that significant?

On September 1st a former astronomer from the satellite service called us: he had seen another mysterious object at 22:00 UT. It was near the zenith and as bright as the American *Echo* satellite. Janine saw one on September 3rd at 20:01 UT. Muller himself has recorded several unknowns with his staff. He classifies them as "aircraft" or "mistake." Later they get lost.

Paris. 26 September 1961.

Today a very discreet conference about flying saucers took place at Meudon observatory. I had succeeded in getting the whole staff of the service to attend, except for Muller, of course. I have started to use the IBM computer after hours and on weekends to compute Michel's great circles. Our staff programmer is teaching me the assembly language of the 650 to enable me to encode my algorithm into this antique machine. This is a formidable engine with a 2,000-word drum memory, and it takes up an entire room. It looks like a locomotive, and it occasionally sounds like one.

A scientist named Pierre Guérin, who is a planetary expert with the Astrophysical Institute and a friend of Aimé Michel, has joined our group. He told us that in his opinion the saucers existed and were probably operated by space visitors. They could be biologically advanced in the evolutionary sense with respect to humans.

Guérin, a tall fellow with jet-black hair whose family comes from the rebellious Vendée, is prone to snap judgments and doesn't like to be challenged. Yet I did question his conclusions about the occupants. I still see all assumptions about "their" superiority as grossly premature.

We debated whether or not it was desirable to seek official recognition for the subject. I am not so sure it would be a good idea. With official recognition would come bureaucratic procedures, lengthy delays, committees about everything. The people entrusted with the power to supervise the research and to control the budget would be the same old scientists who have denied the reality of the problem all along, and have called Aimé Michel a crook. Our research would be emasculated by their lack of creativity and their need to reduce everything to that dull state of uniformity they mistakenly label as "rationalism."

Paris. 27 September 1961.

We have to be very careful now or I could lose my job. Our little gathering yesterday has been noticed and the staff of the observatory is already gossiping about us. The people next door to us belong to a group headed by Dollfus, who does balloon-based astronomy. They know Guérin very well: Dollfus and Guérin are both former students of Gérard de Vaucouleurs, and both are interested in Mars. Now the group wonders, why would Guérin, who is a planetary expert, come to Meudon to talk to satellite trackers? It seems amazing to me that people should find it suspicious and undesirable for scientists of adjacent disciplines to talk to one another. Isn't that what science is all about? The maps I have been drawing are also attracting suspicion. Fortunately for me, nothing looks as much like a satellite orbit as a plot of Aimé Michel's Bayonne-Vichy line!

All this work after hours or on Saturday and Sunday gives me very valuable hands-on experience at the console of the computer, but the learning curve can be very frustrating, because our equipment is notoriously unreliable.

The most beautiful sound I have ever heard is the pure and highly musical hum of the memory drum of the IBM 650 when the computer dies. All power goes out. The motors are still. The console lights stop blinking and of course the program is lost. I become aware of the summer sun in the dusty courtyard behind me. I hear the birds playing and singing. But it takes many minutes for the big drum to slow down to a complete stop. The high pitch gradually turns into a sustained, thrilling note, unnoticeably shifting to a rumble, then just a murmur. Eventually, the drum joins the rest of the computer in death.

This kind of incident happens to us once or twice a day because our power supply often fails. Even when things are working properly this computer is very slow. The satellites go around the earth in ninety minutes, but it takes the machine two hours to compute an orbit, so we are always hopelessly behind. Yet many of our astronomical colleagues consider this computer an example of extravagant waste. They are jealous of us: Think of all the shiny telescopes one could buy with the same money!

Paris. 28 September 1961.

We have now measured with as much precision as we could the positions of the best documented sightings on BAVIC (as I now call Bayonne-Vichy) and a second prominent line. Janine has begun a list of coordinates for all the points mentioned in Aimé Michel's book. We have bought stacks of detailed Michelin maps which cover the whole of France. Unfortunately we do not have the luxury of sending investigators to every location: we have to take the best information we can from the published data: Some errors will be unavoidable. To reduce their impact I am computing the great circles through a least squares fit.

I continue to write science fiction. *Fiction* magazine has just published one of my short stories, *Les Calmars d'Andromède.*

Paris. 14 November 1961.

Another trip to Vanves to see Aimé Michel, who hands me a copy of the world's first sighting catalogue. It was compiled by a man named Guy Quincy, who lives in North Africa. This is a very valuable resource, even if its entries are often sketchy, because it covers over a thousand cases. It is a typescript of which only a handful of copies are in existence.

By plotting the computer data on the map I have made a curious observation: The first three great circles I have computed, which were selected because they linked the most remarkable cases, happen to intersect in a single point, near the town of Cernay in Alsace.

Paris. 25 November 1961.

Janine and I are now doing a study of the correlation between the frequency of sightings and the distance of planet Mars, which comes closest to the earth every twenty-six months. The resulting curve is striking. Guérin tells us he had not expected such a clear-cut relationship.

If the saucers come from space, why don't we see them when they are at a great distance from the earth? Recent experience with artificial satellites has shown that even a small sphere can be detected with the naked eye at an altitude of hundreds of miles. And a simple theodolite enables an observer to detect much fainter objects. Could this mean they don't actually come from outer space? We have no answer to this puzzle.

Paris. 7 December 1961.

Aimé Michel has given me another list by Guy Quincy which covers landing reports. Today I have reviewed all of my notes and I have updated our first catalogue. Janine has been assigned a job as a psychologist with a school near *Gare de l'Est.* After work she helps me by putting on index cards every single sighting we can find. Aimé Michel believes that daily alignments of sightings are characteristic of saucer waves but he agrees with me there is nothing magical about a twenty-four-hour interval. In fact he once thought he had discovered a twenty-three-day period between the most important cases, but he could never confirm it.

Janine and I are tired. She is afraid that I will fall asleep at the wheel, so some nights she drives with me all the way to work in Meudon.

Pontoise. 31 December 1961.

Taking advantage of the holidays I have plotted on a world map all the landings prior to 1954 and all the significant observations before 1947. There is no pattern, no law at all. Now I am trying to find out what researchers in other countries may have done on this problem.

Last week, while browsing along the Seine among the boxes of the *bouquinistes,* I was able to purchase French editions of a book by Major Keyhoe and of *Flying Saucers Have Landed* by Leslie and Adamski. The latter is utter fantasy, with fine fake pictures of a "Venusian saucer" that is nothing more than a dining room lampshade or some such ordinary object photographed at close range.

Seen through Keyhoe's book the attitude of American scientists and military men is quite disturbing. They behave like a well-organized insect colony whose life is suddenly disturbed by an unforeseen event. The Air Force's inability to think about the world in terms of anything other than the Air Force itself strikes me as particularly curious. For example, when they decided to study the flying disks they created a commission composed of rocket experts! And their idea of active "research" is to chase the objects with their jet fighters in the hope of shooting one down....

In the whole history of the problem Aimé Michel has perhaps been the only man who looked at the evidence humbly and calmly. To me this phenomenon is not simply something that should be investigated, ***it is a psychological test:*** the first great collective intelligence test to which mankind has been subjected. The sightings put into question both the structure of our society and the laws of our physics. Naturally we are free to run away from this test, as our scientists are currently doing.

Paris. 21 January 1962.

I have revised the flow-chart of my great circle program. Guérin has given me some useful advice for an article about the Mars relationship. He suggested I send a copy to his former mentor Gérard de Vaucouleurs, a French astronomer now established in Texas, and to a man named Allen Hynek who is the scientific consultant to the U.S. Air Force on the problem. Both are open-minded, he assured me: they visited Aimé Michel with Guérin in Paris four years ago and they were impressed with his data. I have agreed to put my real name on the article, even if it ends up being published in a magazine like the British *Flying Saucer Review.* I should have the guts to take a stand.

Since the first of the year I am no longer working in Meudon. I resigned, fed up with the pettiness of French astronomy in general and with the narrow-mindedness of Muller in particular. It is a small incident that finally drove me to quit. One evening, driving back from visiting my mother in Pontoise, I saw that the sky was quickly getting overcast and that satellite observations would be unlikely at best that night. I went to bed instead of reporting to the station. The next day Muller was very upset.

"But the sky was overcast," I said. "I thought..."

He interrupted me:

"That's exactly your problem, Vallee, *you think too much.*"

Perhaps that was a compliment? In any case I have looked at other job opportunities and I have quickly discovered that I could earn twice as much as a research engineer with an electronics firm. I now find myself with more advanced computing tools, and more stimulating people who are in touch with reality.

A scene that would be amusing if it was not a tragic illustration of French scientific bureaucracy took place after I handed in my resignation. Muller made several valiant efforts to get me to return, stressing the fact that a retroactive raise had recently been given to astronomers on the government payroll to correct the blatant injustice of their low salaries. Naturally, this raise was due to me even if I maintained my resignation. But the administration denied it. I had the audacity to take my complaint all the way to the head accountant who had his office in the main building of Paris Observatory.

It was my very first visit to that august institution. I was awed by the impeccably polished hardwood over which the great Leverrier had walked and where a line in the floor materialized the Paris meridian. I admired the shining copper of the antique instruments on display, the magnificent rooms with their period furniture and their dignified portraits. In these glorious surroundings the accountant turned out to be a cowering bureaucrat who was not used to being challenged. He tried in vain to argue against my retroactive raise. I left the observatory with a modest check in my hand and a very bad taste in my mouth.

Paris. 24 January 1962.

Guérin has sent a draft of my Mars correlation article to Gérard de Vaucouleurs. Aimé Michel has mentioned it to Yves Rocard, who is director of the physics laboratory at *Ecole Normale Supérieure.* Rocard appeared to be intrigued. He recommended that it be published quickly. Aimé Michel's health is better: His worry about a brain tumor has turned out to be unwarranted.

He also told me the funny story of how he had met Guérin. After his book came out in 1958 he received a very irate letter from him, lamenting the shortcomings of his methodology. He called Guérin and came to his lab, ready for a confrontation.

"Why did you send me a letter full of insults?" He asked.

"What insults?" replied Guérin, genuinely surprised. "I wrote to congratulate you!"

Paris. 26 January 1962.

Janine has introduced me to a statistician who is her teacher at the Psychology Institute. We need advice to evaluate the significance of the correlation we have found between the frequency of unexplained sightings and the proximity of Mars. At first, however, she ran into some strange reactions on the statistician's part.

"Where does the data come from?" He wanted to know.

"That doesn't matter," she said cautiously, since I had lectured her about keeping our study very quiet. The gossiping that surrounded us at Meudon has made me very nervous, and we don't want to get Guérin into trouble. She added: "This is a purely theoretical question."

"Not at all," insisted our friend, "these numbers look to me like astronomical parameters. Are you sure this doesn't have anything to do with Mars?"

It turns out that he has a colleague named Michel Gauquelin who has been doing his own clandestine study, an attempt to disprove astrology while demonstrating to his students the power of the statistical method. He did expose for them the myth of the zodiac, but the rest of the calculation backfired: he found a strong, unexpected correlation between certain positions of the planets at birth and the professional destiny of the individuals in his sample. There is an unexplained "Mars effect" which affects many people born with the planet just above the horizon or just beyond the zenith.

We soon met Gauquelin, and we laughed when he and his charming wife Françoise revealed their secret to us. They laughed too, when we told them where our own data came from.

Michel Gauquelin agreed that Janine's correlation computations were on the right track. He has given us some suggestions regarding the type of statistical test we should apply.

Paris. 28 January 1962.

For the last two days I have been making a copy (by hand) of Aimé Michel's card file, which holds a thousand cases. Each card is a reference to a raw document, so it is hard to estimate how many actual cases the file represents. There is a huge amount of work ahead of us. On the same card I sometimes find four or five observations. Most of the time he had only noted the date, the time, the place and a reference number, so I am left very hungry and frustrated: "Bloemfontein, most interesting!"

Later the same day.

I have just finished copying the cards for 1958, 59 and 60, some 400 cases in all. Since there was almost no overlap with the Guy Quincy catalogue I started worrying about the stability of our correlation with Mars, so I ran a comparison between the two sets. The results were beyond my expectations: Here are two researchers taking their data from independent sources, in different countries. Three out of four cases in Michel's files are not known to Quincy, while the latter has many cases from North Africa and from America. Remarkably, not only do the peaks coincide, but the details of the distributions match too. Could this mean that the percentage of errors in each file is actually very low? Are geographic and population factors playing a weaker role than we thought? Is Aimé Michel right when he says there would still be a noticeable wave in the Fall of 1954 if we threw away all the French cases?

In spite of the late hour I went out again among the produce trucks and the streetwalkers of *Les Halles* to find a public phone at the Reaumur subway. I called Aimé Michel to give him the monthly sighting figures and to ask him to think about their meaning.

Paris. 31 January 1962.

This afternoon I went back to see Guérin in his office on Boulevard Arago. I have a lot of freedom in my new work, as well as a measure of intellectual independence that was denied to me at the Observatory. Guérin works in a long, dark, dusty office in the basement of the Astrophysical Institute. Gérard de Vaucouleurs used to work in that room, which he left long ago for the wide open spaces of Texas. By the way, says Guérin, he is looking for an assistant, preferably someone who understands computers. If I was ever tempted to move to the United States, he could be a point of contact....

We have to keep our voices low when we discuss our forbidden subject: the walls are thin. He likes my article about Mars. At no point do I argue that the correlation is a "proof" of anything. We are simply attempting to analyze a phenomenon.

What the saucers are doing on earth eludes us completely. The type of observation that troubles me most is what Aimé Michel calls the "medusa": an object comes down from the sky, hovers a dozen feet above the ground and drops a small probe that touches the earth and goes back up. This suggests that they are taking measurements, but it doesn't correlate to anything we know. As for the landings, we do not have any good explanation either: could it be that our visitors are simply trying to get used to earth's gravity?

Paris. 7 February 1962.

Aimé Michel has given me the address of Dr. Hynek, who is director of Dearborn Observatory at Northwestern University in Evanston, near Chicago. I have mailed my Mars correlation article to him, to de Vaucouleurs and also to Richard Hall, of the civilian research group NICAP.[13] We have finished copying Aimé Michel's cards for all cases after 1954.

Through a friend who is doing his military service at the French Air Force headquarters, I have had access to the saucer files they have been quietly maintaining since the early fifties. They are full of well-documented sightings, including a remarkable incident in which an object was tracked on radar over Morocco.

Paris. 9 February 1962.

A demonstration against the extreme right turned ugly last night. Mad with hate, the cops charged the crowd, forcing it to retreat down the stairs of the subway station at Charonne. The iron gates had been drawn shut. Still the policemen, many of whom belong to neo-fascist groups, charged viciously, beating people with clubs and rifle butts, pressing them against the bars. In the panic several men and women died, crushed under the weight.

I was deeply shocked as I drove home through the area of the Bastille on my way back from work a few hours after these atrocities were committed. Today the whole area remains stunned, in deep mourning. There is a palpable, terrible sense of catastrophe, of despair, a feeling that reminds me of the dark days of the German occupation as I was aware of them as a child. I am fed up with such violence, fed up with this country and its absurd political intrigues. How could anybody ever build something of value here?

Paris. 29 February 1962.

I went to see Robert Kanters in the offices of Denoël on rue Amélie this afternoon. He is an affable man who heads up their science-fiction collection, *Presence of the Future.* He likes *Dark Satellite,* but he wants me to rework the end of the novel.

It is a strange experience to write science fiction here. The area where we live is still the Paris of Baudelaire, or of Villon. It has not changed much in centuries. On our street a dozen pretty girls stroll along the sidewalk by day and by night, wearing the shortest possible skirts and whispering sweet suggestions into my ears every time I pass them on my way home. We have agreed, Janine and I, that we would never "own" each other; that we would keep the wild freedom of love that gives a unique meaning to life in general, and to our life together. I must confess that I take advantage of this freedom: I would lie if I said that I was always deaf to my neighbors' suggestive invitations.

The streets are throbbing with tempting pleasures. Every night, all around us *Les Halles* celebrate life. Huge trucks bring the produce from the four corners of France. Workers unload the beef and the salad crates, the cheese and the pallets full of strawberries, lining them up on the sidewalk among the vagrants waiting for a chance to steal an apple, the poor students looking for a night job, the bag ladies rummaging in the trash, the glittering whores parading before the truck drivers, displaying their Parisian style before the farmers from Périgueux or from Brest. I watch the whole picturesque scene with lusty amusement.

Paris. 15 March 1962.

Yesterday I brought back all his index cards to Aimé Michel. The more we make progress in this research, the more upset I feel with the constraints under which we have to do this work, outside of normal hours, on weekends and evenings. In spite of all the improvisation, the computer programs to reduce the observations and compute the great circles are now ready, and we have a sizeable catalogue at our disposal. Isn't it time to place our work on a solid footing, instead of this suffocating, childish clandestine business? We need a stable structure within which we could do some real science.

"I've decided to move to the United States," I told Aimé Michel yesterday. "Even at the electronics company where I work, I am simply doing technical research on problems whose solutions have already been discovered across the Atlantic."

When I paid a visit to the U.S. Embassy, however, I came away discouraged by all their forms, by their haughty handling of us "aliens," by their bureaucracy. I hope this is not the real face of America.

Again Aimé Michel promised to mention my ideas to Yves Rocard, who will perhaps find a way to keep this work in France. Why can't we find a discreet laboratory somewhere where I could shelter my research?

When I think about it I am fairly sure that no one in France, not even Professor Rocard, can do anything. He may be one of the "bosses" of French physics, a leader of the atom bomb program which is dear to De Gaulle, yet his rationalist colleagues might hurt him professionally if they found out he has an open mind on subjects like dowsing and flying saucers. I find this especially ironic, when the man who directs nuclear tests from the flagship of the Pacific fleet isn't allowed to spend a few francs to explore the frontiers of science with a handful of his friends.

We also speculate about the work going on in America. Aimé Michel suspects they have not even gone as far as we have. He bases this opinion on the fact that "Project Blue Book"[14] has a staff of only three people. Their files are almost useless, judging from the overly complicated questionnaire designed for them by a team of psychologists, which breaks down every case into meaningless details. As Dr. Hynek told Aimé Michel when he saw his files: "You're luckier than I am—at least when you read your documents you can find out what actually happened!"

Paris. 24 March 1962.

I have obtained permission to use the computer all day, and everything is now ready for the calculation of Aimé Michel's largest "network" of alignments. I wonder what will come out of this massive effort.

In the evening, when we do not go out for a walk or a movie, Janine and I listen to the marvellous songs of Jacques Douai:

> Nos plus beaux souvenirs fleurissent sur l'étang
> Dans un lointain château d'une lointaine Espagne:

Ils nous disent le temps perdu, ô ma Compagne!
Et ce blanc nénuphar, c'est ton coeur de vingt ans...

(Our finest memories are blooming on the pond
In a far-away castle in far-away Spain:
They speak to us of all the time we lost, my Darling!
And this white water-lily, your twenty-year-old heart...)

Paris. 26 March 1962.

Nothing of any significance has come out of the computer. I find no remarkable pattern so far in the distribution of the great circles, so our research is stalled.

My employer has taken delivery of a new machine from IBM, the first 1620 model they have shipped to France. Our engineers are astonished: it is entirely built of removable printed circuit cards bearing transistors instead of the traditional tubes; even more remarkable, many connections are wire-wrapped rather than soldered. When something goes wrong the IBM maintenance man runs a diagnostic program, removes the bad card and replaces it with a new one without troubling to identify and fix the specific device that has failed.

I picked up one of these cards from the trash, put it in my pocket, and went to see Aimé Michel.

"I want you to imagine that we live in the fifteenth century. You are the learned abbott of the monastery, you have read all the available literature of Antiquity and you are familiar with the full extent of science. I am a farmer who has just found an object in his field, and I bring it to you to find out what I should do with it. Here it is, Father."

He took the card between his fingers, turned it around carefully this way and that, and finally gave it back to me:

"Burn it, my son, it is the work of the Devil!"

Indeed, what could a fifteenth-century scientist do with such a piece of technology? Even if he had been such a great chemist as to analyze the composition of the transistors, and had had the formidable insight to guess at the existence of germanium, would he have detected the *impurities* in the germanium? And if by some miracle he did, would he have concluded that these impurities had been put there *on purpose*, to create the effect of semiconductivity? And wouldn't we be in the same position with respect to a piece of flying saucer hardware, if it happened to fall on the earth?

Paris. 5 April 1962.

We met with Cristian Vogt yesterday. He is one of the founders of the Argentine research group CODOVNI, based in Buenos Aires. He told us that our current research was far ahead of anything he had heard about, including official American studies. Apparently the U.S. Air Force gathers a lot of data but doesn't really do any analysis. It only looks at the sightings case by case. He also told us about some cases in Argentina that appeared to fall on my extension of the BAVIC line.

Today Guérin called me at work with an exciting observation. He had noticed that the three great circles I had computed divided the equator according to a defined scale, with a basic unit of 12.4 degrees which turns out to be related to the Martian mean time. Could we have found a genuine law? The elegant geometric pattern formed by these three great circles does seem to be more than the product of coincidence. Guérin also asked me how the contact was going with the University of Texas. I told him I was waiting for a firm job offer.

Paris. 16 April 1962.

Taking the bull by the horns I have decided to punch the totality of our data into IBM cards. Such a file will be essential if a serious professional study is undertaken some day. Janine is helping me set up a card format and a code for the sightings. We keep accumulating new cases from every book we can find.

We are always afraid of being overheard when I call Guérin over the telephone from the office. Our colleagues would be intrigued if they listened to us as we say things like: "There has been a landing in Normandy and a vertical cigar over North Africa . . ." Therefore I have devised a code which also turns out to be a good classification system for the sightings. I designate the landings as Type I, the vertical cigars as Type II, and so on.

If we moved to Texas we wouldn't have to hide this work.

Paris. 21 April 1962.

Instead of going to the movies or to the corner café we spend every evening punching cards. After work we now make it a practice to go to the magnificent facilities of IBM France on *Place Vendôme*, where the key-punch units are freely available to IBM customers. There we are able to type, sort and print to our heart's content.

The De Gaulle government, trying to protect the French language, has made it a rule that all high-level computer codes must now be in French. Fortran programmers cannot write "Go To 103" any more: They must express it as "Aller a 103." This is fine, except that there are very few experts at IBM France who are of a high enough level to understand the intricacies of the compilers (the special programs which translate the languages themselves into the code which is executed by the machine), and they now spend all their time correcting the numerous new versions that keep arriving from the United States. They do their best to translate the instructions themselves, but they don't have time to translate all the error messages.

Last night at IBM we met a disoriented Frenchman who was staring at his output without understanding.

"Look," he told us, "this says: CORRECT SOURCE PROGRAM, yet the program wasn't executed!"

We had to explain to him that the English wording definitely did not mean that his program was "correct." On the contrary it was an urgent invitation *to correct it....*

When we come out at midnight we find Paris calm and clean, washed out by the Spring rains. We drive quietly from light to light, in Janine's blue Renault Dauphine. This highly civilized city will wake up again tomorrow in its well-ordered world. From the patient stars has come a sign we are trying to decipher. At the end of our own night, we do not know what kind of awakening we will find.

Paris. 17 June 1962.

Yesterday I finished computing all the alignments in Aimé Michel's book. I still find no pattern. So much for the idea that the great circles might generate a regularly spaced set of intervals at the earth's equator, as we had speculated with Guérin. But we have not given up the search for patterns.

After eliminating from our data the cases that are explainable by conventional causes we now have 2,437 entries on file. We are going ahead with the keypunching of 500 selected observations for which we are computing precise coordinates. In the meantime the U.S. Embassy has begun to process our visa application.

Marigny. 10 August 1962.

A few days' rest at the home of Janine's parents in Normandy, not far from Bayeux. Through the window I see the vegetable garden, the cabbage plants and the neighbors' rabbit cages. Beyond that, a few hedges, some bushes blurry with rain, and the open fields. To my left the quiet village climbs the hillside. As I sort out my notes, new ideas come to mind. Now I think of publishing everything, even if we don't have a perfect key that will unlock the mystery. At least we can vindicate Aimé Michel by proving that his alignments do exist, that he did not make up the data. But what do they mean? And could they be only the product of chance?

At last the letter we had been waiting for has arrived from Gérard de Vaucouleurs in Texas. He confirms that he is offering me a position as a research associate in Austin. Also a letter from Antonio Ribera in Barcelona. Janine is compiling new index cards based on the voluminous files of Charles Garreau, which Aimé Michel first used as the basis for his book. They give us one more way to calibrate and extend his work.

Fat with the rain, this little town sleeps in blissful ignorance of our scientific puzzlement. Marigny will be my last real memory of France.

I will leave without much regret. Since the Charonne episode of last February I have felt disgusted. In vain have I expected any indication that De Gaulle would take disciplinary action against the thugs in his police force.[15] Wouldn't the opposition, communists and socialists, commit similar crimes if they had the chance to get into power? The hypocrisy of French society has become obvious to me, under the veneer of the convoluted commentaries in *Le Monde* or the snobbish style of *L'Express*.

In the countryside near Marigny dogs howl, furiously foraging in the hedges. Behind their enclosures, cows complain noisily. A romantic moon spreads its light over the fields. We are already thinking of what it will be like to sail to New York on the Queen Mary.

Pontoise. 15 September 1962.

From the little plywood dome on top of the roof I spent an hour observing the planets, aware that this may be the last time I looked at the starry sky from my old house. I followed Jupiter and Saturn with my telescope. The night was clear, the moon very bright. For old time's sake I tracked a satellite for one minute on a nearly horizontal trajectory from the West to the Southwest.

Robert Kanters tells me that Denoël will publish *Dark Satellite* in November.

Paris. 18 September 1962.

Pierre Guérin and I have paid a visit to Paul Misraki in his beautiful Paris apartment. An excellent musician, well-known for many popular songs (most notably *"Tout va très bien, Madame la Marquise"*), he is also a deeply reflective man and something of a religious scholar. He has just published a book entitled *Les Extraterrestres*, in which he argues that some religious miracles, notably Fatima, could have the same cause as modern saucer sightings. And what about the vision of Ezekiel? He showed his manuscript to a Catholic bishop, who told him that he was presenting a perfectly valid interpretation of the facts. Although the Church had a different interpretation, this was not a matter of dogma, and he did not require the *Imprimatur.*

Pontoise. 13 October 1962.

All our books and personal effects are now locked inside three large green footlockers over which I have painted our destination in big white letters: *Austin, Texas.* Amidst my parents' old furniture, in the house on Saint-Jean street, how incongruous these words seem!

My mother has helped me dismantle my old observation dome.

Fiction has published a second short story of mine, *L'Oeil du Sgal.* Robert Kanters tells me that *Dark Satellite* will be in the bookstores next month. I won't be in Paris to see it.

Part Two

BLUE BOOK

6

Austin. 29 November 1962.

On the way to Texas we spent two days in New York City to rest from the rough sea journey on the *Queen Mary.* We stayed at the Taft Hotel near Times Square, feeling very much like tourists crushed by an alien city. We took the obligatory trip to the Empire State Building and Greenwich Village and tried to forget the uncomfortable entrance into New York harbor. We were unimpressed by the first human contact with America. Contemptuous bureaucrats kept all passengers waiting for hours, like mere cattle, at the Customs Office.

We flew from Newark to Texas on a four-propeller plane that stopped in Washington and Dallas before landing here. We have found a temporary place to rent. People smile at our halting, imperfect use of English. But the weather is magnificent, and I have become acquainted with the University campus.

Now the movers have delivered the boxes containing all our sighting files, and we are busy reorganizing them. Here I will be free to do my own research in my spare time, and to use the computer facilities whenever I want to, so we have made great plans to extend our catalogues and to punch more data into IBM cards. We don't have to hide anymore. Janine argues rightly that our coding system is too limited: it drops some important features of the sightings.

We are going to turn one room into a study, complete with file cabinets and space for specialized magazines. I have ordered several books in English to expand our library.

Austin. 10 December 1962.

For $800 in cash we have bought a real American automobile, a 1958 Buick Roadmaster. It is black, huge, a real gangster car, with a red leather interior. There are switches for everything: raising, pushing back and tilting the seats, putting up the radio antenna. The starter is located under the gas pedal, a fact which confused us at first. Janine has already found a job as a research associate in the Testing and Counselling center. After one look at her Sorbonne diplomas the University granted her the equivalence with a Master's degree in psychology. She is auditing graduate courses in statistics to familiarize herself with American terms. De Vaucouleurs is planning to send me to MacDonald observatory in a few weeks, and there is talk of my attending a conference on the exploration of Mars in Colorado next June.

MacDonald Observatory. 15 February 1963.

We just drove up from Austin to the snow-covered top of Mount Locke for a series of observing sessions. My main task is to take a series of galactic spectra, starting with NGC 1741. We are using the prime focus of the 82-inch telescope, a huge monster lodged inside an enormous dome. The night is windy, fairly clear. Our target galaxy was discovered by Stefan at Marseille observatory, Gérard tells me. He even recalls that Stefan was using a Foucault telescope with a wooden mount.

Back in France another short story of mine has appeared in *Fiction.* It is entitled *The Planets Downstream.*[1] I was surprised to see that the same issue contained a review of *Dark Satellite* by Gérard Klein, who calls it "one of the most interesting French science-fiction novels we have been able to read in recent time," and he goes on:

> The description of the Paris of the future he sets up in a few pages is a little masterpiece, as well as his precise but too fleeting images of foreign worlds, of alien machines. It seems that the language pulls the author forward, forcing him to venture to the edge of the comprehensible. His wild rush launches him into a maze where time and space and the multiple, misleading appearances of parallel worlds are mixed together.

After analyzing the various characters and situations in the novel, which he compares to *Sub-Space,* he concludes:

> One could attempt a finer analysis of the themes of *Sub-Space* and *Dark Satellite* to try and elucidate (the author). But I believe he will do it soon enough himself when he emerges from his own poetic torrent, from his own formalism. He will then place the

nearly unique explosion of his spontaneous style at the service of a surer, more acute awareness.

Boulder. 24 February 1963.

I am spending two weeks in Colorado at the National Bureau of Standards to discuss problems of two-dimensional data analysis with a French-born senior scientist and his staff, and to learn about computational techniques which we can apply to two of our current problems: the mapping of Mars and the reduction of background data in galactic photographs. Janine came along with me.

We drove through the Sangre de Cristo mountains which were covered with snow, and offered an awesome vision of extraordinary beauty.

Yesterday I witnessed an animated discussion about Vietnam at a party attended by many staff members. After a few drinks the French scientist cautioned his American friends about getting embroiled into a full-scale war: The Vietnamese will push you out, he said, just like they threw us into the sea.

They responded angrily, suspiciously, nearly accusing him of being in league with international communism. He tried to explain:

"Don't give me that garbage, I fought in the Mekong delta long before you did, I was a French officer," he told them angrily, turning red in the face. "You don't understand the situation. You can't even imagine the conditions there."

They wouldn't listen to him. America is so much more powerful, they said with quiet arrogance, "you can't compare what happened to the French with what would happen if we went in there." These people have blinders over their eyes, dangerous blinders.

MacDonald Observatory. 2 March 1963.

We are back at the observatory, in a little house that looks down into the deep valley. In the evening several deer come to our door, begging for bread and apples. On our way back from the Bureau of Standards we drove through those high points of saucer history, Amarillo and Lubbock, where a celebrated photograph of lights in formation was taken in the fifties. We stopped in the latter town to mail postcards to Aimé Michel and to Pierre Guérin.

MacDonald Observatory. 4 March 1963.

Last night we succeeded in taking two spectra, respectively of IC 4296 and NGC 4258. I worked with Gérard and Antoinette, again using the prime focus, but tonight we suffer from poor visibility. There were low clouds to the North of us, and the moon was bright in the sky. It was 9:30 p.m. when we were able to start on our main objective, NGC 4258. The temperature on the telescope bridge was barely above freezing. The sky became overcast again, so we had to stop half an hour after midnight, after less than three hours of exposure, a duration which is too short for the deep-space objects we are trying to capture with the spectrograph. From our position high inside the dome I had a fine view of the Texas landscape through the wide opening, all the way down the mountain.

MacDonald Observatory. 6 March 1963.

We are aiming the telescope at NGC 4258 again. It is a galaxy where Courtès has reported a hydrogen spiral we are trying to confirm by taking a series of spectra under different angles. We had to stop after only a one-hour exposure, but we returned to it later and remained on target until 5 a.m.

After I close the dome and park the telescope I go into the darkroom to develop the film, having rehearsed every gesture so many times I am able to execute it in complete darkness. The result of all this work is a tiny blur striated with spectral lines, half the size of my fingernail.

Austin. 30 March 1963.

I am expanding my library, reading UFO books in English by Keyhoe and by Waveney Girvan. In spite of my interest in the subject, I find this kind of literature incredibly boring. The style is atrocious, many of the ideas are childish. The data alone may have some value. As for Charles Fort, author of *The Book of the Damned*, he is witty and well documented but sometimes he tries to prove too much.

Austin. 9 April 1963.

Along with the staff of the Astronomy Department I recently attended a seminar on models of the universe. The approach to the problem is fairly simple: you begin by assuming that the universe is uniform and isotropic (with the same properties in all directions), and from this hypothesis you begin to write down the equations of its past, present and future behavior. Then you consider the body of observations that "prove" that the whole thing is expanding. By drawing the light cone which represents the path of the radiation that reaches the earth you can show that an infinite region of the universe is not observable. Next you "infer" that the entire thing is rotating, although you have just demonstrated mathematically that you were observing approximately zero percent of it. What kind of a joke is this?

The astronomer who gave the lecture was profoundly Christian and saw in cosmology the ultimate proof of the existence of God.

I came away very unsatisfied. Wouldn't the universe be large enough to have some parts that expand while others contract? Aren't we confusing local statistics and universal laws? And what about those fundamental, everyday phenomena that are still a puzzle to our current science, such as gravitation and consciousness? Shouldn't we try to find a reasonable way to account for them before we generalize? Until human thought has a theory of itself, what is the value of cosmology? The answer is that it is a useful and necessary exercise but it may not tell us very much about the real nature of the space around us.

The same criticism, of course, applies to the very idea of God. From some unexplained local phenomena that were reported at the dawn of history (the burning bush, the pillar of fire) people have erected the convenient model of an omnipotent creator. This is silly. The correct conclusion, in my opinion, would be to acknowledge that an unrecognized form of life and consciousness exists close to our earth. It is not necessarily cosmic in nature any more than our astronomical observations are necessarily telling us the shape of the entire universe. In fact, if there is a God, it is an information anomaly that would prevent the universe from being isotropic: It would violate the premises of the whole argument.

Austin. 11 April 1963.

A letter has arrived from Guérin, who is talking about spending his vacation in Alsace, at the point where three of our great circles intersect. Aimé Michel wants to send his own Martian statistics to astronomer Shklovski in Moscow, who is rumored to be interested in extraterrestrial life. I don't think that will do any good. Tired of the term "flying saucer," which sounds overly sensational and already implies a certain answer, I now refer to the problem as "the Arnold Phenomenon" after a celebrated witness, businessman Kenneth Arnold.

I am drafting a complete report on our work. I wonder what impact its publication would have. I also realize now that I need a doctorate if I want fellow scientists to listen to me, whether I talk about UFOs or about computer science, which I now find more interesting than astronomy.

Austin. 15 May 1963.

To what insights can higher consciousness take us? In my attempts to see more clearly where traditional teachings lead I find that the Rosicrucian tradition is still the most attractive source, the one to which I return again and again, perhaps because it is so humble and does not pretend to have all the answers.

What is that feeling, that attraction of an unseen presence that seems to be speaking to us across centuries of darkness? The elements of esoteric history I found in the book by Sedir given to me years ago by a friend at the Sorbonne continue to resonate for me. Behind all the jargon and the lofty words there is something to be discovered, a truth about ourselves that reveals its grandeur in the glow of candlelight, in the vague nuances of the night reflected by mirrors.

This is not the silly black magic that invokes devils or deities to gain riches and power. Nor is it the enchantment that comes with orgiastic dance and possession. It is a more subtle and demanding search; a personal, solitary investigation through one's own life, a confrontation with the very mind of the world.

As I study these texts it is also becoming clear to me that whatever else these old hermeticists were doing, they should be credited as the real founders of modern thought. In spite of its occasional obscurity this material represents a profound body of work. It may be the only significant and successful effort to preserve the intense spirit of earlier ages within the knowledge of modern science, even if I find it simplistic to invoke a "Supreme Guide of the Universe." Anyone who has looked at remote galaxies, as I did, through a deep-space telescope, should realize how far we are from understanding the higher levels of consciousness, how low we stand on the scale of cosmic things.

Boulder. 3 June 1963.

We are back in Boulder for a conference on the exploration of Mars, where two kinds of scientists who do not speak the same language are trying to argue with each other: the astronomers speak of the reflectivity of the green areas and the spectral features of the red areas, while the space engineers want to know if the green areas are edible and if the red areas will corrode the tracks of their vehicles.

I have just heard a talk by Dr. Franklin Roach at the National Bureau of Standards. He has recently served as a consultant to NASA in the scientific training of astronauts. Given the narrow field of view visible from the Mercury capsule, scientific opportunities are severely limited, he told us. It is even difficult for the pilots to recognize any constellations.

Roach also mentioned that Gordon Cooper was the first astronaut to see the Milky Way from space. He observed the zodiacal light, but he was unable to see the solar corona because of the scattering caused by the thick windows.

He also mentioned several phenomena reported by the astronauts, which did not immediately have explanations. Thus John Glenn saw some strange particles outside the cabin. It turned out they were attached to the capsule: He could make them go away by hitting the window with his fist. An observation made by Gordon Cooper has not been explained yet. It turns out that during the day the astronauts see the earth surrounded with a blue halo. At sunset, for a brief instant Cooper saw an intense orange luminosity which is not fully understood. The appearance of the sky, too, poses some problems. Cooper saw something like "haze" underneath his capsule.

Psychologically speaking, the worst fears of the experts about the negative effects of space have not been verified. Cooper kept his good humor throughout his flight, taking frequent short naps. The dreaded "split-off" effect (it was speculated that the astronauts might not want to come back to earth) did not take place. What did happen is a harmless illusion called "capsulo-centricity": the astronaut has the impression that he can make the whole universe move around his cabin and that he can bring the earth into view just by pushing a button, while he remains perfectly motionless in space.

Austin. 10 June 1963.

Central Texas is a paradise. There are woods and hills here, big oak trees, huge butterflies and endless miles of shoreline along the artificial lake where the Astronomy Department holds its Sunday picnics. People are relaxed and friendly, they seem to have all the time in the world. A small white wooden house on the edge of campus shelters our love. Squirrels run around the shaded roof, playing hide and seek with us. There is white dust on the dry roads. Behind us a wooden bridge goes over a creek.

Janine tells me she is pregnant. She is radiant, more beautiful than ever, calm and sweet. We miss nothing of France. If I had remained in Paris I might still be spending my nights tracking satellites, and I would have to watch every word I said.

The science work, however, is not as exciting as I had hoped, although I greatly admire Gérard de Vaucouleurs. He left France in 1946, incensed at the rigid attitudes of the astronomers of the time. He wanted to study planets and galaxies. The former were too close, and the latter too far away for traditional French astronomy, which concerned itself primarily with the fixed stars. When he was told that nothing new could be found about galaxies, since "Hubble had done it!," he became infuriated at such an absurd statement: Hubble, a great American pioneer of galactic research, was the first to acknowledge he had barely scratched the surface of the immense new domain he had discovered. Gérard went away from France, first to England and Australia (where he specialized in the little-known galaxies of the Southern sky) and later to Texas. He works very hard with his wife Antoinette, but there is no future for me here without a Ph.D., although I am busy with several large and interesting projects, including the preparation of their catalogue of galaxies for publication.

I was hoping to work on planetary problems, but this expectation has not materialized, except for a few efforts related to the Mars mapping project we have undertaken for NASA. Washington provides little support. It is as if planetary scientists were afraid of seriously tackling the real questions before them.

With the approaching Summer comes a stifling, muggy heat that slows down the thinking process.

Austin. 14 July 1963.

We went over to de Vaucouleurs' house for lunch today. We spent a pleasant afternoon with him and his wife. He brought the conversation to the subject of flying saucers. We talked about the questionable quality of the photographic data and about the Air Force's equivocal position. Clearly the problem is not as frozen here as it is in Europe. I remain confused, however, about de Vaucouleurs' own outlook. He is open and encouraging to me in his comments, but he always approaches the subject from a skeptical viewpoint, and he has not drawn out my deeper thoughts. I am reluctant to remain in Austin, in spite of his support and the friendly hospitality Antoinette and he have extended to us. I am anxious to move on with my research and perhaps to pursue a Ph.D. elsewhere. I have decided to contact J. Allen Hynek, in search of a more definite approach to the observational material.

The folks from *Fiction* have written to me: They will be printing one of my stories, *The Artificial Satellite*, in their special Anthology of French science fiction for 1963.

Austin. 4 August 1963.

We are awaiting Hynek's answer to my suggestion of a meeting in Evanston. In the meantime I am printing out our entire catalogue. I miss the good long talks I had in Paris with Aimé Michel, that extraordinary mind, and with Pierre Guérin.

Austin. 27 August 1963.

We went to Mexico City for a week of vacation, and upon our return we did find an encouraging letter from Allen Hynek. I am answering him with the suggestion that we travel to Evanston to see him on September 7th.

This week away from Austin has given me a new perspective. We should not stay in Texas, beautiful as it is. Professional astronomy is a field with only a small number of scientists who hate one another and fight over tiny budgets, ignoring the giant forces that are reshaping the world around them. I admire de Vaucouleurs' insight and dedication, and I do not have any quarrel with the older generation of astronomers who did so much pioneering work, but I am discouraged at the attitudes of many among the younger ones. They show little interest in other scientific disciplines, and they have no exposure to the wider industrial reality around them. They underestimate the real potential of computers, a fast-changing technology to which I am increasingly attracted.

I think it is criminal not to put more emphasis on planetary studies. In this respect I find it ironic to be working with de Vaucouleurs on the Map of Mars project for NASA. Our map will be ten times more accurate than any previous one, suitable for the reduction of the photographs that will soon be taken by the Mariner probes.[2] But in the process we are part of the last chapter in the history of classical Martian astronomy. From now on, all the important studies of Mars will be based on space probes. The great tradition of earth-based Mars observations which comes to us from Galileo and Kepler through Schiaparelli, Antoniadi, Maggini and de Vaucouleurs himself, ends in our machine and in this printer spitting out the coordinates of hundreds of fuzzy spots at 400 lines per minute.

Austin. 18 September 1963.

Last Friday we flew away from Austin. We landed in Chicago at night, after a stop in Dallas. Janine was especially happy to make this trip: since her childhood she has been fascinated by the very name "Chicago." She has long had the intuition she would visit the city some day. Chicago and Samarkand.

On Saturday Hynek called us at our hotel and we had quite a conversation for a first meeting. It lasted all day and into the whole of Sunday: we had so much to talk about! He is a warm and yet a deeply scholarly man, with much energy and a great sense of humor, an open mind, and a deep culture that comes from the sophistication of his Czech ancestry.[3] It is hard not to be impressed by his sharp ideas and his eagerness for action. He has a lively face where piercing eyes are softened by a little goatee that makes it hard to take him completely seriously. He took us on a tour of Dearborn observatory, a charming old building among the trees of the Northwestern campus, at the very edge of Lake Michigan.

I had planned to take time out to look for a job in Chicago, but Janine and I were both tired on Monday morning and found no energy to go downtown. However, a new possibility has come up: Hynek has found out there is an opening for a systems programmer at the Technological Institute, right at Northwestern University. I will look into it seriously.

After reviewing all our results in detail for two days Hynek said it was imperative to publish them as soon as possible. This implies a series of decisions. First I will need advice to find a publisher. Perhaps my French literary agent will be able to give me the name of an American house he knows. It will be easier to pursue these contacts from Chicago rather than from Austin, which is outside the mainstream. Second, I realize that this book will be an unprecedented statement by an astronomer arguing for the reality of the saucer phenomenon. If I remain in Austin professional pressures are inevitable, since the department works closely with Harvard Observatory which is directed by arch-skeptic Donald Menzel, who has just published his second debunking book, *The World of Flying Saucers*. I do not want to put my friends there into such an awkward position.

Above all, I want to work closely with Hynek. There was an immediate bond between us. I am struck by the balance he has achieved between theory and practice; he understands solitary research but he is also a very good team builder. He has a genuine understanding of science but his sense of culture saves him from the pitfalls of specialization. His house in Evanston is full of books on all kinds of subjects. There are classical records everywhere, and current issues of the cartoon-filled *New Yorker* on the coffee table. He is an ethical man, with a realistic view of the politics of science and the role of the military. After our very first meeting he tried to call Colonel Friend to suggest that he join us in Evanston right away. I like this impulse, it comes from a man of action who can make quick judgments about people. Warmly, his wife Mimi offered to help Janine when the baby comes.

Austin. 21 September 1963.

We are packing again, ready for Chicago where I did get the systems programming job. I will leave astronomy behind without much sorrow, to start a new career in computer science. A Pakistani friend is just returning from a tour of East Coast observatories, including Washington and Harvard. He shares my impression that the field is becoming dull: "All the people I saw," he told me, "seemed engaged in pedestrian work, without much passion, without fervor, without the driving ideas that could lift their lives above the routine drudgery of academia."

We will take the big black Buick, driving through Oklahoma, across the Mississippi and the plains of Illinois. Janine is six months pregnant, as beautiful as ever, as enthusiastic as I am about moving to another unknown city, another life.

7

Chicago. 19 October 1963.

Barely one month after our first meeting with Hynek, we are settled in this furnished three-bedroom flat we are renting on the North side, halfway between Evanston and the Loop, as people call the downtown area. We have a large living room with wood everywhere, a bay window, and bookcases with glass doors on either side of the fireplace. Next to the living room, my little office looks over Bryn Mawr Avenue. At the other end of the apartment are the large kitchen, the porch and the back stairs. There are people living on the floor above us, about whom we know nothing. The owners, a cheerful Greek couple and their children, live downstairs.

On the Northwestern University campus in Evanston I audit Hynek's astronomy class when I am not developing computer programs for the Biomedical Department. My work at the Technological Institute—a large modern building which faces Lake Michigan—gives me access to the main campus computer and a more comfortable salary than the measly $600 a month I was earning in Texas. Yet I have not recaptured my spiritual balance. I count on this weekend to renew contact with my friends in Europe. An exciting letter arrived this morning from Aimé Michel: he is giving up free-lance journalism for a job with the French radio research center, where he will work under musical genius Pierre Schaeffer.

Chicago. 2 November 1963.

Hynek and his wife came over for dinner Monday night. They were surprised, they said, at our fast transition from Texas. Things are also moving rapidly on the research front. During his class on Thursday Hynek mentioned his relationship with the Air Force: just the previous day he had visited the Project Blue Book office in Dayton. He was going to need some research assistants, he said, to analyze the accumulated files and to test the investigators' conclusions. Having said this, he stroked his goatee thoughtfully and added that he had already discussed the idea with the Air Force, and that it had been approved.

I asked if citizenship or a security clearance would be required, since I am still a French citizen. Hynek answered that not only were the files unclassified, they might well end up in the trash if some general suddenly decided that the whole phenomenon was unworthy of study. He welcomed my involvement. Another member of the class, Nancy Van Etten, asked a pragmatic question: she wanted to know how much we would be paid. Not much, answered Hynek; and he gave her a copy of *Special Report #14* to read.[4]

Chicago. 9 November 1963.

My major concern now is to create a small elite team around which a real scientific effort could get organized. This brings my thoughts back to Michel and to Guérin. Aimé is a remarkable and dangerous man. His imagination, coupled with a sharp sense of humor and a powerful brain often propel him too fast. Guérin is very different. He is an island of intense thinking, but he has to preserve an image of conformity inside the old astronomical citadel where he works. He cannot afford to let his colleagues guess what lofty thoughts are burning through his mind when he passes them in the hall. Actually, very few of them put a high priority on real science. They worry much more about their pensions, who will get elected to the Labor Union Committee, whose budget will be slashed next year.

Chicago. 15 November 1963.

The French manuscript of my monograph about the "Arnold Phenomenon" is on its way to a Paris literary agent who has promised to look actively for a publisher. Hynek has suggested that we hold the first meeting of our new UFO Committee at his house on Sunday. He told me frankly that he counted on me to be the driving force of this group, because he had practically no freedom of action with the Air Force. That is a crafty way to use me without his taking any personal risk, but I am willing to play the game. I have resolved to come to every meeting with a written list of specific points to be discussed. My objective is to pull the problem out of the quagmire where it is stuck. I want to try and convince the Air Force that this is a serious scientific question, that it isn't limited to the United States, and that it isn't necessarily just the business of the military to investigate it. Current American attitudes about the issue are based on shaky statistics whose only strength is the appearance of authority of the electronic machine that produced them.

As I become familiar with the files, I am struck by their backwardness, compared to the level of sophistication we had reached in Europe. There is not even a general index of Air Force data, and no compilation of cases comparable to Guy Quincy's list. Even Hynek still talks as if landing reports did not exist at all.

Chicago. 17 November 1963.

We have just had our first meeting of the UFO Committee at Hynek's house in Evanston. Janine was there as well as Nancy Van Etten. We spent a great deal of time on organizational issues and on the definition of the problem itself. I still would like us to call it the "Arnold Phenomenon," or any such term that would break with the sensational and somewhat pejorative term "flying saucer," or the overly used and abused acronym "UFO." But the Air Force terminology is hard to change, and we are stuck with those awkward words "Unidentified Flying Object."

Mimi Hynek, who thinks the whole subject is utter nonsense, argued emotionally against me. She did not believe there would ever be a time when the topic would be seriously studied at a major university under the term UFO, Arnold Phenomenon or any other name. Hynek watched us fight, cleaned his pipe and refilled it in silence. Wisely, he avoided getting into the dispute.

My next objective is to develop a formal classification of sightings. Hynek finds the idea interesting, so I have started to read and to analyze the Air Force files, working from his own carbons of the Dayton documents. I extract the significant data, and I put them into our own file after translation into French. This will enable us to compute new, independent statistics.

Chicago. 19 November 1963.

It is practically on a full-time basis that I now work on the saucer problem. Janine, who is resting at home since her delivery is approaching, does a huge amount of work calculating the longitudes and the latitudes of the best sightings. This often leads her to a critical re-appraisal of certain cases and to correcting old errors in our documents. Now we need to test the very existence of Michel's *orthoteny*. The question is no longer to verify that his alignments exist, since I have already validated them. But I need to test whether or not they might simply appear as a result of chance. I have convinced myself that the appropriate methodology for this is simulation, sprinkling fictitious sighting points over a geometrical shape representing France, and systematically looking for all possible alignments. Without a computer, of course, this would be out of the question.

While my programs are running I spend long hours at the library at Dearborn Observatory, catching up with the recent literature on the physics of the solar system or making notes for our UFO study. I often hit theoretical snags, as I did in the correlation between the frequency of sightings and the distance of Mars, and I need to discuss such issues with Hynek. Fortunately his office has two doors, one of which opens up into the library itself. Thus I see him practically every day, even when we do not have a formal meeting scheduled.

Chicago. 23 November 1963.

The assassination of President Kennedy yesterday is one of those events history uses in order to show us how little we have progressed away from barbarism. It deprives us of a sincere man who gave the world a remarkable lesson in genuine democracy. Beyond this it puts a tragic halt to the acceptance of new social concepts, from civil rights to the conquest of space. Certainly the forward march of the Western world will have to be resumed sooner or later—science itself makes this unavoidable—but who will expand the framework of our political system, as Kennedy had started to do?

I discover with surprise to what extent I had judged and admired my new country through Kennedy's actions. This brutal death reminds me of the existence of a volcano of violent realities underneath the orderly unfolding of our best plans.

The real Beyond is not that which follows death but that which stretches underneath life itself. In this sense the saucers are a potential source of cultural and strategic upheaval, just like yesterday's killing in Dallas. Just as in Dallas, we are dreadfully unprepared.

Chicago. 28 November 1963.

The computer simulations have now produced some remarkable and unexpected results. Every evening I have been bringing home the statistics generated by my latest simulation program, to study them with Janine. There is no longer any doubt about the distressing fact they demonstrate: Aimé Michel's networks *can indeed appear as a result of chance, with the same precision as the actual sightings.* This throws into question everything I have said and written before about orthoteny. The problem with the earlier tests was simply that they did not take into account the peculiar topology, the geometric shape of the French territory.

Accordingly I may have just destroyed much of what I was going to publish. I will know it for sure when I recompute the totality of the patterns Aimé Michel had constructed by hand with his maps of France. I now expect to see the whole theory collapse. Of course a few exceptional alignments will remain . . . and also the saucers themselves! The key to the mystery escapes beyond our reach once more. However I refuse to be discouraged. Perhaps it is even a good thing for us to go back to the proverbial drawing board. I am anxious to finish these calculations and to take a trip to France in a few months to discuss the results with Aimé Michel himself.

Mimi Hynek, whose son Ross was born just a few months ago, has been very generous with us, giving Janine some practical lessons in the handling of babies. They have four other children: Paul who is about two years old, Joel who is a teen-ager, and two older kids who have already left home: Scott and Roxanne.

Chicago. 3 December 1963.

Tonight Janine sleeps at Grant Hospital, our son Olivier next to her. He was born at 6:48 this morning. They are both doing very well. While waiting for them to come home, I continue to read the UFO cases from the Air Force, some of which are as fascinating as the most sensational cases in the published literature. Captain Hector Quintanilla, the man who currently heads up Project Blue Book, has promised to visit us next month.

Chicago. 4 December 1963.

This afternoon I saw Janine and my son at the hospital again. She is happily resting. We are both recovering from the tough period before her delivery: the move from Texas, the uncertainty, the trauma and the psychological upheaval that accompanied Kennedy's assassination, our intense work on the UFO files, and her increasingly disabling pregnancy. Our son is an angel: I am going to be crazy about this little fellow. Janine has a very pleasant, cheerful room: she is impressed with the way the hospital runs.

Our doctor, who was recommended by our landlady, is a Greek. So are most of the staff and the patients at the hospital. Since Janine's black hair gives her a Mediterranean look the nurses frequently burst into her room happily giggling in Greek, and they wonder what is wrong with her when she doesn't respond.

The first snow has fallen on Chicago. There is a festive mood about everything. The world is fresh and clean. In Evanston the snow-laden branches around the quaint old building of Dearborn Observatory give a picture of perfect peace.

In the used bookstores of Clark Street I have found five more volumes about flying saucers. I am now thinking of restructuring my manuscript into a more complete overview of the problem.

Chicago. 22 December 1963.

Janine is home and we are getting used to life with a baby. I admit I am fairly helpless when Olivier cries in the middle of the night. As soon as I brought him back to the house with Janine, our landlady put honey on his lips, according to Greek tradition, "so he will be sweet all his life."

Hynek flew to Dayton again this week to visit the Air Technical Intelligence Center (ATIC) that has responsibility for Project Blue Book.[5] He reported on the results of his visit during our sixth meeting, which was attended by Stanley Roy, a new astronomy student. The staff of ATIC, he says, regards us very pragmatically as a group of nice young enthusiastic scientists who might eventually provide them with a convenient alibi if Congress, or the Air Force brass, ever ask pointed questions about the handling of the problem. By using us to verify their statistics they will be able to appear open-minded and genuinely concerned with proper scientific methodology. At the same time, however, they did not count on the amount of real work we have already accomplished. Captain Quintanilla and another officer will be here on January 16, and they have invited us to visit Wright Field. They told Hynek they would be delighted if the phenomenon turned out to be a massive, genuine scientific mystery: the Air Force would rush before Congress to get a multi-million-dollar boost in its budget!

During our meeting I stressed the need to inject new methodology into the field. Sooner or later I believe the Air Force will be forced to make such a change. Janine and I now find ourselves in a unique position: no one else has had access to the private documents in Europe and to the French and the U.S. military files as well. There are still many holes in our catalogue, of course, but the trip to France we are planning to take in March will give us a chance to complete the study of Aimé Michel's files, which will fill most of the gaps.

Chicago. 23 December 1963.

In recent discussions with Hynek, I pointed out that the saucer question may well be part of a complex series of scientific realities, but it also plunges deep into mystical and psychic theories. I found him very receptive to this idea. We must also ask ourselves if an extraterrestrial intervention might have been a factor in man's early history, specifically in the early development of civilization and of biblical events. As Paul Misraki has shown in his book, the immense machinery of the Angels and the divine messages delivered by Jehovah amidst lightning and thunder could be interpreted as a celestial manifestation rather than a divine one. The return of such phenomena today could be explained by the need for some "unknown superiors" to boost our religious vaccinations. Some super-scientific group of cosmic origin, considering mankind as its own creation and seeking to experiment on us, or to guide us benignly towards galactic status, might behave as the saucer operators do.

Another question then arises: Has the future spiritual state of man already been achieved by some individuals? Have certain gifted men already achieved contact, on some plane, with those who may be guiding our psychic evolution?

Chicago. 1 January 1964.

The latest book by Serge Hutin, *Journeys to Elsewhere*,[6] does not fulfill any high expectations of occult revelations. In fact I am fairly disgusted by all the esoteric apostles he quotes, along with mystics of the left-hand, right-hand or middle path. I am growing out of my earlier fascination with the hermetic literature, where the truly important ideas are obscured by occult mumbo-jumbo. Every fragment of every bizarre document left by earlier ages is eagerly seized upon and commented on, with assorted hints of deep and mysterious meanings that are never brought out to be critically examined in the harsh light of day. Hutin's book is replete with such false wisdom, throwing into the same stew the science fiction of Maurice Limat and the esoteric writing of Zozimus!

What about alchemy? Its Adepts are said to have reached a total, definitive vision of the world. Let us set aside the minor questions of information theory this would imply. Hutin gives the names of medieval occultists and philosophers who indeed left their mark on the culture of their time, but they certainly did not reveal to us the final truth about the cosmos: why didn't they write, if they were so smart, about the structure of galaxies, or the atmospheric pressure at the surface of Venus?

Such mastery one attains exclusively through perfect asceticism and total abstinence, according to Serge Hutin, but wait! It can also be reached by sex, through tantric union, possibly combined with orgiastic ecstasy. Or simply by leading a healthy, sane life. All these recipes are contradictory, yet their advocates present each one as the only choice, out of which there can be no salvation.

There are hundreds of manuscripts and grimoires that are said to be filled with the details of the ultimate secret. They describe the Philosopher's Stone, the preparations for eternal life, the potions that enable man to lead multiple simultaneous lives, and to see God in all His glory. But all these published revelations contain deeply hidden secrets: In other words, you will need a Master to reveal what they mean. But no two Masters ever agree on anything.... Now consider the Adepts, who could change the course of history merely by raising their little finger, but who refrain from such intervention. They could live blissfully in eternal cosmic beatitude, yet we are told that they choose to be among men to appease their suffering. If they are so powerful, why don't they use their powers to break the misery of everyday life, hunger, illness, war and slavery?

My impression is that what we have here is simply an interesting assemblage of ideas, images, visions and dreams gathered from disparate sources. Their authors extrapolate from a few interesting facts and interpret them hastily into eternal truths. To me there is indeed a fundamental reality of hermetic science, but it is not based on such visions. Everyone must find his or her own expression of it. *The spiritual path I have chosen is that of intelligence tempered by the fire of love, but always applied to accessible, solid, consistent, calibrated facts.*

Another concern for me is the degeneration of the esoteric domain into a variety of branches and sub-fields that stand in glaring contradiction to each other: Thus Alchemy is said to be a branch of Tantra, which is widely sub-divided. There are some twenty kinds of magic. The Kabala contains many domains, while astrology is split between several disciplines and methods that have nothing in common. All the fundamental principles are said to be absolute, inviolate, pure and true of all eternity, but they are also perfectly incompatible with each other and impossible to test against everyday reality. Thus Hutin tells us that everything comes from the Egyptians, who got it from the Tiahuanaco man, who was an Atlantean. Good. We finally have a solid base. But wait, if you go to the North Pole you will find a race of telepathic blond giants who also have all knowledge, while the truly supreme magicians reside in Tibet, where they hide inside the mountains, as everybody knows. The Ultimate Truth, in the meantime, is buried in the Amazon, while the Extraterrestrials mingle among humans on Main Street, going in and out of their base inside Mount Shasta. Is this clear? In the end such books read like the brochures of a travel agency, not genuine mystical texts.

I am willing to accept the hypothesis that there is a higher plane of perception, because that is confirmed by my own everyday experience. The "great initiates" are exceptional men who are driven by excellent intuition, but they do not have total understanding of the world and are subject to the same weaknesses as other people. A true Adept, asked about his knowledge, should only say that he is a "mere student." Those who brag about their wisdom are crooks and phoneys. As far as Unknown Superiors, Superior Ancestors and other orders of Superiors in the cosmic barracks, their existence is not proven by any real facts. It does seem that the history of Antiquity contains instances of contact between men and other races or alien cultures. But the knowledge about this contact may have been misinterpreted by priests and self-styled masters down through the ages.

Hermeticism contains many beautiful truths, so I would go too far if I said it is worthless and irrelevant to the modern world. What bothers me is that esoteric knowledge gets lost in its contradictions and in the contempt it implies for the concrete observations of modern science, the very source from which it should draw support.

Chicago. 2 January 1964.

When we go back to France I am afraid we may find nothing but an old people out of breath, with a wily old man at its helm, an elderly general who lives only for History. For me the recent history of France remains summarized in the Charonne massacre: what I saw there after the police riot will always be in my memory. The worst hours of Stalin, the worst crimes of the Nazis had a name: Tyranny. The killings at Charonne have no name.

Charonne is worse than the deepest horrors in the tales of Lovecraft. Mad policemen threw dozens of men, women and innocent children down the stairs against the steel gates of the locked subway station, and they beat them up with their weapons. Eight people died. The culprits were never punished. We cannot escape that simple fact. Such is the France that claims to hold up to the world the torch of freedom and civilization: What right does it have to speak of such things?

Chicago. 5 January 1964.

The seventh meeting of our UFO Committee took place this afternoon at Hynek's house. I am beginning to understand the expectations and also some of the frustrations of the Air Force. I also think it may be possible to get them to create a new scientific commission to review the cases. It is clearly towards the creation of such a commission that I should work. The first step is to convince Hynek of its utility.

The absence of research seems attributable to lack of interest among scientists as much as to censorship by the military. We cannot seriously accuse the Air Force of neglecting scientific research, since that isn't their job. It is the scientific community which is guilty of neglecting its duty to the public, by refusing to consider the observations for what they are, namely sincere, genuine reports of unexplained facts.

Chicago. 6 January 1964.

My manuscript is finished but I despair of having it published by such academic houses as Dover in the U.S. or Dunod in France. Those people will not take a chance on a controversial subject, even with such a technical approach to the problem: What we are doing is *Forbidden Science.*

Chicago. 12 January 1964.

At the eighth meeting of our UFO Committee we re-investigated the Mitchell case of 1963, when two pilots saw an object over Nebraska. We called up the witnesses on the phone to get first-hand data. Their sighting is similar to that of the BOAC "Centaurus" that flew for nearly an hour in view of a cluster of dark objects. Genuine team spirit is building up among us. Hynek himself is now drawn into it. The major new fact is the realization that we are in a unique situation, as the only civilian scientists who have had access to the basic documents.

Chicago. 16 January 1964.

We have just had an excellent meeting with the officers in charge of Blue Book, Captain Hector Quintanilla and Sergeant Moody. I come away from it with a clearer sense of how they perceive their role. Neither one of them has any serious training in science, and they make it plain that knowledge is not their business:

"The mission of the Air Force is *to identify, intercept and destroy* any unauthorized object that violates U.S. air space," Quintanilla told me. "In other words, if an unknown object is detected by ground radar or by a pilot we ask it to identify itself. If it doesn't we chase it. And if it tries to escape interception we shoot it down. It's as simple as that."

"What about the global nature of this phenomenon?" I asked. I brought up the French cases, the landings, and the humanoid sightings. They were not interested.

"It's none of our business if a Martian shakes hands with a baker in Brittany. Our responsibility is limited to reports from U.S. citizens. What we are looking for? Enemy prototypes, spy craft, anything unusual that we can understand in terms of technology."

He gave me two examples from Blue Book files:

"About 1951 several technical observers and ground radar personnel saw a strange object that flew very high and did not match any published pattern. It was reported as a UFO. The sighting was immediately classified, because it was realized that the observers had only seen a U-2 on its way to the Soviet Union. Another time, a simple fisherman saw a strange object in the ocean close to shore. The report was checked as a matter of routine: the thing in question turned out to be a Russian submarine!"

In their minds every report must have a classic explanation. Accordingly they see no reason to deny us free access to all the files: we might even find interesting data on some rare forms of globular lightning. But not one American scientist has ever suggested there might be anything there, other than purely natural phenomena.

"Besides," pointed out the Captain, "we brought together five top scientists in 1953 to review the whole problem. They concluded that all the cases had conventional explanations."

"Wait a minute!" said Hynek, "I attended that meeting of the so-called Robertson Panel.[7] These men had been taken away from their regular work for a few hours only, they examined a handful of cases selected by the military itself. What could they do except to re-affirm their beliefs?"

I jumped in: "It would have been a lot more meaningful for the Air Force to bring together scientists who had witnessed the phenomenon themselves, such as Clyde Tombaugh or Seymour Hess."

"We always call experts," answered Quintanilla. "For example, we send all the mirage cases to Menzel, all the meteors to Dr. Olivier. . . ."

I felt disgusted with that comment: "With that kind of approach, how do you expect to ever learn anything new? You will always find that the phenomenon is composed of meteors and mirages! What about the unexplained residue, which you claim is only a few percent? Who will be your expert for the residue?"

"Perhaps the residue can be explained in classical terms, too."

"What about the Loch Raven case?"

They agreed they couldn't explain Loch Raven, which they still carry in their own files as unidentified. Or the Levelland case.[8] Yet nobody is following up on such observations, which are unexplained although they are very detailed and come from multiple, reliable observers. Since they do not fall into any of the explanation categories, these cases are closed. In the triumphant words of Sergeant Moody: "We have identified them as Unknowns!"

Chicago. 17 January 1964.

For the second day I met Captain Quintanilla, Sergeant Moody, Hynek and Nancy at Dearborn Observatory. We went to a restaurant in Skokie for lunch. The conversation revolved around the role of scientists in this problem. Hynek remains very prudent. He is clearly afraid of antagonizing the Air Force and of losing his contract, hence access to the files, if he makes any waves. I have no such reservation, so I speak up against the Blue Book approach.

"You underestimate the level of interest which exists among scientists in private," I tell the two officers. "Even if such scientists deny it in public."

"Carl Sagan himself is more interested than he would admit to his colleagues," adds Hynek. "At a recent astronomy meeting he walked up to me and told me privately that he had learned of my association with the Air Force."

Grumbling among our guests. But there is more.

"Why is Blue Book rejecting all the landing reports?" I go on. "Why ignore Aimé Michel's well-documented accounts of humanoids simply because they seem fantastic? That is not a scientific criterion. Comets seemed utterly fantastic in the Middle Ages. Artists have left us engravings that show comets as the hand of God holding a bloody sword in the sky. Yet competent scientists took the trouble to study them. If they had rejected them just because the report sounded weird, where would astronomy be today? A scientist is supposed to be able to go beyond the report to the phenomenon itself."

Hynek is impressed by this argument, but I can see I am not getting through to Quintanilla, who has made it clear he wasn't concerned with science, or to Moody, who keeps looking at his watch.

Chicago. 20 January 1964.

A New York house is reading the manuscript of my book, which I will call *Challenge to Science*, and my Paris agent has given me the address of one Mr. Henry Regnery, a publisher in Chicago, advising a visit. He met him many years ago, he says, and he was impressed with his business sense. Aimé Michel writes that he has mentioned my work to Frank Salisbury, a plant pathologist and Mars expert from Colorado.

Chicago. 25 January 1964.

Waveney Girvan, who edits the *Flying Saucer Review* in London, sends me a text by Menzel that responds critically to one of my articles,[9] although it is written on a strict scientific level, and Menzel's tone remains civil throughout. He is the director of Harvard College Observatory and his only interest in flying saucers is to deny their existence. I appreciate the fact that he does not "pull rank," he does not use intimidation and scorn against me, as a senior French scientist, a Danjon or a Schatzmann, would undoubtedly have done. After all, I do not even have a doctorate yet. I am small potatoes next to Menzel.

Chicago. 26 January 1964.

Now I am reading the book by Gray Barker, *They Knew Too Much about Flying Saucers.* It describes the alleged intimidation of American investigators by mysterious "men in black," or MIBs. If such terrible beings existed, wouldn't they have neutralized Hynek a long time ago?

We only have vague theories about the nature and origin of the saucers. One could speculate that they may be coming from a temporal rather than a spatial source. Some of the objects change shape. Others disappear on the spot. The saucers observed on the ground do not seem adapted to long-term interstellar flight as we understand it. These are the facts that the believers like Keyhoe are sweeping under the rug because they contradict their preconceived ideas about UFOs. I begin to appreciate the predicament in which the Air Force finds itself. Furthermore, the objects are often invisible to radar.

Chicago. 31 January 1964.

Eleventh meeting of the UFO Committee. Stanley Roy did not attend. I showed Hynek the FSR article by Menzel, together with my answer. This is a touchy situation, since Menzel was Hynek's mentor at Harvard during the artificial satellite days. We have begun the systematic review of last year's cases.

Chicago. 1 February 1964.

Doubts. I am tired and upset, not seeing clearly where I am headed. I need to talk to someone about the data and what it means, but Hynek always remains on the fence, cautiously.

"All this is very interesting, but there isn't any evidence here that I could take before the National Academy of Sciences," he keeps telling me.

"What about these patterns? The time distribution, the behavior classes?"

"That's not as good as physical evidence. A genuine photograph, a piece of hardware: That's what we need. We have to wait for a really good case to show up."

For many years he has been watching, waiting for the one incontrovertible case that would make a sudden dramatic difference—the single big event that no one would be able to deny because the evidence would be overwhelming. That is the way the French Academy of Sciences was forced to admit that meteorites existed, when so many fell at once on a single little town that it became impossible to deny any longer that "stones fell from the sky." But most scientific discoveries do not happen that way.

I try to convince him that we may not recognize that "really good case" if and when it comes, unless we begin in-depth research, right now. My arguments move him but he still does not want to challenge authority. It is not so much that he needs the money he gets from the Air Force. Rather, he is afraid of losing access to the data, of being cut off. Secretly I also suspect him of enjoying the mystery of it all, the opportunity to fly periodically to Wright Field as a scientific consultant, while his astronomical colleagues are stuck in boring, routine jobs. Blue Book puts adventure and intrigue in his life.

Vaguely, I do think that Menzel is wrong, that there are flying saucers out there. The earth is only a dark cave far from the exciting places of the universe. We live in ignorance of the Champs Elysées and the Broadways that are glittering all around us in the cosmos.

All great changes have come through incredible facts, through the fantastic: the lives of Tesla, Newton, Kepler, Paracelsus are examples of it. I believe in the higher dimensions of the mind.

The wind blows over Chicago. The sleepy town listens to the rumble of an engine that drones on somewhere to the North. The deserted streets are wet and mindless. The world dies, the world lives. The city is sad and heart-wrenching.

Chicago. 28 February 1964.

Meredith Press is sending back *Challenge to Science* with a rejection notice: "We appreciate the chance to read your manuscript but we do not see how to sell it profitably on the American market." Back to square one. Should I publish in England? To issue the book in France only would mean to doom my work to oblivion: Aimé Michel's friends would surely applaud, the rationalists would reject it summarily, and nothing more would ever happen.

It is sad to realize that most Americans ignore the richness of life in Europe: the Spanish spirit, the Italian soul, the splendor of medieval mysticism in every French country church, the agony and the creative force of the intellectuals confronted with a changing world from Berlin to Dublin and from Stockholm to Prague. But what is Europe itself doing with all that wealth?

Some eras of history seem completely colorless to me: the eighteenth and nineteenth century in particular, except for the American revolution, leave me cold. The Sun King was boring, his Versailles a display of empty vanity only relieved by Molière's gentle wit. The Philosophers created a form of rationalism that put humanity into a bureaucratic cage for two centuries in the name of "progress." Where is the great enlightenment they promised?

The last twenty years, on the contrary, have seen a remarkable shift of all our cultural systems under this rigid framework of historical rules. There is a powerful, dangerous, exhilarating feeling in the air. Today culture must meet reality, and vision must perish unless it leads to practical, effective actions. All of this was beautifully captured by the anonymous American schoolboy who wrote: "Schubert earned very little money; if he hadn't composed music we would never have heard of him."

Chicago. 29 February 1964.

Those achievements that are beyond the capabilities of a single country like France could be within reach of a unified Europe. The reason for this is simply mathematical: a single equation with ten variables has no solution, but if you construct a system of ten equations with ten variables there is a single well-defined answer. If there had been a real chance for the emergence of Europe we would never have left Paris for Texas and Chicago. To create this great economic empire, however, will take more than merging the holdings of a few conglomerates under the control of big banks. It will have to be a human adventure, not a purely financial or political one. The current apathy of Europe is not making this happen. In France political parties and special interest movements are only yearning for the illusions of yesterday's stability and privileges, denying the need for change as long as possible.

The Left could have become the driver of Europe as early as 1950. On the contrary, its lack of imagination has made it a passive, idle mass. The problem is men, courage, will. On the other side of the political spectrum De Gaulle has launched France on a bizarre journey towards what he calls Grandeur, building symbolic monuments like the *Plan Calcul*, an effort to build computers for the captive Government market which will concentrate all the decisions about French electronics in the hands of a few technocrats[10] instead of capitalizing on the real potential of the young people. The country will find it hard to come back to reality.

Chicago. 1 March 1964.

Fourteenth meeting of the committee. I am so fed up with the fanciful "explanations" given in the Air Force files that I am tempted to write a satirical book entitled *The Universe According to Sergeant Moody*. This would be a universe in which globular lightning lasts over twenty minutes, where meteors routinely make ninety-degree turns and where Venus rises in the North! Hynek agrees that such conclusions are an insult to scientific truth, and he continues to give me support, but when it comes to the Air Force he always takes an opportunistic course of action rather than staking out a scientific position and defending it.

It is at his request, however, that the Air Force has now declassified the final report of the old *Project Sign*[11] to enable us to review it. I find the document remarkable and open-minded. After such a fine beginning, why did the research sink into the quagmire we see now? Hynek is happy to be able to read the conclusions of this report to which he contributed, but which he was never cleared to read: it takes twelve years for such a report to be declassified, he told us.

Chicago. 7 March 1964.

André Gide wrote about religious belief in *Les Faux Monnayeurs:*

> As a soul sinks into devotion it loses the sense, the taste, the need, the love of reality. . . . As for me, I care for nothing as much as seeing clearly, I stand in astonishment before the thickness of the lies in which the true believers love to wallow.

The true believers I am thinking of are not the bigots he was attacking but another sort of zealots, the UFO groups which clutter the intellectual landscape, eagerly awaiting their Venusian friends. Their influence is far from negligible: they release bubbles of irrational imagery through the social fabric.

Chicago. 12 March 1964.

Frank Salisbury, the botany expert from Colorado who is a correspondent of Aimé Michel, called me up on Tuesday evening: He is preparing a new study of the Mars saucer correlation. I promised to send him my data. Janine and I are now moving fast in screening the Air Force cases. During the day she checks all the index cards we have selected together the previous evening.

At Northwestern I have begun work on an interesting pattern recognition problem, getting the computer to "learn" letters of the alphabet even when their shape is altered. Some day computers will be able to read printed text and even handwriting. Naturally if we can do it for letters we could extend the program to other figures. This is the fourteenth computer program I have written for the Biomedical Department since last November. The main problem I have worked on is the analysis of respiratory cycles, with automatic identification of physiologically significant pauses. That program is now in production. I work closely with an Australian scientist named John Welch, who collects the data from human patients at the hospital.

Chicago. 15 March 1964.

Yesterday we had our sixteenth meeting of the committee in my office at the Technological Institute, after a quick dinner together at the cafeteria. Hynek told me he wanted me to join the staff of the observatory to manage the computer applications. Indeed I am tempted to return to astronomy, but I cannot leave my current job right away.

Our committee works hard on last year's cases, some of which are remarkable. The attitude of the Air Force in the face of the phenomenon remains consistent: open and motionless, like a lazy schoolboy yawning in the back of the class.

Janine, who has returned to full activity after Olivier's birth, continues her statistical work. We began our catalogue with 1,062 cases from Guy Quincy and we added forty-five reports from the French military files. After sifting through the documents accumulated by Aimé Michel and Charles Garreau and after rejecting all the marginal reports we reached 2,864 cases in August of 1962.

Since then we have added another 700 cards based on the reading of various books and magazines. The Air Force cases we are screening now will be merged with this body of data: out of 3,318 Blue Book reports made between 1957 and 1962, we are only keeping 726 sightings as truly unexplained.

Not only has the catalogue expanded, but it has greatly gained in quality. New sources often overlay the old, doubtful cases as well as contributing new entries. Thus we have wiped out almost a third of the Guy Quincy catalogue from which we started. Many cases mentioned by Leslie were only bad quotations of Charles Fort: in such cases, naturally, we always return to the original source.

Chicago. 17 March 1964.

The Air Force files range from the dramatic to the grotesque and the utterly comical. Thus an old lady has reported a luminous thing in the sky. She was sure it was a spaceship. An investigator pointed out to her she had been looking at the planet Jupiter. Undeterred she wondered, as a good American should: "Does it belong to us, or to the Russians?"

Another witness came out of the building where he had his office and saw a spinning luminous disk in the sky. The remarkable fact is that the man was an eye specialist, who had been in utter darkness prior to his sighting, examining a patient's retina with an instrument that uses a spinning, luminous disk....

Next week we will be seeing France again, for the first time since we boarded the *Queen Mary.* We will breathe the air along the Seine, the fragrance of oranges in *Les Halles*, the taste of the Chateaubriand at Doucet's. Aimé Michel is likely to take us to some impossible restaurant in an unknown corner of Paris. In Pontoise, long-stemmed blue flowers will be covered with dew on Easter morning. Maman writes that life in France has become even more complicated than before, that prices are rising everywhere. The newspapers she sends me tell the story of the Bull computer company, mired in a series of absurd decisions in the name of a mythical plan for greatness: De Gaulle is killing innovation in electronics while trying to promote it.

It is difficult to make a choice between the two cultures. America is tough, greedy, egotistical, and truly magnificent in spite of all its faults. If only it were softened with a little humility, a little tenderness, it would easily reach to genius. It could become a beacon, a blessing for humanity everywhere. But I see mostly cold efficiency around me, and little recognition of those demanding desires described by Marcel Aymé in *La Jument Verte:*

> In my imagination the call of luxuriousness raised heavy burning dreams, a priapic tumult.

American midwestern society does without such embarrassments as "painful concupiscence" and priapic tumult. The passionate man has no place, it seems, in this hypocritical, puritanical land. Feelings of indignation, anger or desire are regarded as abnormalities of behavior meant to be repressed, hidden away, banished to the couch of the Freudian psychiatrist. Americans strive to show themselves as cool, virtuous citizens. But in everyday life this attitude often denies the warm and tumultuous reality of the human heart.

8

Pontoise. 24 March 1964.

The *Mens Magna* is an absurd machine I built years ago to amuse Eric and Denis, my nephews. It is a black tower of blinking lights and whistling sounds, impressive and useless. Yet I found it again with pleasure, here in the attic where I played as a child and where my brother's children now run their games. Pontoise is marvellous. Downstairs Olivier laughs with his grandmother, babbling happily. This attic is my watchtower. Up here every memory is a magic mirror.

France is like the *Mens Magna:* useless and pretty, absurd and surrealistic. Yet I am tempted to live here again. This temptation is like a recurring death wish.

Marigny. 29 March 1964.

We are back in fat, rainy, luscious Normandy, laughing with Janine's sister Annick, playing soccer with her little boy and eagerly reviewing the files of Aimé Michel which cover the period from 1946 to 1954. They consist of press articles, notes typed by physicist René Hardy, a pile of letters from Aimé's readers and a series of observations volunteered by readers of *Le Parisien Libéré*, of more mediocre interest. These documents have never been screened systematically.

Marigny. 30 March 1964.

France looks gray and complicated. For the first few days of our stay the Latin Quarter seemed strange to us, as if the proportions were suddenly all wrong. We must have become used to the vertical lines and the bold architecture of Chicago.

Guérin told us bluntly that what was left of French science had become a nest of snakes. There is so much jealousy and acrimony among his colleagues that some astronomers do not dare go away, even to use their hard-earned observing periods at the major telescope of Haute Provence, for fear that their budget might be cut back and their office reassigned while they were away. Guérin himself has many enemies, not only because he is interested in our forbidden subjects, but because he does not follow the Marxist party line which has become *de rigueur* in French science.

Paris. 2 April 1964.

We left Olivier in Pontoise and we came back here to make copies of Aimé Michel's documents prior to returning them to him. I continue to feel depressed by what I see here. The people met in the street seem exhausted, disoriented. The weather doesn't help. Everything seems gray, the rain is cold and miserable. People bark at each other in the streets, the shops, the subway.

I have met with Guérin again. He has read the manuscript of *Challenge to Science* and made a list of points to be worked on. I am too tired to attack them tonight. I will wait for Aimé Michel's own reactions. Unfortunately he has inexplicably skipped town to spend a week in the Alps, where he owns a retreat away from the world.

Pontoise. 4 April 1964.

Still no news of Aimé Michel. Yesterday morning we met with my agent, a retired gentleman who keeps a hand in literary matters by representing a few authors. He owns a grand apartment in an expensive section of the city. We discussed the French title. I decided on *Phénomènes Insolites de l'Espace.* He encouraged me again to pay a visit to Mr. Regnery in Chicago. Regnery's family, like his own, has roots in Alsace. Regnery understands the European mind, he argues.

Everything I see here leads me to think that France is headed towards economic and intellectual bankruptcy. The wind has gone out of the sails of scientific research. Publishing only survives by issuing masses of translations from the American and by jealously protecting a few islands, a few intellectual circles that are little more than tiny cults around people like Sartre, or the *Nouvelle Vague.*

Paris. 6 April 1964.

Occasionally I see and feel another world. It is a realm of colors, shapes and beings outside of us, entities foreign to our lives; a world most men cannot or will not recognize. It is our own universe I see, but through the eyes of another time.

In those moments I also perceive (more like a memory than a perception) the shadow of a vast new wisdom that has nothing to do with today's science. I dream of a body of knowledge that would encompass our emotions. Current thought is narrowly based on cognition, with the assumption that perceptual errors must be subtracted from the observed universe to yield the "true" picture. To my mind, without the perceptual errors that inject emotion and meaning into life there would not be an observed universe. Thought, consciousness, science itself: those are only big names thrown around by high priests trying to impress the common citizenry. They are mere labels stuck on the tiny emerging portion of a giant octopus which goes on swimming, undetected, unrecognized by our great thinkers.

I have dived below the surface. I have glimpsed the eyes of the octopus.

Paris. 7 April 1964.

Another day lost to science and knowledge, and dedicated entirely to happy, idle walks through Paris. "Flâner," "promenade," are two French terms that have no accurate translation in the English language. "Flâner" gets translated into "loitering," which almost suggests criminal intent. In the United States one can actually get arrested just for walking aimlessly around.

Today, then, I have loitered from one end of Paris to the other. Along the Quais I bought *Letters on Astronomy* by Albert de Montémont, and Humboldt's *Cosmos*. I found Koestler's *Sleepwalkers* next to *Les Diaboliques* by Barbey d'Aurevilly. I happily confess I indulged in the mixture of scholarship and lust which is the only authentic Parisian pleasure.

The sun has finally come out, putting a more pleasant face on the irritations of the city. There is a thrilling, superficial force rising in the air, but it is not genuine change. I suspect that it comes from the mere passing from one fashion to another, from one set of meaningless words to another, words to which no deep idea is attached, giving rise to endless symbolic battles that have no contents.

I hear the television set blaring next door. For the last half-hour two idiots have been arguing about their definitions of poetry.

"The poet is, above all, an erudite..." yells one expert.

The other fool disagrees:

"Poetry is a spark which..."

Such useless fights, such pointless life.

Pontoise. 9 April 1964.

The temptation remains strong to engage in comparisons, to measure Europe by American standards. I must reconcile myself to the fact that we are dealing with two different planets, among which such comparisons are meaningless. Yet I find the urge irresistible.

Americans are disappointing when they fall into dull conformity. At Northwestern the students themselves refuse to consider any idea that isn't blessed by a book or stamped by some "authority." Within our own committee for example, Nancy and Stanley can't understand why I feel the need to invent a classification system for UFO sightings. It is as if I were asking them to break some terrible taboo, as if such new knowledge might make them different from the bland mass consciousness to which they aspire to belong.

Hynek shares my frustration: young men and women come to the campus expecting to find all knowledge in the books, he says. Their parents are in real estate, in accounting or law. The kids conform to the expectations of their families. In their eyes, various academic degrees simply measure one's level of acquaintance with the library. As soon as you set aside the books to ask them about the future, about new ways to observe the stars, about the possibility of life in space, they feel lost, scared, threatened to the very core of their being.

Aimé Michel has come back to Paris. He is experimenting with telepathy, a domain where he argues that everything remains to be discovered. As he hobbles along, forcefully hitting his cane on the uneven stones of the streets in Vanves, he tells me:

"Even hypnosis has never been studied seriously. About thirty or forty years ago a few people—amateurs!—did some interesting experiments, then little more was said about it, the major issues were never resolved. We are sitting on masses of explosive discoveries that could go off at any moment."

Unfortunately an idea for which nobody has the courage to fight cannot win by itself. Europe is full of new ideas that die, stifled by the bureaucracy, by the intellectual mediocrity of its self-appointed cultural elite.

Last night in the house of Serge Hutin, my fellow Seeker of Truth under the Rose+Croix, I had the feeling of having reached one of those points of singularity from which an entire spiritual universe can be probed and explored. He showed us admirable paintings by the shy Belgian artist René-Maria Rener, who has no "name" in the art world, and seems happy to remain hidden in obscurity, pursuing and perfecting his vision. His work is a vast tapestry of humanoid beings floating in blue space, directing hordes of humans who carry spheres on their backs, or staring across infinity from the top of large boulders. They are as majestic as Mesopotamian priests. Their women have large, beautiful eyes on the sides of their heads, and lips on their nipples. They smile inscrutably as if reminiscing about pleasures unknown to mere humans.

What good is our technology, all our space research, if it does not encompass that experience of the heart, that inspiration, that hope of worlds beyond? And who are the strange beings under Rener's fantastic brush?

Chicago. 14 April 1964.

Our mailbox was overflowing when we returned to the United States. Frank Salisbury has discovered a remarkably close correlation between our data and his new sighting frequency curves, based on the files of APRO, the Aerial Phenomena Research Organization in Arizona.[12] No one is arguing that this correlation proves that the phenomenon comes from Mars, but the internal consistency of the data is very striking. Another letter has arrived from Waveney Girvan, the editor of the *Flying Saucer Review.* He proposed one of his colleagues, a man named Gordon Creighton, as a translator for *Phénoménes Insolites.* Creighton is an accomplished linguist, retired from the Foreign Service, a great traveller, a former British consul in Brazil and expert on China.

Seventeenth meeting of our committee: more work on the Blue Book cases, which we continue to discuss in detail, one by one.

Following up on my agent's advice I went to see Mr. Regnery in his downtown office. We had an amusing exchange.

"What's your book about?" he asked me.

"Unidentified Flying Objects," I replied, already on the defensive.

"You mean Flying Saucers?"

I objected: "I am a scientist, Sir. The correct term is UFO."

He seemed to accept that, then he asked me bluntly:

"Do you ever ride in them?"

I was shocked. "I beg your pardon?"

"I mean, do you fly to Mars and Venus like George Adamski?"

"Certainly not!"

Now he seemed genuinely puzzled.

"Let's see, you're a scientist. So you must explain them away: Mirages, clouds, atmospherics?"

"No Sir, I don't explain them either."

"Then that's final: You don't have a book about Flying Saucers!"

It took me another half-hour to convince him that there was a third way to think about the problem, the way of science which looks at both sides of the evidence and tries to weigh it. In terms of publishing he is absolutely right, of course: the only books on the subject that have been successful with the public at large were either written by contactees like Adamski or by arch-skeptics like Menzel. There has not been a serious hardcover book on the subject since 1957, when *The Great Flying Saucer Hoax* was published by APRO leaders Jim and Coral Lorenzen. Even that book was issued by a small regional press, with very limited distribution and no impact on the public at large. As far as publishers are concerned the subject has been dead for ten years.

"Well, that offers us the opportunity to re-open it," I argued valiantly.

Mr. Regnery remained skeptical, but he has a daughter who speaks French: He will give her my manuscript to read.

Chicago. 19 April 1964.

How could I help my friends in France get more support for their work on the borders of science? The truth is, they do very little to break out of their routine. Serge Hutin is a wise and quiet man who lives in Fontenay-aux-Roses with his aging mother. He "awaits the decisions of the Cosmos." Michel and Françoise have a small apartment behind the Panthéon, with no heat in the place, and they write on a freelance basis. Aimé Michel may be the happiest of the group, reading the Greek classics in the original text in his little village high up in the Alps.

I don't have such luxuries, but I am acquiring solid computer experience at Northwestern, where I now make my living as a programmer for the Physiology Department. In one typical experiment we run simulations of the cardiovascular system. Columns of numbers express ventricles pumping blood, vessels expanding, lungs breathing inside the mathematical "model" that tabulates blood pressure and volume throughout the system. The doctor with whom I work puts on his white coat and begins his operation. I remain at the console and I keep my hand on the switch that will feed the real data into the machine.

Above the racks of electronic equipment is a glass panel. On the other side of the glass is an operating room with a surgical table. I see the doctor walking and talking with an aide who connects a tube with a measuring device to the aorta of our "patient." On the table, restrained by leather straps, is a large dog with its chest open. The instruments provide readings that are converted into digital impulses. Cables go through the partition and into my side of the facility. A hundred times a second, pressure and volume data from the dog's heart and lungs are fed into the program and compared with the model, which recomputes all the parameters in time for the next data point, one hundredth of a second later. The project is designed to uncover new properties of the human cardiovascular system. Dogs, in this respect, are very close to humans, since the physiology of their circulation and respiration approximates ours.

Chicago. 25 April 1964.

Hynek and I have just spent two days at Wright-Patterson Air Force Base with the Blue Book staff. It was a curious, eye-opening trip. Quintanilla and his wife met us at the airport as we landed in Dayton on Thursday night (April 23rd). The consensus was that it was too late to begin work. Accordingly the evening was devoted to drinking and dancing at the officers' club. I didn't feel like joining them, so I spent a few hours slowly getting mad, staring at my coffee cup and wondering what this festive beginning was designed to prove, or to hide. I had made the trip to work on one of the world's greatest research puzzles, not to listen to a syrupy record in some stuffy night club.

I did get into a funny argument with Quintanilla and his wife about "today's youth" and the strange music kids were listening to, notably the Beatles, which they find objectionable and violent.

"What do you mean, violent?" I asked. "Their songs say things like *I want to hold your hand*... You call that violent?"

"Well, you know what I mean, they encourage the wrong kind of behavior."

We never resolved our disagreement. We went to bed at one in the morning and I woke up painfully to begin work on the Base by 8:00 a.m. Only then were we introduced to the rest of the team, which occupies a few offices at FTD, the Foreign Technology Division.

The FTD building is a long aluminum structure with no windows. A model of a MiG fighter hangs down from the ceiling in the lobby. In the bathrooms are prominent signs, reminding the staff that classified conversations are prohibited in all public areas. Whenever we needed to use the facilities a staff member came with us and stood outside the cubicle to make sure we did not engage in some act of espionage.

Quintanilla proudly showed me the UFO files and several metal cabinets full of "evidence" which had turned out to be chaff, random pieces of metal and ordinary junk. These are the items which the Air Force chooses to display in front of visitors who ask questions about the status of Project Blue Book. No wonder Congressmen and scientists always go away convinced that the whole subject is garbage! They have never been shown any of the really puzzling data. The whole thing is a joke, or a charade.

We had lunch at the fancy officers' mess, after which Quintanilla took us on a tour of the Base, notably to the edge of the very impressive section that belongs to the Strategic Air Command. Back in the office we continued to study selected files until 4:30 p.m.

The evening was spent at Quintanilla's house on the Base. Soon the lights were turned down, and dancing resumed. By the time the music stopped we had missed our plane, but never mind, there was another one later. The whole group agreed we should go into town for dinner. The women demanded more cocktails and more music, in an atmosphere of forced gaiety which drove me crazy as it became more and more dull and dreary, a parody of carefree happiness. I had expected to find many things at the Foreign Technology Division, but certainly not this absurd form of giddy entertainment.

It all ended with a race to the airport. We nearly had an accident on the way and naturally we missed the plane again. That new delay provided an opportunity to visit a downtown Dayton night club for more drinks and more dancing.

As I watched the entire staff of Project Blue Book wasting this precious time, my mind went back over the other things I had seen that day: the immense empty runways of the Strategic Air Command, with the rows of huge B-52s ready to take off on either side of the tarmac, and the openings of ominous tunnels nearby, where the crews were waiting day and night for a signal indicating that World War III had begun.

What kind of life is this? What have we done with the human soul? I thought about France: What did De Gaulle's *Grandeur* mean for these bombers whose wings were trailing menacingly in the grass like giant night moths? To them France was little more than a few minutes of flight time over a few acres of terrain. I did feel a breakthrough, a new understanding of the scale of things. When our party finally staggered out of the night club I had to look up at the sky, at the stars, to find a little reassurance.

Chicago. 28 April 1964.

This morning I was working at the downtown medical facility when a technician from the Physiology Department ran to fetch me: Janine was trying to reach me on the telephone. There was an emergency, but no bad news, he said right away. I rushed to the nearest phone. Mimi Hynek had called Janine to relay a message from Allen: *A flying saucer had landed in New Mexico!* Captain Quintanilla had called him and he had jumped into the first available plane.

I turned over control of the computer to one of the assistants and drove home along the Lakeshore Drive as fast as I could. The observatory staff only knew that Hynek had flown out to Albuquerque on the 12:45 flight. Finally I reached Mimi: A saucer had indeed landed in full view of a policeman. It rested on tripod legs, she said, leaving traces on the ground. Hynek wanted me to join him at Kirtland Air Force Base, where he had alerted Security to expect me. However he also said he would be back in Chicago tomorrow, so there was really no point in my flying there. Although the case only came to Hynek's attention this morning, the landing happened four days ago, on April 24th, last Friday at 5:45 p.m., *when Hynek and I were just leaving Project Blue Book headquarters.* If this observation can be verified, it certainly comes at the right point to support the proposal for new research I presented to Quintanilla when we were all in Dayton.

Chicago. 30 April 1964.

Are we at the beginning of a new wave of sightings? Newspapers are widely commenting on the New Mexico events and on Hynek's trip there. The witness in Socorro is a very sincere man, a highway patrolman named Lonnie Zamora. Hynek tells me he did not want to be interviewed before talking to a priest, because he thought he might have seen something diabolical. Now another landing is reported in Montana.

Chicago. 8 May 1964.

Things have calmed down a bit. On Tuesday night we convened the UFO Committee. Hynek described the Socorro landing scene in detail for us. There is no doubt in his mind that the policeman is telling the truth, but we still argue about the nature of the object. An electronics engineer named Bill Powers, who works for the observatory, has taken an interest in the ground trace measurements made by the police and the FBI. If he is right the egg-shaped object seen by Lonnie Zamora rested on a very sophisticated landing gear with four legs (not a tripod as initially reported) which was capable of equalizing pressure on its points of contact with any rough terrain.

I have developed a mathematical formula that tracks the probability of being visited by a space civilization as a function of its distance from us. The most probable origin of such visitors would be a solar system only a few hundred light-years away from us. That optimum distance is a compromise between the number of civilizations expected to exist (which grows like the cube of the distance from earth) and the likelihood that they can acquire knowledge of our existence. Such knowledge is expressed by an integral function which decreases as one goes farther and farther away from our Sun.

Chicago. 10 May 1964.

The committee is scheduled to meet today for the twentieth time, at Dearborn Observatory. Hynek is writing his report on the Socorro landing and he wants us to critique it. He can no longer deny that there are indeed unexplained landings, accompanied by humanoid sightings. The Socorro case reads like an incident right out of the French files of 1954, except for the insignia seen on the craft. Our discussions on this point are intense. Unfortunately the Air Force analysis of the Socorro soil samples is not turning up anything of interest, other than eliminating rocket or jet propulsion as a flying mechanism for the strange craft.

I have given Hynek a copy of an article I wrote about the 1946 Swedish reports of "ghost rockets," a series of events of which he was unaware. He said he would like to quote it in his forthcoming interview with CBS on Tuesday.

Chicago. 24 May 1964.

Plenty of new ideas have been inspired by the Socorro landing and its sequels. I save them for a new book I have undertaken, which will be called *Anatomy of a Phenomenon.* I will write this one directly in English, because I do not want to repeat the mistake of having a manuscript languishing in the drawers of publishers in Paris, at the mercy of their capricious decisions, and then having to go through the agonizing process of yet another translation.

Gérard de Vaucouleurs wants me to go back to Texas in July as a consultant on the NASA Mars project. In Chicago the weather is magnificent. Janine and I often walk through the park or along the shore of Lake Michigan, pushing little Olivier in his carriage. I feel very happy. On weekends we find a spot in the grass for a picnic in the park, we play and laugh together.

Chicago. 28 May 1964.

Screening Aimé Michel's old files I stumble across these wonderful words of professor André Danjon, the pompous director of Paris observatory: "Our good old earth is made exactly for us and it would be very wrong to hope to find something better elsewhere."[13]

Every day my work still takes me to the Northwestern medical school where I continue programming our model of the cardiovascular system.

Chicago. 4 June 1964.

Bill Powers is becoming a close friend. He is a tall, energetic fellow with a fine sense of humor, a maverick cybernetician trained in psychology, a great reader of science fiction. In recent discussions with him and with Hynek I stressed that our UFO Committee had come to the limits of its usefulness. They concurred: we need a new study group, a real working team, staffed by scientists rather than by students. We have just had an informal meeting at his house to consider the idea further. I brought Janine and two of my own colleagues, an engineer and a mathematician named Carl De Vito. A couple of astronomers and Bill Powers completed the group.

Chicago. 28 June 1964.

Janine, who has been staying home with our son for the last six months, starts work again tomorrow at a downtown data processing consulting firm. We have hired a warm and homely girl named Hélène to come from France and take care of Olivier. Next to my typewriter a pile of pages is mounting: The first draft of *Anatomy of a Phenomenon* is practically finished. Gordon Creighton writes from England that he has already translated half of my earlier French manuscript. I am beginning to understand that I will never be able to run my own research projects here if I do not obtain a doctorate, so I have reluctantly decided to embark on a Ph.D. program at Northwestern, although that means going back to school and taking again some mathematics and engineering courses I passed years ago. Chicago is stifling hot.

Chicago. 10 July 1964.

I no longer find the calm retreat into myself or the concentration I need in order to write. Since our return from France I have gone through a period of dry, helpless disappointment, even though Chicago is as magnificent next to its big open lake as Paris is frustrating and bureaucratic.

In a couple of weeks I will go back to Texas on a consulting assignment. It should be interesting to see how the place has changed since last year.

Chicago. 18 July 1964.

Since I finished writing the first draft of *Anatomy of a Phenomenon* my mind seems to have dried up. I float in little eddies at the foot of the waterfall. What will happen when both of my books are published? I have to expect many misunderstandings. The believers in spaceships and cosmic brothers will hail me as a supporter of their bigotry, and the Rationalists will attack me without taking the time to examine my arguments. My only long-term hope is to nudge Hynek and Blue Book towards some fundamental changes and more research.

The clarity of vision which the true seeker gains through solitary work also applies to spiritual attitudes. It endows him with the right to go forward in life without fear. While he gives up the gratification offered by membership in some recognized movement or church, he gains a far greater privilege, the freedom to inquire into the nature of his own soul.

The notion of the "good yet frightful God" of the Bible and of the Gospels seems like a swindle to me: it is the biggest, most cruel confidence game in history. Are we really supposed to be scared of some plaster god surrounded with little blue angels? Simple human dignity should make us reject all that with indignation. That does not mean we should be ashamed to kneel on the earth and cry like children when we contemplate the evils of mankind and our pathetic weakness.

It is man I fear, his lack of respect for himself and his lack of reverence for everything sacred around him. I do not fear him as the creature of some unknowable God but as an unreliable entity, biased and unpredictable. I fear man's stupidity and violence.

Austin. 25 July 1964.

Nothing has changed here. I spent a happy evening in the poorer part of town with a friend who grew up in this sleepy capital of Texas. Old cars drove by, kicking the dust from the dirt streets. It settled on the trees and the dry bushes. The glowing twilight lingered. Doors were slammed, people sang, phones rang. Thankfully, a breeze was felt throughout the wooden house. As we made love, drapes shuddered in the hot night.

Austin. 26 July 1964.

Between two computer runs, I daydream about things to come. I have fun writing a novel of our future life in my head. It would begin in a blue bedroom, sealed from the world by blue drapes. The bed would be in one corner, next to a purple chest of drawers and a blue lamp. The carpet would be gray, with vulgar white furs carelessly thrown over it. A big sofa of black leather would face two tiny chairs, lilac and blue, near a small table where we would take our coffee and our ease.

In an oval-shaped golden frame set on a bright red background I see our private map of Tenderness. Mirrors, oil paintings, drawings, small fantastic motifs in black and purple are everywhere. Black shelves too, with golden nails along the edges, supporting a vial in which a bloody heart swirls ominously.

In another corner of my vision is a statue of a green faun wearing a hat of fire. Around his neck is a collar adorned with rectangular red stones.

Like the Genie of the Bastille he is stepping up on one foot, holding an orange ball in his upraised hand.

The entire house is dripping with sumptuous things. Along the staircase foxes hang by their tails. As an additional note of bad taste, the upstairs rooms are bathed in red light. A huge color television set brings a note of stark reality into this oppressive atmosphere. A large window opens on the vast expanse of the attic, made visible at that level by the subtle play of mirrors. Twenty suits of armor stiffly stand at attention there, under the ceiling beams. They resound with an astounding noise when they are hit by all the bats who live there.

Stravinsky's *Petrouchka* and Rachmaninoff's *Second Concerto* are heard over this tumult. Occasionally a blond girl of great talent but little virtue is brought over. She takes off all her clothes, keeping only one ornament, a wide belt which bears the Arms of Zagreb University. Sitting on the skin of a rare blue tiger, she improvises on the harp, inspired by the random motion of two Chinese fishes swimming in a large sphere which is lowered from the ceiling for this express purpose.

In the depths of our park three alchemists spend their time in prayers and dangerous experiments in a large brick house. Silent assistants, their faces hidden behind masks of black velvet, use marble stretchers to bring to their masters a steady supply of cadavers. Ignoring their sinister traffic, we wallow for hours in the delights of intimate conversation. And woe to the mischievious angel who would steal a single strand of your hair, woe to the bird who fails to sing, woe to the toad who doesn't jump out of the pond to greet the dawn with us, as we prepare for another long day of dreaming.

Chicago. 1 August 1964.

From the *Guide of Mysterious France*, published by Tchou, under the heading "Jasseron":

> In Tharlet Wood some spirits called *Sénégouges* used to gather around an oak tree. One day a farmer fired at them with his hunting gun, and he heard a human cry. The *Sénégouges* have seldom been seen since 1815.

Chicago. 13 August 1964.

Harvey Plotnick, a young editor with Regnery, called me last week at the Technological Institute. He had read my manuscript, liked it, and wanted to discuss publication. We have reached an agreement: They want to issue it as *Anatomy of a Phenomenon* (the title was Harvey Plotnick's idea) but I must restructure the book. Yet I cannot work on it until I complete my report on the Map of Mars.

Chicago. 22 August 1964.

I am practically rewriting the whole book from the beginning, sorting out the figures and eliminating the passages that are too technical. I have improved my style as I compiled the book: "All you have to do now is to rewrite the beginning like you wrote the end," says Harvey with humor, remarking on my progress.

We are living through a curious period. We have never had so much freedom to organize our life. We have never been so happy. But we both feel very tired. Janine works long hours downtown while I drive to Evanston every day. We have temporarily solved our financial problems thanks to the money from my consulting project in Austin, but we are still unsettled. Life in this country creates such an energy imbalance that it is imperative to define wide cycles of consciousness in one's own life, or be swept away.

Chicago. 24 August 1964.

Dr. Hynek and I agreed over lunch today that I would start working for him at the observatory in September. I gave him a draft copy of *Anatomy* to read, asking him to keep it confidential for the time being.

Chicago. 27 August 1964.

I had a shock when I received the latest copy of the *Flying Saucer Review* from London yesterday. The cover bore a big red title: MENZEL versus VALLEE! This made me very angry at the editors. And today, a letter arrived from Menzel himself, dictated from Portugal. The tone is not so triumphant any more, since I have shown he was wrong by a factor of two when he tried to disprove Aimé Michel's alignments. I am gaining some respect for Menzel through this exchange. He does prove what he states, namely that many UFO reports are explainable. We already knew that. But his books push the evidence too far. They are like those manuals on Natural History, as Jonathan Swift says, where the elephant is always much smaller than reality, and the flea much bigger!

Chicago. 5 September 1964.

As I negotiate with Regnery, I have managed to keep the presentation of the book on a sober and straightforward level. Before UFOs can become a valid object of research for science, a wide public needs to get used to the idea that the phenomenon is the product of an intelligent force.

I do want to call attention to the facts, but I want to do it with dignity, and I want to remain behind the scenes. Regnery would like me to create a media sensation, to promote the book aggressively. That would be a mistake. I am not an entertainer, or a missionary calling for some new crusade. Perhaps it is my childhood in wartime Europe that has convinced me that the most important changes occur in secret, behind the curtain, undetected by superficial observers. Also, I continue to question my own objectivity. I am finishing *Anatomy* with an odd mixture of joyous anticipation and fearful trepidation. Koestler found the right word when he spoke of scientific researchers as "Sleepwalkers."

Chicago. 9 September 1964.

All joy has disappeared. What remains is fear, unsettling solitude. My arms try to reach for shapes that vanish. I have slipped my moorings. Something has broken. I drift away.

The fervent hopes of my search are tainted by low discouragement, a vague notion that all desirable things are suddenly beyond reach. Life sometimes appears unmanageable. The hot nights make things worse. I feel angry, without knowing why. Janine comes over with a kiss, she wants me happy: everybody wishes me well. But I feel like a jet fighter helplessly grounded in a beautiful field of violets and primroses from which it cannot take off.

Chicago. 18 September 1964.

My work is changing. The Astronomy Department will now cover a quarter of my salary under a new project on stellar evolution, Hynek's scientific specialty. At the biomedical center I continue the computations of respiratory mechanics I have started with John Welch.

Last Sunday I finished retyping the manuscript of *Anatomy*, and this morning I had another argument with Mr. Regnery and with Harvey Plotnick over control of the book's presentation. To begin with, I do not want the dust jacket to mention that I work at Northwestern University.

"How will the public know that you're not just another crackpot like George Adamski?" asks Plotnick.

He is a smart fellow with a very abrasive way to ask direct questions.

"If you understand my book you must know the answer."

I am probably too careful, too conservative, but I do like to stay behind the scenes. Regnery, on the other hand, has to sell books, so I understand their motivation and I feel like a fool.

As a result of our work together, Hynek's position is finally beginning to change. He took my draft manuscript with him when he went on vacation in Canada, as he does every summer. He liked it and sent me a note of encouragement. Today he read to me a Foreword he had composed while reading the book, but eventually he decided not to release it: He is still afraid it might compromise his position with the Air Force and his colleagues. But the gesture was significant: Hynek had declined to write a Foreword for Menzel's book.

There is almost a mystical atmosphere on this little campus in Evanston, in the sweet rains of Autumn.

Chicago. 24 September 1964.

Today is the tenth anniversary of the sightings along the now-famous Bayonne-Vichy line, and my twenty-fifth birthday. Long gone is the time when I wrote, after reading Aimé Michel's book, "I hope the events will wait for me..."

Chicago. 27 September 1964.

Lunch at the Orrington Hotel today, with old Mr. Regnery and Harvey Plotnick, to resolve the last marketing issues. I introduced them to Hynek. He was impressed, as he told me later, by my firm stand against any fancy promotion.

Menzel has just written to Hynek about Socorro, in his usual absurd way. "It sounds like a hoax, or perhaps a hallucination." In the meantime the Air Force continues to look into a curious fact I have uncovered: the insignia seen by patrolman Zamora looks very much like the logo of *Astropower*, a subsidiary of the Douglas Aircraft Corporation. I found the logo in an ad they recently published in an engineering journal. I am suspicious of this aspect of the sighting. To my knowledge there has never been a genuine report of a saucer with an insignia painted on the side. Could the Socorro object be a military prototype?

Chicago. 29 September 1964.

Hynek and I eagerly discuss Socorro at every opportunity. We disagree about the relevance of the landing, which he often rejects in frustration by saying:

"What can we do with it? It's a single-witness case. How do they say in Latin?"

"*Testis unus, testis nullus.* Single witness, no witness."

He laughs good-heartedly. "I always thought it meant having one testicle is like having no testicles!"

"If we fail to follow up the single-witness cases, how are we ever going to discover if there were in fact other confirming witnesses?"

He brushes my objections aside. "I can't exactly rush over to Carl Sagan or to Donald Menzel with this case in my hand and tell them, look fellows, here is the proof you've been waiting for!"

He would like to be able to explain away Socorro because of its implication that the diminutive pilots are a real factor in the phenomenon. It is true that there is something absurd, even ludicrous, about the humanoids. Yet they are real.

For me, I would also be tempted to discount Socorro as an isolated observation of some secret prototype, perhaps a test of a moon landing module, if it weren't for the fact that a genuine wave of sightings is actually under way, all over the world. Hynek called me again yesterday: he had just received a package of recent sightings sent to him by Blue Book, showing a significant rise in the number of reports. We will have lunch tomorrow to compare our data.

Chicago. 11 October 1964.

The Hyneks came over for dinner at Bryn Mawr on Friday night. We had a very relaxed evening. I showed them the illustrations I plan to use for the book.

Yesterday, Bill Powers and I drove to Wisconsin with Mimi and Allen for a meeting of midwestern astronomers. I had noticed that the witnesses in a sighting that took place in Monticello were living in Madison, so I insisted that we should take the time to meet with them in their home.

This excellent case may force Mimi Hynek to change her mind about the phenomenon. She continues to be firmly opposed to my pleas that UFOs should be considered in the mainstream of science. A strong-willed person, she is a militant member of the League of Women Voters, and an ardent follower of Chicago politics. She considers herself a realistic modern woman, and realistic modern women don't believe in all that UFO nonsense.

I do think the witnesses impressed her: they are a young couple working in the anthropology department of the University of Wisconsin. The wife's mother and sister were with them as they drove between Monticello and Argyle. They clearly saw a huge formation of lights that simply cannot be explained as natural phenomena.

Chicago. 18 October 1964.

Hynek has mentioned my forthcoming book to Quintanilla. He will take the manuscript to Dayton on his next trip. I want the Air Force to read it before publication, out of courtesy if nothing else. Yesterday the mail brought Gordon Creighton's translation of *Phénomènes Insolites.*

The end of the week has seen some remarkable world events: Nikita Khrushchev has relinquished power in the Soviet Union, while China tested its first atom bomb. I spent this Sunday afternoon and the whole evening thoroughly reorganizing my files in preparation for the voluminous correspondence and the new data my book is sure to bring. I am entering a new phase of simplification, of alchemical distillation. I feel winter coming near, my winter.

9

Chicago. 24 October 1964.

What is happening on the moon? Aimé Michel has written to me that some of the unpublished photographs from the Ranger missions showed well-defined formations which looked like large artificial domes. His information was based on the fact that Dollfus, the current president of the Planetary Commission of the International Astronomical Union, has been urgently called into conference by Kuiper, the astronomer in charge of lunar photo analysis for NASA. The purpose of the meeting was to discuss, as he put it, "whether or not it is appropriate to release the photographs." The press has been kept in the dark, and so have the Soviet members of the commission. What does it all mean?

For the last few days Hynek was in Boston, attending a conference on balloon astronomy. I caught up with him on Wednesday. We had lunch at his house.

"Do you know who I met in Boston?" he asked. "None other than Audouin Dollfus. I was frankly surprised at his demonstrations of friendship. He showered me with his attentions."

"Perhaps there is something to those stories of mysterious formations on the moon, and he is having second thoughts about the possibility of UFOs in the solar system . . ." I suggested.

Hynek shook his head. He was not as excited about the lunar imagery as Michel and Guérin, he said. He reminded me that Bill Powers had already noted some peculiar formations on the photographs that had been published much earlier. They contradicted the common geological theories about the moon, but they certainly did not show any spacecraft or any artificial structures.

Now comes a letter from Guérin. He writes that Boyer, a Pic-du-Midi astronomer, was at Meudon observatory when Dollfus received Kuiper's letter. He was literally "decomposed" when he read it. Before Guérin could learn anything more Boyer had gone back to his eagle's nest in the Pyrenées and Dollfus was on his way to the United States. In Boston we find Dollfus suddenly being very friendly towards Hynek, who wonders about this abrupt change of attitude. From Boston Dollfus flies to Tucson, Arizona, for reasons unknown.

Armed with this latest information from Guérin, Hynek acted quickly. He called Dick Lewis, a reporter friend of his at the *Sun-Times:* "Find out why the press is being kept away from the most recent Ranger findings," he told him. "The photographs may be censored." Lewis got excited, picked up his telephone and called Shoemaker, head of the Ranger project. "I don't have time to speak with you," Shoemaker said, "because I'm flying to Arizona to meet with Professor Kuiper."

This confirmed everything we suspected, so Dick pressed on with his questions. Shoemaker eventually confessed that about twenty photos showed "peculiar formations" similar to those Bill Powers had noticed. They could be just large round rocks. They could also be artificial domes, or even landed spacecraft. In a later conversation with Dick Lewis, Professor Urey stated that he did not have a natural explanation for them.

Chicago. 31 October 1964.

Returning from Wright Field today, Hynek tells me that Quintanilla has made only a few comments after reading *Anatomy.* The Air Force does not want to either endorse or criticize my book officially. That is fine with me, of course. But he conveyed his personal opinion that the work was outstanding. He added that he was impressed by the extensive research it represented.

Guérin and Aimé Michel continue to be very excited about the Ranger affair. They think we may be close to having the proof that there are extraterrestrial intruders in our solar system.

For the last ten days I have been trying to convince Hynek to prepare a research proposal that would bring an exclusive contract to Northwestern for the analysis of UFO files. Under the terms of the contract, we would pool our experience and our documents, making them available for the first time to the scientific community at large.

Chicago. 2 November 1964.

The UFO Committee has completed its work on the Air Force statistics. Stanley and Nancy have gone on to more lucrative pastures. Our new group, composed of scientists and professionals, met today for the first time at our Bryn Mawr apartment. We drank four pots of Janine's coffee, ate a big strawberry cake, and proposed tons of ideas, some of which were new and exciting.

As I kept talking critically about "the Air Force" and its misguided policies, citing recent reports from Major Keyhoe's *National Investigations Committee on Aerial Phenomena*, Hynek laughed and rebuffed me:

"Jacques, you'll have to understand something those UFO believers over at NICAP have never, never wanted to get through their heads: *There is no such thing as 'The Air Force'!* There are various clusters of military brass within the structure of the government, and they have many diverse interests and conflicts among themselves. What you call the 'Air Force policy' is only the attitude of one group which happens to have the upper hand against the others for a little while."

"There must be some authority to which they respond?" I said.

"Everything works as if the only thing the Pentagon reacted to was a little old lady in Nebraska who happened to be the widow of some stockbroker who left her 51% of General Motors and 60% of General Electric when he died. If that woman ever sees anything alien in the sky, or if she reads in the *National Enquirer* that a dozen Martians have just landed in Florida, she gets very scared and she calls her Congressman. Since he counts on her to put up most of the money for his next election campaign, he gets very excited himself and he writes a terse note to the Secretary of the Air Force, demanding to know what they're doing to protect his constituents with all their billions of dollars!"

We all laughed, but Hynek went on more seriously. "The only occasions when I have seen the Blue Book staff working really hard were when they had to answer such inquiries from Capitol Hill. The Congressman in question, if they antagonize him, could well cast a vote that would affect the Air Force budget at the next hearings. So public relations are a big concern, a very high priority for them."

Chicago. 7 November 1964.

The Ranger affair is becoming more clear, thanks to Dick Lewis, who published an article about it in today's *Sun-Times*, cautiously discussing "rock formations that are visible in some craters and whose origin is unexplained." Neither Lewis nor Hynek have been able to confirm the fact that some of the pictures taken with the P-camera of the spacecraft may have been kept secret.

In the midst of all this speculation I feel tired and morose. All of a sudden the idea of going back to France permanently seems less absurd. Here there is no time to think. Once my two books are published perhaps I ought to drop all flying saucer research and try to complete my doctorate in two years. Leave Chicago. Get back to other pursuits. A good computer scientist can find a job anywhere in today's world.

Chicago. 8 November 1964.

All weekend long I have worked on the card file. My mother has mailed to me Aimé Michel's documents for 1957 and 1958, which add to the mass of data. A strange fervor is raised within me when I engage in this slow, painstaking work. I feel like a Benedictine monk translating the Fathers of the Church. Will this be useful to someone, some day?

The UFO problem keeps changing. Many pioneers like Ruppelt, Wilbert Smith and Carl Jung[14] have passed away. Yesterday I learned of the death of Waveney Girvan, editor of *FSR*. I will not forget that he kindly published my first articles and supported me in my early efforts. I am sorry I never met this Celt with the curious, mystical mind. The world is a vast, changing place where I feel lost.

Chicago. 11 November 1964.

On Monday night we had another meeting with Bill Powers, Carl, Hynek, Harry Rymer and his wife, and concert pianist Samuel Randlett. They form an original, stimulating group. They offered me many useful comments on *Anatomy*. At Henry Regnery, Harvey Plotnick has now annotated the first half of the manuscript. He has shown me several mock-ups for the dust jacket and we have begun the tedious process of requesting permissions for all the quotes and figures.

Chicago. 23 November 1964.

This evening my desk is clear and my time is free, for the first time in many months. Bill was right about the moon photos. It turns out they show nothing but natural formations, although Guérin has correctly surmised that new theories of lunar evolution will be needed to account for them.

Janine and I will probably not stay in the United States. I am afraid of something here, a dark force, a profound emotional sickness that I cannot quite define. Our specific decisions will depend on the public reactions to my book, which cannot be predicted yet. I may find myself confronted with a hostile barrage not only from skeptics, but also from the believers in extraterrestrial beings who expect imminent salvation from the stars. American society scares me because it is capable of producing terrifying movements based on lies, religious fanaticism and other aberrations that nothing can stop. I am thoroughly disgusted with the sensational presentation of the UFO subject in those American tabloids that do not hesitate to make up sightings out of whole cloth. Bill Powers keeps reminding me that what I see as flaws in the American mind are really defects and weaknesses of mankind in general: Americans did not invent religious fanaticism, nor do they have a monopoly on yellow journalism.

Chicago. 29 November 1964.

The Boufflioux photograph, which was published by Aimé Michel and by Jimmy Guieu in their books, has turned out to be a hoax. I have finally decided, out of caution, not to use any such alleged saucer photograph in *Anatomy.* A few pictures do seem trustworthy, but I am not able to document their origin with enough confidence.

While perusing the French files I have come across a letter addressed to Raymond Veillith, one of the most dedicated private researchers in Europe. It was written by Colonel Poncet, chief of the 'saucer' bureau of the French Air Force, in answer to Veillith's offer of information sharing:

> The scientific office of the Air Force staff is concerned with much more important subjects. It finds itself forced to leave to amateurs the task of spreading the truth about the interplanetary ships and of greeting our space brothers.

He added: "It would be pointless for you to go on sending us your reports." Thus the door was neatly closed to any disturbing fact.

Chicago. 2 December 1964.

In response to a recent report I sent him on my current work, Aimé Michel describes the trials and tribulations of a French physicist, Olivier

Costa de Beauregard. This man is pursuing top-quality work on the theory of gravitation, yet he is unable to get any support in France. Nobody will take him seriously in Paris until his work is duplicated in the United States. Aimé Michel adds:

> In the mental system of our contemporaries there is no framework for ufology. The compilation of your catalogue, like that of my own files, is already a violation of this mental framework: we are studying a phenomenon without having the definite proof of its existence.

Chicago. 3 December 1964.

Today our son is one year old, already!

I had breakfast with Hynek and I told him, point by point, about Costa's work.

Chicago. 9 December 1964.

Despair. I feel we are making no progress. A strange disenchantment hangs between Janine and me, or is it my own failure to cope with the stress of our relentless work? It seems that nothing encourages me, nothing smiles at me. Torture: I am exhausted at the memory of the happiness which was here, within reach, just a moment ago.

Chicago. 19 December 1964.

Anatomy of a Phenomenon will be ready for the printer next month, thanks to the excellent editing work done by Harvey Plotnick. The dust jacket design shows a fiery meteor tearing through a dark blue grid. The next step is to prepare *Challenge to Science* for publication. Harvey has read Gordon Creighton's translation and likes it. I will have to work on both the American and the French version at the same time. But my first priority now is to bring Costa and his assistant to the United States for a series of lectures and high-level meetings.

Chicago. 20 December 1964.

Returning from a cosmology conference in Texas, Hynek called me tonight to say he had met in Austin with Wennersten, the top physicist for the Air Force. They had no trouble contacting two leading relativists, Papapetrou and Lichnerowitz, to get some references about Costa's credibility and scientific status. What they heard satisfied them. So why shouldn't the Air Force pay his way over here? Costa would be able to see my UFO files (Aimé Michel writes that he is dying to read them) and he would be able to talk seriously with Hynek and me about physics.

This Journal is a constant lesson for me: regret for the mistakes I made, terror before ignorance and human weakness. But those feelings also turn to tenderness and pity. When I close my eyes I often see a country cemetery in Pontoise. It is raining. When my father's coffin was taken to its resting place I walked behind the car with my brother, as we climbed the street that leads towards Gisors. A man in working clothes who was pushing a wheelbarrow put it down as he saw us, removed his beret and respectfully crossed himself.

My mind is full of shadows, my consciousness thrashes a hundred questions I cannot research. I must follow my own way, I must select between bifurcating roads. The choice has always been easy for me. Whenever I found myself facing two paths, one of which was smooth and predictable, the other burning with questions, problems and potential revelations, I have always taken the latter.

Chicago. 22 December 1964.

We had a disappointing meeting at Bryn Mawr yesterday. Bill Powers spent almost an hour on the phone with a witness, telling him about our group and how great we were instead of focusing on his sighting, on what the man had to tell us. I feel we are wasting precious time.

Today Hynek and I had lunch in private. I gave him a complete file of Costa's publications. I told him about a new possibility: should we think of the phenomenon in terms of artificial intelligence? I pointed out to him that many observations, including Socorro, Vauriat, Toledo, showed behavior that resembled that of automata rather than that of intelligent beings.

We just had a fun Christmas party at the Observatory. Dearborn was full of laughing kids. The staff had assembled odd musical instruments in the library. Bill Powers played the accordion, Hynek came in wearing a red hat and a red vest to direct the orchestra with his baton, and Santa Claus arrived mysteriously through the slit in the telescope dome, carrying a big bag full of toys. Olivier watched all this and went on to play in Hynek's office, which was littered with bright boxes.

A fruitful year is coming to a close. The publication of *Anatomy* is imminent. I have fulfilled that deep desire I felt to "respond" to the mystery triggered by my own observation of 1955. I have acknowledged that I saw the problem, that I understood its scope. But when it comes to solving it, as the French expression goes, that is quite another pair of sleeves....

Chicago. 26 December 1964.

We spent a warm Christmas around Olivier; a little work, much play. We think about our families left behind in France. Maman will come over this Summer.

Hynek called me this evening, looking for advice. He was thinking of sending a forceful letter to the Air Force, he said. They have asked his opinion of the annual Blue Book report, in which he is cited as scientific consultant. For the first time he is tempted to rebel. He understands that his signature is only a formality. The work we have done together for the past year has made him aware that there are no scientific contents to the Air Force report: "I don't want my name associated with this piece of trash," he said, "especially when *Anatomy of a Phenomenon* is so close to publication." We discussed the way in which the book might make people aware of the depth of the subject, even awaken scientific opinion. He told me he didn't want to be tied to the Air Force party line any more.

I have received a very interesting letter from an Italian military officer I will call Luciano:

> I am a Captain in the Italian Air Force and I am employed at the Ministry of Aeronautics in Rome.... My interest in the UFO problem began many years ago when I had occasion to speak for the first time with direct witnesses of whose sincerity I had no doubt. Previous to that I was very skeptical on the matter.

He described his sources of documentation, his research, and his files of over 6,000 index cards containing the details of sightings since 1947, about 800 of them from Italy itself. He went on:

> When I saw Aimé Michel in Paris last November I was sincerely surprised that he did not know that after the French "flap" of September-October 1954 a corresponding flap took place in Italy. It was of the same magnitude, with a lot of landings, and much falling of angel hair.

I am answering him right away. Aimé Michel assures me that Luciano is in close touch with military Intelligence in his country. For the time being, I say nothing to Hynek about this correspondence.

Chicago. 1 January 1965.

People speak of "the rapid development of science." What a joke! Most scientists are merely cranking through data to satisfy some bureaucratic ambition. The framework I can see around us is not the pursuit of knowledge, it is an irrational and romantic illusion. Scientific research is only a protective shield, a pink-colored screen to hide the frightening reality of nature.

Our laboratories resemble those American bars bathed in dim light, false conviviality and contrived frivolity, where everything would be seen as blatantly fake if the light of the sun was ever allowed to come in. The great caves of the old earth god where fauns and inebriated satyrs used to chase the laughing nymph have been razed by the bulldozers of Reason. What we have now is a dusty, murky saloon with chrome bar stools, a broken jukebox and a syphilitic whore. Do not pinch the naked women: the staff is paid by the Administration, like the furniture. They may form negative opinions of you if you fall in love, and they will write an unfavorable report on your behavior if you get an erection. Do not steer from the predictable path: Pay your money, get quietly drunk like everybody else, and pursue only those researches whose results are already known. Do not raise unsettling new questions that could bring Theory to its knees. Such is the unfortunate setting for much of modern science.

Chicago. 10 January 1965.

Another letter has come from Luciano in Rome, thanking me for my quick answer and expanding on the state of UFO research in his country. He deplores the fact that the few public advocates of the phenomenon have turned out to be crackpots who brought ridicule to the subject.

> Things changed a little last year when, after a flurry of sightings in the Rome area during the summer months, I was asked to provide my opinion on the matter. I have been appointed a UFO consultant to the Italian Air Force. In some cases I participated in the investigations carried out by our authorities. Naturally I have been given full access to the Air Force files.

In return he asks me what I regard as the most interesting U.S. observations of 1964.

Chicago. 13 January 1965.

I have been working on the text of *Challenge to Science* at the Regnery offices. The title was suggested by Bill Powers at one of our Bryn Mawr meetings. My editor is an energetic young woman named Betty McCurnin. Formerly at the *Encyclopedia Britannica*, she is obsessed with always finding the right word, the short phrase that completely captures an idea. Gordon Creighton, who is retired from Her Majesty's Foreign Service, has translated my long, beautiful French sentences into long English sentences. Betty tells me:

"Jacques, sit down, I want you to listen to something."

She reads to me one of my long sentences, shakes her head as if to scare off some annoying flies, then:

"What does all that mean?"

I give her a few words of explanation, summarizing the arguments, weighing the evidence. She leans back in her chair, holding her blue pencil lightly between her fingers:

"What you just said makes perfect sense. So why didn't you write it that way?"

We cut, we restructure, we trim. She is patiently teaching me to write with the concise precision afforded by the English language, rather than the flowing, evocative, visionary sense of French.

I am adding a whole section to the book, proposing a reorganization of UFO research. It shouldn't be up to the Air Force to decide if the unsolved "residue" in the data contains new scientific evidence. Blue Book should only be a filter, and the unexplained reports should go to a permanent civilian office staffed by scientists in charge of the analysis.

Chicago. 15 January 1965.

Taking me aside in confidence, Hynek has shown me some extraordinary notes he took a few years ago from a classified report written by Colonel Robert Friend, one of the most competent officers ever to be responsible for Blue Book. Friend was still a Major at the time. The report concerns a meeting that took place at a CIA office on Fifth and K street in Washington, on 9 July 1959, under the direction of a man named Arthur Lundahl. Seven CIA officers and one from the Office of Naval Intelligence completed the team.[15]

It appeared from Lundahl's statements that five years earlier, in May 1954, a certain Mrs. Guy Swan who lived in South Berwick (Maine) had contacted the Navy through a retired Admiral. She was able to "channel" outer space entities such as Affa and Crill, she claimed. She would pose a question, relax with a pencil in her hand, and soon an unknown force would take over and provide meaningful answers. A Naval Intelligence officer, Commander Larsen, visited her and tried without success to establish a psychic contact under her guidance. On 6 July 1959 (three days before the meeting related here) he had discussed the case with Lundahl and Neosham at the CIA. They encouraged him to try again and this time he was indeed successful in receiving messages from the Affa entity, who lived on the planet Uranus.

After several exchanges of platitudes typical of psychic communications ("there will not be a third world war, Catholics are not the chosen people," etc.), they requested to see a flying saucer. Affa told them to look outside.

All three rushed to the window, the assembled officers were told. *And suddenly there it was!* Lundahl, Larsen and Neosham had seen what they described as a circular object, the edges lighter than the center. Neosham had called the Washington airport radar and had been informed that electromagnetic signals were unaccountably "blocked" in the direction in question.

Chicago. 18 January 1965.

Hynek is back from Washington, where he has seen many people. The Air Force chief scientist is getting translations made of all the papers by Costa which I had transmitted. There is some new movement in the

UFO field, and I must recognize that this time the credit goes to good old Major Keyhoe, who is stirring things up with his many articles and television appearances. As for Hynek, he keeps telling the Air Force they should undertake a true scientific study—but nobody listens to him.

"What do you think of the rumors that Blue Book is only a cover, that there is another research project somewhere?" I asked him recently.

"Well, *with the enormous budget this country is spending for classified work, it is almost unthinkable that there wouldn't be a secret project on UFOs*. But I wouldn't jump to quick conclusions," he added with a chuckle, "remember what they said about military intelligence being a contradiction in terms!"

"Still, we may be wasting our time, while some group of scientists working behind the fence has the real data...."

Hynek thought about it for a while, and shook his head.

"No, I have a hard time believing this. I would have picked up something along the way somewhere," he said. "If there is a secret project, Ruppelt certainly didn't know about it, and Friend didn't know about it.... Yet it would have to get its data from the same sources we have, wouldn't it?"

"It could have its own detection systems."

He shrugged impatiently.

"Look, I've had a secret clearance since the war. And it is true that what I was working on, the proximity fuse, was kept strictly secret. I didn't even know what the next guy was working on. Sure, the same thing could be going on here. There may be secret projects looking at esoteric propulsion schemes, I'll grant you that. They could be hidden away in the Four Corners, or in the middle of Australia. We would never hear about it. But it's utterly unlikely."

Chicago. 21 January 1965.

Edward Teller gave a lecture at the Technological Institute today. He is a dry, humorless man. His goal is not the betterment of mankind but only science for pure science's sake. He wants to blow up "clean" nuclear bombs. He doesn't speak of understanding nature, only of taming her, of dominating her, of conquering her. His science is a big, precise machine from which man can learn nothing that will enrich his value system, only more facts and more numbers.

Chicago. 27 January 1965.

We saw Major Keyhoe on television last evening. His performance was rather pitiful. A pretentious reporter cracked a few easy jokes before introducing the panel. A skeptic had been recruited among local astronomers. He kept using irrelevant, insulting arguments like "you cannot speak objectively of the reality of the saucers because you make a lot of money with your articles."

The Major gave me the feeling he was a nice, nervous man, way over his head, ill-prepared for this kind of debate. He allowed himself to be easily maneuvered.

Chicago. 28 January 1965.

Serge Hutin has sent me his book *Mysterious Civilizations.* He writes:

> It is said that some spiritual centers exist . . . hidden to profane eyes, protecting the world with their invisible influence. Some locations (like certain parts of California and the city of Lyon) have been magically prepared by great initiates in ancient times to serve as gathering points for qualified magical researchers in centuries and millenaries to come.

I muse that this passage could well apply to Evanston, where a strange convergence of paranormal research is taking place, even though the University shows no interest in supporting it. Not only is the observatory directed by Hynek, a man with mystical insights, but Evanston is home to *Fate Magazine*, that popular standard of occult lore, and the campus seems to attract people with a private passion for higher truths.

Chicago. 4 February 1965.

Major Quintanilla, Sergeant Moody, Hynek and Carl de Vito have just left our Bryn Mawr apartment after three hours of intense discussion. I feel tired but pleased with our progress. Quintanilla now understands how we work. He saw our card file, the original documents, and evidence of Aimé Michel's sources, dispelling his last lingering suspicions that the French reports were vague or had been invented out of thin air. We discussed the goals and the methods of the Air Force in detail. I gave the Major several examples of Blue Book sightings that should have been placed in the *unidentified* category, and for which he had no answer. I have no quarrel with the Air Force and no axe to grind, but I believe I demonstrated to Quintanilla that we should not think in terms of isolated cases any more: instead we ought to think of UFOs as a global phenomenon. This would lead to entirely different methods, new classes of evidence.

Janine created a congenial atmosphere for these talks, a sense of working together even if we disagreed on some points. She put everyone at ease with her charm and her practical hospitality. Thanks to her, Bryn Mawr is warm, comfortable, a pleasant retreat from the world.

Chicago. 6 February 1965.

As Hynek drove our guests back to their hotel, Moody made some dumb remark and Quintanilla said with a sigh: "Sergeant, sometimes I wonder if you understand the problem!" Everybody burst out laughing.

Moody deserves a Nobel Prize in fudging for his bold UFO "explanations." Thus he is the discoverer of a new species of *birds with four blinking lights.* It was also Moody who once decided that a certain observation was without merit because "the reported object did not match any known aerial maneuvering pattern"!

Quintanilla has no scientific curiosity but he shows conscience and integrity. I hope that I have convinced him that he stands before a serious problem, not just some routine military job, and that the reports from the public he handles every day deserve to be treated with some care.

At noon, as Janine and I were about to have lunch, the doorbell rang and a telegram was delivered to us. It came from my French agent, who has been whipped into action by my recent threats of going around him: Table Ronde, a Parisian publisher, is sending us a contract for *Phénomènes Insolites de l'Espace.* If all goes well, both of my books will thus be published within a few months of each other. Many important facts about the phenomenon will be on record at last.

Chicago. 8 February 1965.

This evening we resumed the regular meetings of our research group, which Hynek jokingly calls "The Little Society," by contrast with President Lyndon Johnson's "Great Society." At today's meeting I described our classification system in detail.

Chicago. 10 February 1965.

In preparation for a new biomedical study I recorded my first electro-encephalogram today. I converted all the data to digital form in order to run a computer analysis. In spite of my interest for this work I would be thinking of leaving the University and of returning to industry if it weren't for Hynek, and for the expectation of working with him once the new observatory is built. The positive side is that my department leaves me free to do as much personal research as I want, as long as my normal work gets done.

The Vietnam war is getting worse. The U.S. Army communiqués about recent victories sound eerily identical to the French news I was hearing as a child. They speak of "police operations," of routine rounding up of suspects, of mounting casualties among the "rebels," of their imminent defeat. Speaking of the Geneva Convention about Vietnam, the U.S. Government says there would be no war if the enemy did not violate this agreement. This is like saying that houses could be built without any roof at great savings to everybody if only rain did not fall from the sky. Once again, with Vietnam as with UFOs, American decision-makers seem incapable of thinking in terms of historical and social phenomena, of global strategy. They keep dealing with superficial symptoms and parochial interests. Even Hynek is a conformist at heart. He looks at me with embarrassment and suspicion, and he soon changes the subject if I dare voice my opinion that Vietnam is a stupid war. Other staff members shrug when they hear my arguments: "What happened to France cannot possibly happen to America, we are so much more powerful! There will never be an American Dien-Bien-Phu. The war is only an economic issue, we will drive the cost so high that little North Vietnam will never be able to afford it. We will never fall into the same trap as France did."

Chicago. 11 February 1965.

There must be other levels of consciousness, and other lives than ours. I dream of the transfiguration of man into a conscious being with full access to his world: beyond borders, beyond everything that is absurd and arbitrary in our narrow social, moral, sexual rules.

Some evenings we still manage to break away from the tensions of work. We go out with friends in the bitter cold of the Midwestern winter, for a hot cup of coffee, or a bowl of soup and a movie. The icy weather brings glorious purity to the tall buildings of the Loop, the streetlights reflecting off sheets of white everywhere. Petula Clark seems to explode with all the power of Chicago as she sings "Downtown," lustily expressing what we feel inside: "Forget all your troubles, forget all your cares—Downtown!"

Chicago. 16 February 1965.

Hynek wants to learn about computers: every morning I give him a one-hour programming class. From the conference room in the biomedical department we can see the landfill area where barges are bringing sand from Indiana. A few trees have already been planted there, and trucks have cut a road towards the site of the future Lindheimer observatory that will house a 40-inch telescope.

Every Monday Hynek and I have our regular lunch to compare notes on the status of UFO matters. I have two hours of classes in the afternoon as I build up credits towards the Ph.D. qualification. And in the evening about 6 p.m. the Little Society convenes for dinner at the cafeteria.

When he returned to France after a trip to the United States in 1961 my French astrophysics teacher Vladimir Kourganoff lamented the lack of cultural density from which, he said, he had greatly suffered here. The contrast was even greater when he later travelled to Holland: on one side he only saw American mediocrity on a grand scale, while on the other side was evidence of what he called the warm European soul. Such judgments are biased and superficial. It is true that in Europe every stone has been painted twenty times by classical masters, it has served as a seat for a hundred disconsolate poets, and dozens of philosophers have rested their foot on it to tie their shoelaces. But Kourganoff, like most European tourists, saw nothing of the new culture which is bursting out of the ground everywhere in North America, because it is not made obvious by museums and plaques on prominent buildings. The media only reflect the most vulgar level of life. The casual observer cannot see what I am slowly learning to see: America's deep and secret beauty, its buried emotions.

Chicago. 23 February 1965.

It is becoming very hard for the Air Force to ridicule the UFO subject now that *Anatomy* is circulating in manuscript form and that Hynek has begun to change his tune in public. In fact Moody is in danger of losing his job in the process. His universe is a world where neatly organized catalogues of Rational Events furnish well-behaved Models of Authorized Phenomena. Hynek jokes that the Air Force must have sent him to an elementary psychology class, from which he seems to remember only that the world is made for "normal" people like him and that anyone who reports an unusual phenomenon is simply "nuts."

In December 1964 a huge craft is said to have landed in a field near Harrisonburg, Virginia, after flying over a car whose engine stalled. Some local college teachers measured strong radioactivity at the site. A few days passed, and a report was duly sent to the Air Force. More days went by. Quintanilla eventually sent two sergeants to investigate the case. They came back to Dayton and stuffed the report in the "psychological" category ("the witnesses were nuts").

This neatly explained everything except for the radioactivity readings. One of the college professors, one Dr. Gehman, got angry and mailed his own observations to NICAP. He also reported on the peculiar investigative techniques used by the two sergeants, who had arrived on the scene no less than three weeks after the events. They carried a Geiger counter, but all they did with it was to sweep it over the field, which was now covered with a thick blanket of snow. Yet they seemed to be detecting something. Every time the needle of the counter hit the top, they would reset the calibration, saying reassuringly: "That's all right, it does that all the time! There's no radiation here!"

In his most recent letter to me Aimé Michel is reporting on a curious change he believes he is detecting among some secret services in Europe which seem to have taken an open interest in the subject. Thus a former agent of the British Intelligence Service has leaked the news that Great Britain was now swapping information on UFOs with the Soviets, both having reached the conclusion that the objects were real. Another agent, an American, has assured Michel that the FBI took the whole issue very seriously. Finally, Colonel Clérouin, whom he had not seen for years, suddenly invited him for lunch and told him there was a lot of interest in the subject among the French military staff.

Aimé does not trust any of these shadowy people, whose very business is lying and cheating in the first place. He thinks that all these rumors are manipulated, but by whom? And why? Someone is using us to "snow" somebody else, he thinks. But the sudden renewal of interest in the topic, the fact that they even talk about it, is very curious indeed.

Chicago. 25 February 1965.

Yesterday Hynek and I called Wennersten to discuss Costa's upcoming visit. Today I spent the morning correcting the galleys of *Anatomy* with Harvey Plotnick and Betty. I had lunch with Hynek and Richard Lewis of the *Sun-Times*, whom I have known since the Ranger affair. He is thinking of writing an article based on my book.

Among Aimé Michel's papers I find a remarkable letter from an Englishwoman living in South Africa. Her house sits high on a hill above a wide river valley. One night in February 1957 she saw a "moon" that moved and emitted multiple flashes of light. She got up from her bed, went to a room which looked over the valley and she opened the window. The "moon" came towards the house with a fast zig-zag motion, changed to the shape of a golden football and flew over the roof. As it did so she smelled an odor like that of an overheated radio. At one point the light went behind a tree and part of the object was then visible on either side of the trunk. It disappeared at high speed.

> If it was a saucer, if it had a crew and if they saw me at the window I must say they probably went away with a strange impression of an earth creature: I am blonde, almost six feet tall; on the night in question I was naked because of the heat and the humidity, and my hair was in metal curlers! It is not surprising that they left and did not return....

Chicago. 28 February 1965.

Running through the library of the Technological Institute in Evanston and at the Midwest Exchange Center in the Loop I have found a wealth of old sources mentioning either "intra-mercurial" planets or dark bodies crossing the sun. Such objects were seen repeatedly during the nineteenth century.

I relax from computer work and library investigation by painting a series of fun and wild scenes on the walls of Olivier's bedroom: butterflies and monkeys and animals of every description. We play together every night. He babbles happily. Whenever music is heard in the living room he insists on pushing his playpen into the door frame to sing along with the melody in his own fashion.

Chicago. 4 March 1965.

It is snowing over Chicago today. In the light of State Street or the canyons of Lake Shore Drive my windshield wipers cut through a marvellous whiteness where the world seems delicately suspended. In the glowing night warm happiness passes, so close, so strong! Pools of eternity glitter in the darkness, reflecting other galaxies. Pools, I want to kneel next to you and drink you.

Dream, lead my furious life to the great blue roses of stained glass, to the kingdom near the sea where died the Fair Annabelli.

Chicago. 15 March 1965.

A strange movement is sweeping me along now, like the eruption taking Jules Verne's visitors away from the center of the earth, propelling them towards an unknown surface. I feel new forces awakening; they clearly show me who I must become: a man who is free, without fear, dedicated to the old-fashioned search for naked truth. The planet should not belong to boredom.

Another morning spent revising the galleys as *Anatomy* nears publication. I would have to be stupid not to feel terrified, but this very terror forces me to look for new sources of energy within myself. I have conquered these fears, except for one: I am afraid of being wrong, not so much because it would reflect badly on me, but because I might draw others into my mistakes.

The last line in the book reads: "The sky will never be the same again." But I will never be the same either. I feel like the little boy who had been told that he would be hit by lightning if he ever said naughty words, when one day he happens to get uncontrollably mad, yells "Shit!" at the heavens and discovers that he has miraculously survived after all.

Chicago. 22 March 1965.

We have been leading a very cloistered life that was becoming heavy and drab. At last we left all our various projects behind last week, and we roamed through Chicago with new-found friends who do not belong either to the academic community or to the UFO circles. It was refreshing to be able to talk about ordinary subjects, to discuss movies and life in general, to forget science and its narrow rituals. And it is fun to be guided through this formidable city by people who know it well.

Chicago. 30 March 1965.

Hynek has come back from a trip to New Mexico. In the plane he wrote a statement praising *Anatomy*, and he has authorized me to publish it. At the same time Carl Sagan denies me permission to quote him. As for Quintanilla, he demands that I refrain from mentioning that Hynek is consultant to Blue Book if I publish his statement of support: those are reminders that the fight is far from being won. My book is simply changing the rules of the game towards a more scientific debate where the Air Force now must alter its tactics and its language. Eventually it may even be forced to open the door to new ideas.

Chicago. 9 April 1965.

Anatomy of a Phenomenon will be out in the bookstores within a month. An article by Dick Lewis based on the book has appeared in the *Sun-Times*. It was picked up by other papers from Florida to Maine. Now I am beginning to receive some friendly letters and phone calls, principally from members of NICAP who are open-minded enough to challenge Keyhoe's party line.

Yesterday the contract for *Phénomènes Insolites* arrived from Paris, with an advance on royalties that will barely enable me to pay Gordon Creighton for his translation work.

I had a confidential talk with Hynek today. I briefed him on my correspondence with Luciano, the Italian engineer who analyzes UFO cases behind the scenes for the Italian military, and who has given me permission to disclose our correspondence to him.

Chicago. 13 April 1965.

Life has become too heady, like a strong wine. Janine and I have left our research aside for a while. We are both changing jobs. She could no longer stand the pettiness and the greed of the contracting company where she worked until now. And I have entered the engineering graduate school full-time, planning to take the qualifiers in June.

A funny scene took place today when Regnery sent two men to pick up my model of the Hopkinsville humanoid: they want to put it in the window of Chicago's largest bookstore to promote my book. This model is an exact replica of the goblin seen at the farmhouse in the famous 1955 incident when an entire Kentucky family shot at the intruders until they ran out of ammunition. The creature was four and a half feet tall, dressed in silver, with a large chest, a big head, huge ears, long arms. My model of it is so life-like, with its two red eyes on the sides of its head, that the burly truckdrivers backed off in some apprehension when they first saw it. Realizing that it couldn't bite, they eventually picked it up, folding its ears carefully for the trip down the stairs.

Luciano has sent me information about a near-landing that took place on 20 August 1963 at 9:32 p.m., and which he investigated with a secret service team under a special clearance from the Italian government. The witness was the trusted chauffeur of the Italian President, driving his official car. The site was the hunting reserve of the President, not far from Rome. A disk-shaped object resembling an upside-down saucer with a turret on top hovered at low altitude above the car. The case created quite a stir among the Intelligence services, understandably. The report was communicated to the U.S. authorities in Washington, who never followed up with the Italians but gave assurances they had passed it on to Hynek for evaluation. Yet Hynek has never seen the report, never heard of it! *I have used this case to point out to him again that he didn't see all the reports, that there must be another study somewhere, using Blue Book as a mere front.*

Chicago. 20 April 1965.

I am finishing several tasks related to Hynek's project on stellar evolution. Since the rule at Northwestern demands that I be a full-time student for three consecutive quarters I will receive no pay for my work on the second edition of his Bright Star Catalogue, but I will complete the job anyway.

Hynek has now read my whole correspondence with Luciano, which goes back to December 1964. He emerged from this reading visibly shaken. He had never suspected the existence of a serious UFO project behind the scenes in Italy, a national project whose files were not shared with Blue Book, and now he wonders what else he doesn't know! He has requested information about Luciano from his Intelligence contacts in

Washington. He is also asking the Air Force to find out all available information about his European counterparts: He wants to find all the scientists who might be consulting, as Luciano does, for the various military forces.[16]

Surprisingly, Dick Lewis' article about my book has earned me congratulations from the engineering faculty at Northwestern. This open-minded reaction gives me new hope.

Spring is coming back. All over this little planet men talk and fight, fear and fight, hope and fight and die. The Vietnam war lingers on. Poor folks on both sides are crushed under the crumbling ambitions of stupid leaders. Will the world always stagger blindly from war to war? Nobody seems to care. The level of international news we get from radio and television media in Chicago is pathetic. What little data we glean is ridiculously short and biased, smothered in local politics and "local interest" anecdotes that are meaningless.

Chicago. 4 May 1965.

Hynek has just told me on the phone that Menzel has suffered a heart attack, but survived.

I think I have found the right topic for my doctoral dissertation: how to build an artificial intelligence system capable of answering English questions about a database of bright stars. Such a program would be driven by a linguistic analyzer. My adviser, Professor Krulee, doesn't think the system can be built successfully, but he believes it would be interesting to try anyway. The only question-answering systems in existence are little more than toys, like MIT's *Baseball* program designed for simple-minded inquiries like: "Did the Red Sox beat the Dodgers?" The technique has never been applied to a full-blown scientific problem.

Chicago. 6 May 1965.

Michel Gauquelin and his wife are continuing their pioneering work on their revisionist approach to astrology, with great statistical success. I brought their results to the attention of Regnery, with my recommendation to have their book, *Cosmic Clocks*, translated and published in the United States. Regnery's daughter Susanne is reading it in French.

I spent the morning with Harvey Plotnick reading the early reviews of *Anatomy*. This evening the first box of books arrived from the printer.

Janine picked them up on her way home. The very first copy, of course, is for her. Two others for Hynek (who will forward one to the Air Force) and the fourth for Aimé Michel. Now the rocket is launched.

10

Chicago. 8 May 1965.

A good old Chicago-style heat wave has struck. The muggy weather is draining. I feel exhausted in spite of the soft tepid wind and the blue sky above. Janine has gone shopping with Hélène. My son is supposed to be sleeping, but he is just as uncomfortable as I am and cries in his room. I feel helpless, dispirited, unable to work as hard as I should on the art-work for *Challenge to Science.* Part of my discouragement comes from the irritation of having still more examinations ahead of me. I thought I had left all that behind a long time ago.

Chicago. 17 May 1965.

A curious incident recently took place during a conversation with a Martin Marietta engineer who says he is compiling a reference book on UFOs. As he was spending the evening at Bryn Mawr with our Little Society the conversation came to sightings in the Soviet Union. He told us he had written (in Russian) to their Academy of Sciences, and had received the reply that no study was being made of the subject. He showed us the Russian reply, held in a thick black binder. As the conversation continued the binder was passed around and it came to Sam, who read the letter and innocently turned the page. The engineer leapt out of his chair like a tiger and tore the binder away, tersely spitting out, "the other papers have nothing to do with that!" We were left fairly shocked at the violence of his reaction. Of course we began to wonder what else might be in that binder. There are rumors that major aerospace companies are conducting their own secret studies of UFOs.

Today Hynek invited me to have lunch with a fellow from the Illinois Institute of Technology, an engineer who thinks he knows how to build a flying saucer. It was a waste of time.

Chicago. 21 May 1965.

Last evening I met with Hynek and an NBC television reporter. The network has decided to shoot a documentary on UFOs. I see this as a further sign that my book comes at the right time. However I have not gotten over my discouragement of the last few days, which may correspond to the "decompression" that follows my intense work on the two books. This realization does not make it any less painful: I feel fragile and vulnerable.

Chicago. 26 May 1965.

We had another meeting with the NBC man last night at Hynek's place. Allen greeted us informally, dressed in a sweater and slippers, his face drawn and wrinkled, with signs of age and a great tiredness in his eyes, reminiscent of some pictures of Einstein. Underneath his exhaustion, what we saw was a genuine humanist, caring and vulnerable.

A strong wind was beating against the screen-enclosed porch. It made the lampshade swing over the crude wooden table. Our maps were always threatening to fly away.

When I look at Hynek, at his gentle and ironic way of contemplating the world, at his conscience and his simplicity, I see a rare example of a scholar who is also capable of inspiring great projects, visions of new values, of freedom beyond the ivory towers. Others often consider him a second-rate scientist, because he never tries to impress, he never pretends to know more than he actually does. He loves to share his sense of mystery, of puzzlement.

Chicago. 29 May 1965. Noon

Janine and Hélène have gone out with Olivier. I am working at my desk and my thoughts drift again to France, to the sweet France of the meandering rivers, of the fiery sunsets behind the tall cathedrals, the gentle prairies, where cows walk slowly in the morning dew while the angelus rings from the tower of the country church, and baby-face clouds waste their time in the sky above.

But what about the French people themselves, their vision and their appetites? I see them clearly enough through newspapers and books, through biased news published in perfect good faith and impeccable style. What comes through all that is technical incompetence, an unfocused view of international events, a vaguely liberal self-image that is no longer connected with any genuine love of humanity at large, a pretense of caring for others that is denied at every page by the glossy pictures of luxury cars, extravagant perfume and expensive toys.

It is the ads that count. They show men in their thirties, supremely elegant, handsome as Alain Delon, coming back from the tennis court with some vapid debutante at their side. This generation represents a wave of odious upstarts with empty dreams and vacuous conversation. They were born around the time of a World War that was fought and won by others, in a world invaded by forces their own parents could not fathom, broken and decimated as they had been by the previous war, the "Great War."

When I was seventeen I had many angry arguments with my father and with my uncle around the lunch table every Sunday in Pontoise.

"It's your generation that made all these mistakes," I would tell them. "You went through the First World War and you failed to understand what was happening. You went ahead with your colonies, your so-called French Empire, your old bourgeois values. You created this mess, all this injustice we see in the world."

Perhaps I was right. But what I failed to see was the imminent replacement of what I called the "old bourgeois values" by the young bourgeois values of my own generation. How could I imagine that they would be even worse? More unjust, more ugly, because they would not even rest on any tradition of cultural continuity? In the meantime the most solitary, pitiful generation of men and women is now dying in France, forgotten by all the magazines, ignored by the publicity men and the chroniclers with the elegant pen and incisive style. It is a decimated generation that tried to preserve its obsolete culture amidst the most sordid display of cruelty and terror ever seen by man. The promising young poets, managers, inventors, doctors and statesmen who could have led Europe to greatness were turned into bloody pulp by the cannons and the machine guns and the poison gas. One only needs to take a look at the age pyramid of France and Germany to see where the genius of Europe has gone, why it can't solve its modern problems, why the great leaders and the great thinkers are missing.

When my father was at Verdun as a liaison officer, he was often the only thing that moved in the whole landscape, and the physical and psychological skills he developed to survive in that terrible environment were extraordinary. One day he came back from a mission under the fire of German batteries to find that his whole company—the two hundred men with whom he had lived and fought and faced death in the mud of the trenches—had all been killed or maimed by a direct hit from the enemy guns.

That generation, or rather what little was left of it after the carnage, is now dying. It is being replaced by a wave of arrogant fools who are only concerned with the size of their apartment, their little nest egg in the anonymous coffers of Switzerland, their shiny new Citroën.

Chicago. 8 June 1965.

For five hours I have worked with the NBC television team filming a new documentary. As a result the Toulouse multiple-witness sighting and the case of Vins-sur-Caramy, with its remarkable observation of the vibrating road signs, have been duly put on record for future audiences.[17]

Chicago. 13 June 1965.

Hélène flew back to France yesterday, taking with her the records she is instructed to return to Aimé Michel in person. My mother arrives tomorrow: we have spent two days preparing her room and generally putting the house in order. There is a strange atmosphere here because of the uncertainty in my life. Fortunately Olivier is an angel and lets us work in peace. I am right in the middle of my written examination. Letters from readers of *Anatomy* are flowing in. They express the happy feeling that someone is now working seriously on the UFO problem at last. I wish I could be as enthusiastic....

My bitterness about these silly examinations leads to discouragement. The University is not measuring our scientific merit, only our dogged obstination. They are not selecting the best researchers, but the ones who are the most pig-headed.

Chicago. 14 June 1965.

Maman arrived safely this evening, bringing me another batch of original documents from Aimé Michel. A nice letter came from Guérin, who rightly pointed out one of the flaws in my book: I don't spend enough time on the physical problem itself. After all, relativity does not admit of any speed faster than light, he says, and dematerialization belongs in science-fiction novels. *There is no adequate physical framework for flying saucers.* I am hoping that Costa de Beauregard, the French physicist who is a friend of Aimé Michel and plans to come here next January, will provide us with some new insights.

Today McDivitt and White, the Gemini astronauts, are in Chicago for a big parade. So is Hector Quintanilla, more discreetly, being interviewed for the NBC documentary.

Chicago. 3 July 1965.

Lecky's book, *Rationalism in Europe*, which I am currently reading, is a fascinating study of the changes in public opinion that led to the discarding of medieval ideas about sorcery. I find in it an obvious parallel with our current quarrels. Aimé Michel and Guérin have trouble understanding why scientific minds do not embrace the concept of extraterrestrial life with enthusiasm. But the problem is not one of logical reasoning. Public opinion is evolving in spite of anything we or the skeptics can say and print. That is why the heavy debunking efforts of Menzel have so little effect, and why the impressive "evidence" compiled by NICAP has no impact either.

Regnery has decided to take my advice and to publish the Gauquelins' book on astrology. I called Michel and Françoise to give them the good news, and I told them they should come over to America to continue their research. I had planned to talk to Hynek about their work, but I wanted to wait until there was something definite about their book, since he does not read French. I did tell him about it last Wednesday on the way back from lunch. This devil of a man surprised me again, revealing that he had long been interested in astrology since his early student days, although he didn't buy all the false science that came with it. He had even done some statistical calculations along the lines systematically explored by the Gauquelins. Naturally that is the sort of thing he would never admit in front of his colleagues.

A remarkable sighting has just taken place in the Alps, in a field near the small town of Valensole. A farmer named Maurice Masse saw an oval object land, with two small occupants that stepped outside.

Chicago. 6 July 1965.

Janine's brother Alain has joined us from France. He is a pastry chef who was wasting his time and his health working under abominable conditions in a bakery shop in a Paris suburb. We helped him get a job with a Chicago restaurant, and now he is adapting rapidly, although he speaks almost no English.

I had a long phone conversation today with Gérard de Vaucouleurs, who congratulated me for my book and told me I must not give up this work. These encouragements are important to me. They were quite unexpected, even though I have long understood that de Vaucouleurs, like Hynek, had reached personal conclusions about the subject.

Chicago. 9 July 1965.

Last night Carl de Vito and I discussed mysticism. Lecky does a magnificent job of describing the disappearance of the belief in witchcraft and the obsolescence of the miraculous. He demonstrates the following principle: it is not through a campaign of logical arguments that these ideas evolved in the minds of people. Instead, public opinion changed by a barely detectable "shift" or "drift" which was not reflected in the writings of any scholar of the time. The books about sorcery written by the best thinkers of the seventeenth century deplored the increasingly skeptical attitudes of the masses on the subject. The Bible, after all, gives clear and precise instructions to exterminate sorcerers: "Thou shalt not suffer the witch to live," in obvious contradiction with "thou shalt not kill." Practically no authoritative book dared to argue in the other direction. Yet, according to Lecky:

> In 1660 the majority of educated men still believed in witchcraft, and in 1688, the majority disbelieved it.

Lecky adds that by 1718 those Englishmen who still believed in sorcery had become an insignificant minority. We are probably in the same position today with the belief in alien visitors. The new concept of extraterrestrial intelligence could be studied like a spreading epidemic, using a stochastic model of growth. Citing these facts, I pointed out to Carl that the Church hierarchy had always been forced to follow the movements of society: once public opinion shifted, the fewer witches they burned and the less witchcraft was actually practiced in the countryside.

Carl said he noticed in me a strong desire to go beyond mere scientific knowledge, but he did not think I had a true "mystical" mind. That is a matter of definitions. To me, religion has nothing to do with mysticism. Therefore he is absolutely right if he measures the weakness of my mysticism according to my lack of faith in the common images of God. In any given era faith is determined by opportunistic social factors that have no relationship with reason and intelligence, or with profound mysticism.

If there is a God, then the most important thing he has given us is our brain. I believe he would want us to use it to question him, rather than throwing away our wonderful inquiring abilities to wallow in blind faith in front of him. To question the divine plan would be the greatest compliment one could pay a Creator.

As Lecky demonstrates in his discussion about rationalism, the evolution of ideas seldom takes the form of a visible debate leading to reasonable change. Instead mass thinking slides unnoticed underneath the professional thinkers who firmly believe they are planted on unshakeable axioms and dogmas. To a mystic, space and time are only appearances secreted within the *neighborhood* (in the topological sense) of a given individual and a given epoch. Therefore mysticism is not a doctrine or a belief but only an orientation of consciousness, a direction of thought away from ordinary space-time reality. Monge used to say that in mathematics the shortest path between two propositions in real space often went through the imaginary domain. I think the same is true here.

If an Adept thinks he perceives the shadow of God and his prodigies beyond the plains and the mountains of the fantastic country of consciousness, that is his business. Since the mystical attitude transcends the everyday world it is natural for it to become associated and confused with the idea of a superior being responsible for all creation. But in my view this identification is not necessary.

Chicago. 13 July 1965.

Yesterday I had the bad surprise of receiving a very nasty, threatening letter from a New York attorney. A woman named Isabel Davis who did a study of the Hopkinsville landing a few years ago wanted to sue both me and Regnery because I had quoted two short passages of her report without specific permission. She didn't have a very strong case: short quotations with full credit customarily can be used without written authorization. My good faith is obvious since I printed a full reference to her work in my book, including her name.

Things improved a little today. Hynek, who is spending the summer at Harvard, knows this lady, so he phoned her to find out the reason for her lawsuit. One of her close friends, it turns out, is a lawyer who pushed her to start litigation. Perhaps he thought that my youth and inexperience would bring them both a lot of easy money.

Not only had she not realized that she had everything to gain by being quoted in *Anatomy*, but it has now come to light that her report itself was based entirely on somebody else's work, an investigation conducted by one of Hynek's former assistants, who has now granted me permission to mention his own extensive findings in the case. In fact she had even signed an agreement at the time not to publish anything! Not only did she not suffer any injury or loss, but she is the one who could be sued. Now that this has come out she is anxious to make peace with me. But she has also angered my publisher, so any mention of her is being removed from future printings.

This little affair does illuminate an aspect of American ufology to which I had not yet been exposed. These people may claim they are acting for the greater good of mankind in disinterested fashion, but their private behavior belies that: their bickerings and back-stabbings are unworthy of true researchers.

This little unnecessary crisis has made me lose two days. I have also wasted time in Evanston convincing the University that they owed me $200 in back pay. Eventually I won my case, so that our bank account has gone back up to $249, a veritable fortune.

While we are wasting our time in such fashion, the phenomenon seems increasingly active: the level of sightings which had remained high since Socorro (April 1964) is climbing to a peak. New cases are reported in Chile and in Portugal. Public opinion has become keenly aware of the problem again, and the media are reacting accordingly: *True* magazine has decided to reprint a condensed version of my book in October.

Chicago. 23 July 1965.

For the last two days Chicago has been hot, ugly and muggy. There is news of intensifying war in Vietnam, even rumors about possible mobilization in the U.S. Something is very wrong in Washington. And I have failed the oral qualifying examination. I will have to try again. Janine brings me a cup of coffee and comforts me:

"A life without problems, that would be really boring!" she says with a wonderful smile.

Yet this rotten summer gets to me, this crisis summer. I have always liked Fall and Winter best.

Man is pathetic when he fails to perceive his own nature, when he passes his own true self in the dark on his way to oblivion.

Chicago. 25 July 1965.

I don't find myself here. I long for a smaller world, for a retreat. I want to get down from the merry-go-round for a while. I have had my fun, I have screamed and laughed with all the others. It may soon be time to leave the fireworks of America behind.

I long for prairies that smell of rain. I would like to be among poor simple people again, laborers, the folks of the old country, arguing about the soccer game at the corner bistro. I have had my fill of watching the big American circus, I am fed up with the tinsel on clown costumes, the strong men breaking fake chains with gusto, the trembling feathers stuck in the asshole of the showgirls. Perhaps I will buy another ticket and come back to be amazed by all that again. But for now, give back to me the deep silence of our walks through the woods; the quiet pursuit of science; and the little kisses Janine used to give me when I came home to her.

Chicago. 28 July 1965.

Tonight Janine's brother Alain called me from a Rush Street coffeeshop at 9:45 p.m. He was with five sailors from the French destroyer *Bouvet* which is docked at the entrance of the Chicago River. They had just seen a yellow luminous disk over Lake Michigan, with bright lights around the periphery. It flew South and disappeared in the distance. It had the apparent diameter of the Full Moon.

I called the tower at O'Hare right away. I was told categorically that no advertising plane was flying over the lake. So here is one more case for the unidentified file.

Chicago. 30 July 1965.

We are having difficulty finding a trustworthy person to take care of Olivier. To make matters worse our neighbors are not helping us in this predicament. They resent the fact that Janine works downtown, wearing smart outfits, operating expensive computers. They clearly think her place is at home taking care of the cooking and having more babies instead of using her brain. It is curious how women blame so many of their problems on men, while at the same time they display so much envy and jealousy among themselves.

A friend of my publisher has come forth with a private revelation. He was one of the people in the radar room when the famous 1952 UFO flyover took place above Washington, violating restricted airspace—the same case Donald Menzel explained as "temperature inversions." When the abnormal returns showed up on the screens an officer ordered two men to go out with a camera to take pictures of the source. They soon came back, the photos were developed on the spot and they were immediately confiscated. All men in the room were told to remain silent and never to mention the photographs, which showed perfectly clear luminous objects.

But where did those photographs go? And why doesn't Hynek know anything about this?

Chicago. 5 August 1965.

We took my mother to Evanston today: lunch at the Orrington hotel, then a walk through Shakespeare Gardens and the campus. She was impressed by the site, amused by the squirrels that run everywhere and even cross the busy streets by performing acrobatics on the telephone wires.

Aimé Michel writes that a deluge of sightings is taking place all over Europe. "How come nothing is happening in the United States?" In fact we are witnessing a similar explosion here. There are so many new observations that the Air Force has pulled Hynek away from Harvard to send him to the Southwest, where a wave is in progress.

The *Medical Tribune*, one of the most exclusive papers in the U.S., has published an excellent review of *Anatomy.* This triggered interest by a reporter from the *Sun-Times*, who will interview me tomorrow. I am pleased and surprised at how objective and serious most of the reviews are. The book was initially ignored by the major papers like *The New York Times* which pride themselves on assuming that any phenomena unexplainable to Midwestern peasants would be easily within the rational understanding of the superior minds in Manhattan.... But there is now an avalanche of local reviews, from California to Maine. They show a genuine groundswell of interest which is forcing the big city papers to consider the book seriously. Clifford Simak, writing in the Minneapolis *Star*, called *Anatomy* "fascinating in its detached approach" and the New York *Daily News* mentioned it as "the best book we've seen yet on this intriguing subject, a brilliant effort."

Maman leaves for New York and France on Sunday. The time has come for Janine and me to consider our future. Aimé Michel tells us we would be crazy to go back to Paris.

Chicago. 6 August 1965.

Gérard de Vaucouleurs has sent me a letter of congratulations, together with a curious newspaper clipping indicating that Leonard Marks, director of the U.S. Information Agency, was fascinated by the saucer question and wished to see the government organize a program to study "the probability of intelligent life on other planets."

Chicago. 8 August 1965.

Crazy schedule: We accompanied Maman to the plane bound for New York. The day was rainy, stormy, gray, rumbling, sad as a Midwestern summer. Afterwards I drove Janine and Olivier home. They were both exhausted. I went to Evanston to catch Hynek, who was just coming back from Texas and was leaving again, so that I ended up driving him back to O'Hare airport just to spend more time with him. I am glad I did, because I learned a great many things during our discussion.

To begin with, it is not the Air Force which sent Hynek to Texas. He insisted on going there himself. The Blue Book people were concerned that his trip would give even more unwanted publicity to the numerous local sightings. Generally speaking, the Air Force behaved very shabbily. They issued silly public statements such as "The witnesses must have seen Jupiter *or other such stars*"! One would expect Air Force officers to know that Jupiter is a planet and not a star. Besides, that explanation cannot even remotely explain the reports.

The most interesting fact is that Hynek did not even decide to go to Texas by himself: it is de Vaucouleurs who requested that he make the trip! He called Hynek on the phone and started talking about *Anatomy*, asking what he thought of it. Always the crafty politician, Hynek expressed some reservations at first. De Vaucouleurs countered him: "But it is excellent!" he said. "How can you say otherwise?" Hynek told me that he quickly agreed with Gérard, and they came to the main subject of the local sightings.

It turns out that one of the observations was made by radar operators at Tinker Air Force Base. To get quick help they called . . . the Highway Patrol! Indeed, a number of cops saw the objects independently and another radar tracked it as well.

Two pictures have been taken during the Texas flap. Hynek extracted them with some difficulty from his bulging briefcase to show them to me. They seem genuine. One of them shows a luminous source moving up and down over a constant background where the stars have left characteristic trails.

In Austin itself one of de Vaucouleurs' students saw an object which looked very much like the disk reported to me by Alain over Chicago. Gérard grilled his student for two hours, covering the blackboard with figures in front of an amazed Hynek, whose investigations rarely get into such a level of painstaking detail.

One of the satellite tracking stations set up by Hynek in South America observed three objects and sent the data to him. I wish Muller had done the same at Paris Observatory in 1961.

But the most memorable thing is a conversation Hynek had with a pilot who told him about an incident that had taken place last July 3rd, at about 7:30 p.m. while he was flying over Canada on his way to Montreal. It was already dark on the ground when one of the passengers knocked on the cabin door and asked:

"Do you mind telling me how high a dirigible can fly?"

"Not above 10,000 feet."

"And how high are we flying now?"

"About 35,000 feet." The businessman seemed taken aback, then he pointed out the left window and said:

"In that case, can you tell me what that object is, over there to the left?"

What the pilot saw was an immense cigar-shaped craft about to penetrate a large cloud 10 kilometers or so away. The front of the object emerged from the cloud while the back had not yet entered it. With his binoculars the pilot saw five rows of windows over the entire length of the object, which he estimated at twice the size of an aircraft carrier.

Two Canadian jet fighters came into view at that point, looking very tiny as they climbed towards the craft. The cigar seemed aware of them. It accelerated in a smooth, continuous manner and disappeared in a few seconds with a fantastic display of blinking red lights in multiple rows along its whole body.

The same witness stressed that saucer sightings were common among his colleagues but that nobody would report them for fear of being sent to a psychiatrist and losing his job. "Why don't you conduct a survey of retired pilots?" he suggested.

Hynek confided to me he now realized that the time had come to do something, perhaps by contacting Leonard Marks or even J. Edgar Hoover. As for me I feel rather tired at the moment. I am supposed to prepare another silly examination that covers the history of industrial engineering and the tricks of good management. But I am thoroughly disgusted with this fellow Taylor who had defined cooperation between managers and workers by the fact that "the workers must agree to do everything they are told to do, without asking questions and without making suggestions!"

Janine, Olivier and I have reached a turning point: For the first time in a year we are alone together. This will mean a lot of work for Janine, also much happiness at our intimacy. In spite of my tiredness I am becoming myself again. In a few weeks Autumn will come, bringing delightful rain, the freshness of the air, the rich smell of dead leaves.

Chicago. 16 August 1965.

This year's wave of sightings has already brought two new facts: first, a truly accurate description of a small occupant seen at close range, thanks to the clear and specific account given by Maurice Masse in Valensole.

Second, the fact that the saucers are now allowing themselves to be seen on radar and fly slowly enough to be tracked over populated areas.

Chicago. 19 August 1965.

Hynek has forwarded a copy of a letter from Colonel Spaulding:[18]

> I have recently discussed with Bob Wilson of the National Academy of Sciences the possibility of that organization responding to an Air Force request to evaluate all known information on the subject of UFOs. I would like to have your reaction to this proposal.

Hynek asks me to comment. What's the Air Force up to? I smell a rat.

Chicago. 22 August 1965.

Most of my energy now goes into staying focused, memorizing facts and figures for my next exam. I earn a meager salary by writing computer programs for Children's Memorial Hospital and for the department of Material Science at Northwestern, but I am wasting long hours I could be spending at home with my precious wife and my little boy.

A young man from Chicago has called me after reading *Anatomy*. His name is Donald Hanlon. He wanted to tell me about his research on some old local sightings and on the Fatima miracle. We will be meeting on Tuesday night. I hope he will prove to be a good recruit for the Little Society. There is no hope of finding another Aimé Michel here, but a group of compatible minds, comprised of specialists in different areas, could make faster progress than Michel alone. It is for that reason that I keep the Little Society going.

Hynek acknowledges that our group has already played a major role in allowing him to voice ideas, hypotheses or scenarios that he could never have articulated before a dry and formal scientific gathering. When we meet in the evening around one of Alain's special pastries (Hynek's favorite is strawberry cream cake), a cup of Janine's strong hot coffee in hand, we can let our imagination soar. Allen's current dialogue with Spaulding was ignited as a result. Carl also says that by taking his mind off his dissertation our evening sessions have helped his own research. He has even begun to work with Harry Rymer on the problem of quasi-periodic functions. As for Sam, who is endowed with a photographic memory, he is always ready to supply a mine of amazing bibliographic references.

Chicago. 28 August 1965.

Don Hanlon has surprised and impressed everybody. He turns out to be a twenty-year-old printing worker, with a deep mind, very well-versed in the literature of the paranormal. Not only has he read a lot of books, but he has corresponded with Carl Jung and Ivan Sanderson.[19] His views on Fatima and the apparitions of the so-called "Blessed Virgin" are strikingly similar to those of Aimé Michel. It is for people like him that I wrote *Anatomy.*

Chicago. 2 September 1965.

After lunch I came home, tired and vaguely disgusted, leaving behind the book I am supposed to be studying. It is still on my desk at the Technological Institute, open at the page entitled "Definitions of the Function of the Personnel Director."

Hynek has written to me again to explain his answer to Colonel Spaulding, in which he suggests forming a panel of civilian scientists "for the express purpose of determining whether a major problem really exists." He adds:

> If the preliminary survey of the problem should bear me out, namely that there exists the possibility of new scientific information in the UFO phenomenon, then definitely let the recommendation be made to have the National Academy of Sciences, or some other civilian group of recognized stature, undertake a longer-term study of the problem. I would offer a strong opinion at this juncture: even the preliminary panel should be a working panel.

The success of *Anatomy* has triggered new interest for the topic in the media. The producer of every significant radio or television show in the country has received a copy of the book, at a time when the recent wave of sightings in the Southwest is forcing a reappraisal of the reality of the objects. Many new articles and television appearances by believers are fanning the flames. We speculate that is why Spaulding is now proposing a study by such an august body as the National Academy of Sciences. But are they serious?

Chicago. 6 September 1965.

We stayed up until 3 a.m. to watch the UFO program on the local Norman Ross show entitled "Off the Cuff" in which I participated with Bill Powers, Dick Lewis and several local astronomers. We had asked Don Hanlon to come over for dinner. Our first impressions were strengthened during this long evening. He has a powerful mind. He is intellectually alive. His interests and his passion for learning range over a wide array of topics.

Chicago. 11 September 1965.

A very typical debate has taken place on a local radio station, pitting Menzel against two Chicago NICAP men. The public was invited to ask live questions and these were, by far, the most revealing: The average American now knows as much about the subject as the NICAP members, and displays a better ability for rational thought than Menzel, who was emotional throughout the program. He kept wrapping himself in his professorial toga while the two NICAP representatives fell into his traps "like supporting actors setting up funny lines for a great comedian," as Harry Rymer remarked. All the cases they brought up were ridiculous. For instance, when they mentioned the Brazilian sightings at Trindade, Menzel jumped all over them:

"Now, did you personally research this specific case?"

"Not really . . ." they said.

"Well, I did, and let me tell you what I found out," continued the director of Harvard Observatory, flattening them mercilessly.

Hynek came home yesterday. He leaves again for Great Britain on Tuesday. Since I put him in touch with Gordon Creighton they have become great friends. Creighton is organizing a dinner in his honor next Thursday. Today I will introduce Don Hanlon to him, to give Don an opportunity to show Hynek his excellent research on the 1897 wave. I felt very proud to have unearthed several forgotten cases from that era, which I published in *Anatomy*, trying to re-awaken interest in the UFOs of the American past. But this pride was soon deflated when Don arrived at our house, smiling mischieviously as he presented me with roll upon roll of newspaper facsimiles showing many old sightings of which I had never heard. He has shown me his notes on dozens more.

Three important things have happened since Hynek's response to the Pentagon:

1. Spaulding got very mad at Hynek, yelling that "all his plans were going to go up in smoke." The reason for this may be that Hynek has demanded that the committee be an effective working group, not a phony setup like the Robertson Panel, which worked superficially for only two afternoons and signed a statement which was merely a rubber stamp for the Air Force decision to squelch the whole subject.

2. The Chief Scientist of the Air Force, Dr. Markey of MIT, has made the remark that Hynek could always send a copy of his letter to the Secretary of the Air Force if he didn't like the way it was handled by Spaulding. Of course, he added casually, such a move might lead to ending his contract. Hynek thinks that if he pushes too hard the Pentagon will purely and simply dissolve Blue Book, which appears more and more as an expendable escape valve, not a genuine research project.

3. Hynek had a conversation with Sagan, who suggested the following: If the saucers exist physically they must be detectable by radar. Therefore it should be possible to get information from the aerial defense system to find out if they have such targets on their screens, and how often. Hynek took the suggestion seriously and he forwarded it upstairs as a formal inquiry. This triggered a whole upheaval. There were meetings high up in the Air Force hierarchy, and finally Winston Markey was again sent to tell him that the whole thing was unthinkable. His very words were ***"We advise you to pursue this UFO matter no further!"*** As Hynek remarks with some bitterness, that statement is rather remarkable, coming between the Air Force and an observatory director who has served as its consultant for eighteen years!

Now he wonders if he has not been duped all those years, if there isn't indeed an astute plan afoot to hide the truth from the scientific community and from the world at large. Perhaps Project Blue Book has been a cover from the beginning, operating thanks to the collaboration of a few scientists who go along with anything the military tells them to do.

When Hynek examined the files of photographs gathered by the tracking stations he had set up around the world, he found entire series of UFO observations from all twelve of them, some of which he saved for me. Something very grave is going on, but it is impossible to put a finger on it.

Chicago. 13 September 1965.

A French television team is in town. We have shot a long interview that consumed most of the day. Tonight they took us to dinner. I managed to keep the interview on a scientific level. I even quoted Danjon, the director of Paris observatory, who has said "a subject is only scientific by the manner in which it is approached."

Aimé Michel has sent me a disappointing letter dealing entirely with the sequelae of the Valensole landing. The witness has confessed that he has indeed received from the 'little pilots' a message that was not simply composed of gargle sounds but had a specific meaning. He refuses to disclose what it is when Michel interrogates him. Since I am studying the apparitions of the Virgin Mary at Don Hanlon's suggestion, such contact stories with attending messages do not surprise me any more. We may be dealing with the same class of phenomena, masquerading through various types of entities adapted to each culture.

Chicago. 14 September 1965.

The rain and the fog are back, thank Heavens! Harsh grass is growing over the land reclaimed by the campus from the wide and menacing expanse of Lake Michigan. Furious waves kicked up by the storm crash on the rocks around the site of the new observatory. I love this powerful lake; it gives me a new measure of things.

Chicago. 15 September 1965.

Another good letter from Aimé, who sends me a whole stack of documents. It seems that farmer Maurice Masse has indeed had a contact with the operators, who are supposed to have told him they would come back in October, but he doesn't want to give any details. There are indications that the object was hovering around on the nights preceding the landing, presumably waiting for the propitious moment to make itself seen.

Chicago. 17 September 1965.

Harvey called yesterday morning to report that Ace Books, the mass paperback publisher in New York, is interested in the rights for *Anatomy.* This would raise the distribution of the book to the hundreds of thousands. And last night I found a copy of *True* magazine at the newsstand, with the condensed text in it. I have trouble believing that this is going to be read by three million people.

Chicago. 19 September 1965.

A book I just finished reading seriously at Don's urging raises the possibility of other physical dimensions interacting with consciousness. Under the title *A Woman Clothed with the Sun*, this work by John Delaney is a summary of alleged apparitions of the Blessed Virgin Mary, otherwise known in the literature as the BVM. The reports are extraordinary, very troubling. One is tempted to imagine there is indeed some sort of power that follows, and intervenes into, human affairs. I want to study these observations in more detail, and to compare them with the events related by Charles Fort, whose complete works I am also in the process of reading. I am spending the weekend compiling more sighting index cards. There is so much to learn! The model of previous waves continues to be observed: large cigars, landings, and so on.

Later the same day.

The radio news drones on, spewing out the endless nauseous garbage of human stupidity: some nation "will seek the adoption of a resolution to condemn. . . ." The world is bloody again. This is a miserable planet, mutilated, covered with scabs and bruises. The men with whom I work pretend, as I do, to be virtuous scientists in one of the great civilizations in history. Indeed, tomorrow I will go back to my well-ordered campus, with its nice lawns, the wealth of business well-done, the easy gymnastics of serene brains well-taught. . . . But around us are agonizing masses, fierce beasts. A Marine pilot comments that it's a real pity the Viet Cong don't have airplanes, because it would make the war so much more fun if we could shoot them down! An Indian soldier swears he will die at his post rather than give up one inch of national territory. A Pakistani crowd yells slogans: They will fight to the last man, says the translator, to make Indian blood flow like a river. And the object of all this heroism is a narrow band of useless terrain no one has ever heard of, between two precipitous ridges where nothing can grow.

Chicago. 24 September 1965.

The flu keeps me in bed again and my head is crowded with new ideas. I have spent the week revising my notes for the next examination. To take my mind off my studies I have been reading *L'Express* and it has made me angry at the flood of confused verbiage that keeps coming from France. There is an economist, for instance, who has discovered that "the flexing of the *conjoncture* belongs to the past!" In the meantime *Planète* sends me a brochure written by Louis Pauwels, announcing a new Encyclopedia of Human Civilization in twenty-three volumes, no less, whose complete collection "shall reproduce the spectrum of light, the rainbow, the Ark of the Covenant." As if this were not silly enough the great thinkers at *L'Express*, whose intelligence cannot function unless it quotes other great thinkers, mention "that fine observation by Louis Armand: *Culture is the Actualization of Heritage.*"

There is something awesome and admirable about this stream of stupidity which weaves important words into meaningless sentences. It has an effect on the mind which is not unlike a powerful anesthetic. I am almost tempted to go back to that France now governed by the Angel of the Bizarre, that France where I am ready to bet there is really a Monsieur Louis Armand. I imagine he sits in a well-appointed office with fancy carpets, surrounded by a team of bright young economists.

Whenever we see the appearance of a repetition in history, that is a sure sign that the true significant events have passed unnoticed. When a nation puts at the helm in 1958 the same man, no matter how charismatic or brilliant, who ran it twelve years before, as the French have done with De Gaulle, that is a sure sign of decay.

A mass of polar air has descended over my stately, sturdy Chicago. In the icy wind workers continue to assemble the new observatory at the edge of the lake. As it rises it does not look like a pioneer fort any more, rather like a great warship. Many ideas are brewing within our group. Bill Powers is writing to Costa and has started some experiments based on his work, but he has not been able to exhibit a non-classical effect on gravity.

Chicago. 26 September 1965.

Yesterday morning I took the examination again, staggering around with a high fever. I think my answers were adequate. My head is a bit more clear now. I am studying the Air Force files for 1952, and I am surprised to discover how little the American amateurs have understood about the nature of the phenomenon in spite of all the documents they had available from the beginning. Keyhoe's NICAP is particularly naive when it distinguishes six "phases" in the history of the sightings since 1947, with the landings a characteristic of the last phase only: "In 1954, writes one of their reports, there were a few alleged landings but no proof!" APRO fares little better: They now see the sightings as a systematic exploration of American strategic sites. Those alleged "phases" in the phenomenon only reflect the biased framework of the beholder.

By far the two aspects that fascinate me most are the landings and the ancient sightings, especially when I find these two aspects converging towards the issue of "contact," as they do in the apparition-projection cases that involve suggestion. In this sense the addition of Don Hanlon to our group is timely: observations like those of the Virgin at Knock, to which he called my attention, are indeed amazing: they throw into question any research on the phenomenon conducted purely on the classical scientific level.

I am considering more seriously Aimé Michel's notion that some higher power is manipulating us. Contact with that power or its messengers (in Greek the word for "Messengers" is "Angeloi") may have given mankind its first concepts of the universe. What is upsetting is the modern recurrence of these contacts and the absurdity of the "messages." What is the role played here by the coarseness of the receiver, or by the noise in the system? Is there a deeper, real meaning in the message? What is the possible role of directed hypnosis, of spontaneous suggestion? We know so little about the nature of our own consciousness that we cannot answer these questions.

Aimé Michel believes "they" have brains superior to the human brain. I still question this belief today as I did when I first wrote to him in 1958, seven years ago. What right do we have to talk about the human brain, when we know it so poorly? Perhaps we function like a powerful computer which is barely aware of its own abilities while somebody else, outside, is programming it to use its full power towards an unthinkable goal, or simply pushing the buttons of our unconscious. "Pray more," the apparitions keep saying. Is there a race of beings who suck up the psychic energy of such prayers? Perhaps our major value in "their" eyes is our very stupidity. Everybody likes a nice clean fat pig. But a pig who asks questions about the square of the hypotenuse is an embarrassment to his own species, and an annoyance to his masters.

Perhaps our visitors are not divine or diabolical but just stupid. Bill Powers joked the other day: "Maybe they are some sort of missionaries, who think they do all this for our own good . . ." In any case we are manipulated. The thought is not pleasant.

Chicago. 2 October 1965.

Hynek is back from yet another trip, in time for my thesis committee to meet. Now I am seriously starting work on my dissertation. The coming year will be filled with classes, problems, committees, reviews, seminars and lectures. I have selected a very challenging topic in artificial intelligence that interests me and justifies this heavy time investment, namely the unsolved problem of real-time interrogation of computer data-bases in natural English.

During his trip to England Hynek met the major British ufologists, men like Gordon Creighton, Charles Bowen and Le Poer Trench. Most of them, he says, are literary minds who "seem ready to believe anything." All this remains at the level of a general discussion around the table. There was no genuine scientist in the whole bunch, he told me. This makes our own role and responsibility here all the more significant.

There were two important facts this year: on the one hand, the loss of public confidence in the Air Force, which was precipitated by the publication of *Anatomy,* because my book openly showed how much could have been done. And on the other hand, a few clear examples of cover-up on the part of government agencies: witness the Washington case last January.

Chicago. 3 October 1965.

We have just watched the documentary released by NBC. How fragile and irrational we look, Allen and I! In a way the most solid character, stable as the rock of his own strong inner conviction, turned out to be Joe

Simonton, the old Wisconsin country boy who saw a flying saucer land in his backyard. Three occupants came out of it and he is absolutely sure they gave him some pancakes. Besides, he still has the pancakes to prove it! Quintanilla, too, looked very solid as he stood at attention in front of the camera and recited the Officers Handbook, open at the page: "Undesirable Phenomena." But this doesn't fool anybody. Naturally NBC put the emphasis on silly local sightings that proved nothing. They did not use the Toulouse case. They also cut out Vins, which I had spent so much time documenting for them in an effort to show the scientific relevance of the observation. There is an important lesson to be drawn from all this: These people are just entertainers. I was a fool to believe they would seriously try to show how this problem transcended our knowledge. That is not what the public wants to hear. After watching this documentary I measure better the extent of our loneliness, of our vulnerability.

Chicago. 7 October 1965.

The Little Society met last night, minus the Rymers and Carl, who is now married and ready to go teach mathematics at DePaul University. We were joined by a University of Chicago psychology graduate named Michael Cszentmihalyi. (We quickly decided to just call him "Mike.") He teaches sociology in Lake Forest. His dissertation topic was creativity in art, which clearly qualifies him for our group. Don was with us too. We reviewed the files of the July-August period. Hynek told us the Air Force had now asked its scientific advisory board to review Project Blue Book. Therefore we may soon have to select our best cases in order to present them to the committee.

I am in a strange mood again. I have some free time to study, but I lack the passion that would make me dive ahead. How could I say that I don't have every happiness? Indeed there is nothing I would change in my life. Deep inside me is the joy and love of Janine, the fullness of our existence together. Yet something else looks for expression. An orchestra inside me is playing Wagner and it won't stop. All doors closed, no sound is filtering outside to scare the neighbors while the Valkyries sweep down from the sky to take my soul away. I try to remain calm and cool while they drag me by the ear into their fog-filled caverns.

Chicago. 23 October 1965.

In a single hour yesterday afternoon, Don and I have located three more excellent observations of the 1897 airship. We are systematically going through the collection of the old *Chicago Herald* at the Midwest Inter-Library Exchange Center, a collection so ancient the yellowed pages crumble between our fingers.

In the evening Bill joined us at the Hilton for a meeting with Jim Moseley, a UFO researcher who edits a magazine called *Saucer News*, almost entirely devoted to amusing personal quarrels and petty gossip. He was waiting for us with three of his friends. They tried to capture our conversation on a hidden tape recorder, but we managed to elude this obvious trap. We failed completely in our attempts to convince Moseley that his magazine could fulfill a sorely needed role if he could only give it a more serious orientation, but I liked his sense of humor.

Chicago. 30 October 1965.

Bill Powers and I are becoming close friends. I have begun to study the original psychological theories which he is developing, based on cybernetics. He used to work in the psychology department at the University of Chicago, but he was getting no recognition for lack of a doctorate. His sense of humor, too, was a liability among professors who took themselves very seriously. He is remembered there as the inventor of the Diesel-powered Pogo stick, among other zany contributions.

Hynek called me today to tell me that the UFO question was coming before the Air Force's scientific advisory board on November 5th. If they decide to launch a special investigation, as he is hoping they will, Hynek expects to be called to Washington. Things could move very fast, he said with excitement in his voice.

Last Wednesday we had a full group meeting at Bryn Mawr. We had invited over a man who had written to the Air Force, insisting that he had witnessed UFOs over Chicago and knew the whole secret behind flying saucers. He brought with him amazing rolls of paper covered with diagrams, and he proceeded to give us a grotesque demonstration of his obsession. It was painful to watch him. He believes there is a machine in the sky, hovering over Lake Shore Drive, and it controls him. It is as large as a city and shaped like a baker's rolling pin. It is filled with washing machines and staffed with dead people.

He also believes that each of us is connected to the machine in the sky by a sort of psychic umbilical cord which should never be tangled or knotted, and this fact naturally places our orbiting astronauts in extreme danger. Indeed, we observed that he was very careful not to tangle up his own cord when he moved around the apartment on his way out. He retraced all his steps very precisely when we managed to convince him it was time to leave. He went away happy in the feeling he had enlightened us, and he drove home while we wondered what would happen if the big machine in the sky came to get him while he was driving in traffic at 70 mph.

"I'll think about this guy, every time I take the Lake Shore Drive," said Bill.

That was not a very pleasant thought.

Chicago. 14 November 1965.

I am doing a survey of all the UFO groups I know around the world, and I am using the weekend to bring my correspondence up-to-date with France and England. I corrected the galleys of the French text and I found time to prepare a seminar in mathematical logic based on Novikov's book.

Last night I walked through the "Old City." Chicago is a prodigy, built by humorless giants who have now faded into history, leaving behind deserted, dangerous streets. The only thing alive is the great North wind, howling in despair, looking in vain for the passions and the excesses of yesteryear.

This is not a sad city, but it is buried so deep in its own wealth, or its private misery, that it has no use for such luxuries as public squares with benches, corner coffee shops, or a common language. It feels nothing and cares about nothing. It lies squarely in the hands of corrupt cops. An outside pretense of Morality rules even what is left of the old Mafia. Chicago is a middle-aged whore struck by a mystical crisis, who thinks of entering a convent. It is a suburban matron counting her money, and a little black boy who cries for the toys he will never have, unless he steals them. The great common denominator is the wind.

Chicago. 15 November 1965.

With the imminent publication of *Challenge* in French and of *Anatomy* in paperback, my year of residence at Northwestern almost completed, I should have the right to catch up on my sleep. However, I do work hard and with much pleasure on my thesis research; the program modules are falling into place one by one, and I read everything I can find about artificial intelligence. I look forward to earning some money again when all this is over. I have considered jobs at the University of Chicago or at the Illinois Institute of Technology but that would mean leaving Hynek. He wants me to stay at Northwestern.

Chicago. 26 November 1965.

Life has become a bit slower and easier, although we are utterly broke again. Last week a French television reporter came to see us on the recommendation of Aimé Michel. He has been assigned the general topic of scientific trends in the U.S. I tried to give him some orientation. He was also bringing us some concerns about Aimé, who suffers from sudden fever attacks and once again fears he may have a brain tumor.

Janine, Olivier and I spent Thanksgiving with the Hyneks, as we have done in previous years. At their house we met Chinese astronomer Su Shu Huang, who is joining the astronomy department faculty. He is a sincere and pleasant man. He looks like a perfect example of the kind, brilliant cartoon theoretician, like Tintin's brilliant and zany *Professor Calculus.* As we discussed the question of possible visits by interstellar travellers, I showed him the probability function I had drawn in the appendix to *Anatomy.* As we were about to sit down at the table where Mimi was serving the turkey he tapped the book, smiled and said, with a heavy Chinese accent: "If I had been asked to draw this curve, I would have guessed it was exactly the way you did it here!"

Chicago. 28 November 1965.

The wind was icy and pure when we visited the construction site for the new observatory today. Everything had turned to pastel colors: the beach, the grass, the sand, the sky.

It has been three weeks since the Air Force scientific advisory board met in Dallas. Hynek has not been asked to join them. Repeated phone calls from him and from Quintanilla have not provided any information about what was discussed, or any subsequent decisions that may have taken place.

Finally Hynek did hear something last week: Colonel Spaulding had met one of the Board members in the men's room, and he was told that "the Board thinks it has found a way to approach the problem."

So this is how science works! Our future research depends on a few words some military fellow overhears while pissing in the latrines of the Pentagon. That is a fine image of this society, and of the way it makes decisions.

While all this intense intellectual work was going on in Washington we did not remain idle. Instead we are stubbornly plowing ahead with the study of both old and recent sightings. On Thursday Hynek and I called the witnesses of the Cuernavaca case, where a massive power failure followed the sighting of an aerial object. We have to do this privately because the Air Force does not want any more "good" cases submitted to them, *for fear their statistics would deteriorate.* It would be a disaster for Blue Book to have to report to Congress any increase in the percentage of unknowns.

I am still patiently sifting through Aimé Michel's original files. I have found a new package of data about the landings of 1954. What an amazing period.... And what a shock it would be if these facts were ever presented publicly in the light they truly deserve! Among our group only Don Hanlon has understood the full scope of that wave.

In Aimé's files there are many letters from witnesses and from his readers. I am surprised and impressed at how serious and articulate they are. They show a sober quality, a high level of intellect, far above that of the journalists and the scientists who comment on the cases without bothering to study them.

Chicago. 3 December 1965.

Over lunch Bill Powers and I have been talking about the operators of the craft.

"In some cases," I said, "it almost seems that they are not real beings, but artificial humanoids."

"Yes," he replied, warming up to the subject, "they could be noticing machines, with fast pattern-recognition abilities! In a few minutes on the ground they could gather reams of data about us, couldn't they?"

Any rational definition of thought implies the theoretical possibility of building an automaton that will exhibit that activity. I am studying Markov's theory of algorithms, and of course Alan Turing, looking for clues. The problem of UFOs boils down to two major issues: first, the theory of time and gravitation, then the question of the nature of human intelligence.

Today we are happily celebrating Olivier's second birthday.

Chicago. 5 January 1966.

Classes started again this morning under a magnificent, pure sky. Perhaps it is a symbol indicating that the end of this agony is in sight for me. I have passed the qualifying examination this time. I have registered again for my last mathematics courses, although this means no gainful employment for another quarter, and penny-pinching until next summer. This month again Janine had to go tell our Greek landlady we would be a little late with the rent money. But the computer part of my research is progressing well into uncharted territory: The problem I am attacking is the automatic translation of English-like, natural language questions into computer programs that consult a data-base of ten thousand stars.

Chicago. 10 January 1966.

Some enormous machinations are under way in Washington. Were they caused by the series of new sightings that took place last summer, and are still unexplained? More likely it was a combination of several factors, including these new sightings, NICAP's public statements and the repercussions of my book, if I judge from the letters and the phone calls I am receiving. In any case, someone has become concerned and the huge wheels of the Pentagon have started turning. Hynek has been pushed aside and Project Blue Book was ignored as a minor pawn in the new game.

It is practically in secrecy that the Advisory Board met in Dallas. After a long silence we have learned that a physicist, Brian O'Brien, had been placed at the head of a special scientific committee charged with making new recommendations. He had *carte blanche* to staff his group. He began by selecting Carl Sagan[20] who is an avowed adversary of the reality of the phenomenon. From the choice of members it is easy to deduce what the final decision is likely to be. After all, said Hynek, "Carl won't become a full professor at Harvard by taking flying saucers seriously. And he knows that!"

Hynek had already been told "pursue this UFO question no further!" And hints had been given that he would lose his Air Force contract if he displayed any real independence. Now it is clear that the O'Brien committee wants to see the files; ***but it will ignore the advice and the contributions of those who are in a position to know about those things which are not in the files.***

What foolish illusions we had! And what a lesson in science! Scientific honesty and objectivity appear as a travesty. It was absurd of us to think we could make a difference. Research under such conditions is a meaningless ritual, since truly important issues can so easily be treated behind closed doors by groups already committed to burying the truth.

Chicago. 17 January 1966.

Hynek calls me from New Orleans: Winston Markey, the chief scientist of the Air Force, has just told him of a UFO observation made by Roger Woodbury, associate director of the MIT instrumentation lab. The story is simple: His son rushed into the house screaming there was a flying saucer in the sky; the professor came out and saw it. Now Markey may be forced to take the subject seriously.

Chicago. 22 January 1966.

The Little Society just had a meeting attended by Hynek, Bill, Sam, Don, Janine and me. Hynek wanted to talk about everything that has taken place since the publication of *Anatomy*, including the Woodbury case and the sightings at Exeter that writer John Fuller has recently reported in the *Saturday Review.*

I have received an invitation to participate in a television debate in New York with Hynek and Menzel, to be chaired by President Johnson's cultural adviser. I declined, because I just have too much work ahead of me. Besides, trying to change Menzel's viewpoint is a big waste of time, in my opinion.

For several months I have been trying to convince Hynek that we should prepare a formal research proposal to centralize observation files at Northwestern, with the establishment of a technically competent office to analyze the significant reports. He has agreed that the time has indeed come to develop such a proposal.

Janine and I are slowly emerging from another period of deep exhaustion. We have been working like dogs all week. Wednesday night, after the meeting with Hynek, we wrote computer programs until two in the morning. In the last two weeks I have also drafted two scientific papers, one to be co-authored with my adviser Gil Krulee and the other with Hynek.[21]

Chicago. Wednesday 26 January 1966.

For the last few evenings we have been having intimate dinners in the kitchen, just the two of us, by candlelight. Olivier likes to tell long, complicated stories to the butterflies, the cats and the birds I have painted on his bedroom walls. Janine radiates sheer, sweet beauty. Our nights are splendid. The snow that falls outside seems to insulate our love from the whole world. When she smiles at me all difficulties vanish. But if she is serious or concerned, I suddenly fail to see anything ahead. If she frowns, if she doesn't come over to kiss me often enough I lose my place on the charts, I drift aimlessly for hours like a lost raft.

Chicago. Thursday 27 January 1966.

My two thesis advisers had lunch together today to discuss my research findings and the theory behind the new system I call Altair ("Automated Language Translation and Information Retrieval").

To their surprise it is now running on the Northwestern computer, a Control Data 6400. It is hard for anyone to question the theory once they have eaten the pudding. *Altair* takes English-like questions like "What is the proportion of spectroscopic binaries among stars bluer than K2?" and it converts them into internal formulas that enable it to search through the entire catalogue of bright stars to give a numerical answer in a few seconds.

By contrast, the same problem previously required writing a special program and running it each time against the catalogue tape, with all the attending chances of programmer or operator errors, and a twenty-four-hour delay to get the results.

The significance of this is that a similar program based on the *Altair* framework could follow stock market quotes at the Exchange and answer questions such as: "Which specific companies have shown a growth in earnings greater than ten percent in one year among Utility firms with price-earning ratios lower than twelve?" At Children's Memorial Hospital, a doctor who knows nothing about programming could walk over to a terminal and type: "What is the percentage of infants who develop complications that include a high fever within ten days after an operation for cardiac malformations?"

Following publication of *Phénomènes Insolites* I am getting reactions from my frustrated friends in France. Guérin wants to be vindicated. Aimé's reaction is different. In the United States he would have achieved the recognition he deserves a long time ago, but France has nothing but contempt for people who are ahead of their time. Now he is bitter and dreams of revenge. Both Aimé and Guérin tell me, "You're not going to convince anybody. Your book says clearly that the calculations do not lead to firm results. People will take that as an excuse to reject the whole problem."

Fine. That is their right. I cannot prove to them that this phenomenon exists. Indeed, our theories about UFOs are still obscure and contradictory. But one could say the same thing about many fields of science, like elementary particles or cancer research.

Chicago. Friday 28 January 1966.

A haughty academic, Professor David Park, of the Thompson Physical Laboratory at Williams College, has written an irate letter to my publisher to complain about *Anatomy.* He does not like my quoting the astronomer who said, a few months before *Sputnik*, that "Space Travel is Utter Bilge."

> To choose a distinguished scientist who is now dead and unlikely to make a fuss, and quote his words in a context which makes it appear that he was an idiot, deserves an appreciative chuckle from any huckster. I will not detain you with an explanation of what Spencer Jones was talking about; probably whoever found the quotation can enlighten you.

I answered immediately:

> Before accusing me of quoting such a distinguished scientist out of context, I believe you should check your own sources a little more

> carefully. If Spencer Jones said anything about space travel being utter bilge, I am not aware of it. I was making allusion to a statement by Sir Richard Van Der Riet Woolley, Astronomer Royal and former director of the Commonwealth Observatory at Mount Stromlo, a statement which he *repeated* in a speech to the Press Association annual meeting on June 15, 1960: "I said space travel was utter bilge when I arrived here four years ago. It remains utter bilge."

I thought it would be useful to add:

> To the best of my knowledge Sir Richard Woolley is still living, and probably still of the opinion that space travel is utter bilge (. . .) This is certainly an illustration of the dangers of the Principle of Authority which has so much contributed to the decline of European science. I believe I detect in your letter much contempt for those who dare question the words of the so-called "great" scientists. Let me remind you that the tasks of the Astronomer Royal were defined by King Charles II of England as follows: "The Astronomer Royal shall apply his care and activity to the rectification of the tables of celestial motions and the positions of fixed stars, in order to be able to determine the much-needed longitude of terrestrial locations and thereby increase the perfection of the Art of Navigation." The King said nothing about space travel. Sincerely yours, etc.

I do not expect to hear from Professor Park again.

Chicago. Sunday 30 January 1966.

Costa de Beauregard and his assistant have arrived in the U.S. at our invitation. They landed at O'Hare this afternoon. We greeted them and took them to Bryn Mawr without delay. The temperature was 25 below zero, but Janine warmed us up with a wonderful French dinner. Costa's lecture on Tuesday will be entitled *Modern Ideas in Field Theory*. This will be followed by a slide presentation of their experiment. Dr. Chapman, a physicist from MIT, will be in attendance.

Chicago. Wednesday 2 February 1966.

The lectures are over and we have had ample time to discuss our favorite subjects in private. At the latest meeting of the Little Society, attended by Costa as guest of honor, Hynek conducted a telephone interview of a New Jersey policeman. He had seen an object similar to that observed by Woodbury. However no one, not even Costa, has any new theory to offer to explain the physical effects of UFOs, and I find this very disappointing.

Costa did tell us a few anecdotes of his work with Louis de Broglie. They do not agree on the subject of relativity any more, and this has created a rift in their long-standing friendship.

"It's really a pity to get to that point," comments Costa seriously. "There are so many other things in life we could discuss, and agree on. Quantum mechanics, for example!"

Chicago. Friday 4 February 1966.

It turns out the Brian O'Brien panel has recommended that the Air Force spend about $250,000 a year to scientifically document a number of selected sightings. At least that is what Hynek has heard, and he is going to propose that the work be conducted at Northwestern.

Aimé Michel is feeling better. He is encouraged, he says, by the good reactions our book is enjoying in France.

This afternoon I briefed Costa on the features of the sightings with special physical interest. He listened carefully, and encouraged me to compile a natural history of the phenomenon. He compared my research to what physicists were doing in the early days of electrostatics, which appeared just as absurd as UFOs do today. He believes the objects could be looked at from the point of view of magnetism (are they magnetic monopoles?) and of artificial gravity.

There was a strange incident when Costa and his assistant were at Bryn Mawr. The bell rang and I went downstairs to open the door. A jovial man I had never seen before shook my hand in a funny way. He proceeded to explain with much effusion that he was from the local Masonic Lodge and wanted to give his brotherly greetings to Costa's companion.

I invited him in. The man openly explained his rank and his role in the organization, much to our guest's embarrassment since Masonry, in France, attempts to preserve its status as a Secret Order. In America, of course, it has become little more than a social club.

Later Costa told us: "That's funny, I have worked with this fellow for many years, but I had to come to Chicago to discover that my assistant was a Mason!"

Chicago. Sunday 6 February 1966.

Jim Lorenzen called me this morning to say that Dr. Olavo Fontes, the best expert on Brazilian cases, would be changing planes in Chicago this afternoon. Since we were driving our two French physicists to O'Hare anyway, Don and I waited around for Fontes, who was travelling with his wife and three children.

Dr. Fontes is a small-built, energetic man with intense gray eyes, dynamic and serious, sincerely worried about UFOs. He finds certain correlations in the data very disquieting. For instance, the apparent relationship between the objects and power failures in large urban centers during the most recent wave bothers him: he wonders if control of energy is not the ultimate phase of "their" plan.

I asked him to clarify a statement he had made, that the phenomenon might be hostile.

"They are not necessarily hostile by nature," he said, "only they don't allow us to interfere with their plans; they use violence if necessary, whenever a witness stands in their way."

Chicago. Saturday 12 February 1966.

To a great extent Americans are right when they claim their country enjoys the most freedom in the whole world. But they would be horrified if they could measure how quickly the very expansion of their society is eroding that freedom. I love the United States, the flowering of ideas, the genius in organization and in management that exists here. Having seen this, one can only feel contempt for those nations that only have small problems and still succeed in drowning in them. But I don't like what the U.S. is becoming.

Chicago. Sunday 20 February 1966.

Pursuing the detailed study of the landings of 1954, which represents a massive amount of work, I discover an inverse correlation between the number of reports and population density. As I plow ahead through piles of documents more and more details are accumulating: dates, times, descriptions of the craft and their pilots. But the most important key is still eluding us.

Chicago. Tuesday 22 February 1966.

A strange surprise: I just found my picture in a Japanese magazine, in the middle of an article I naturally cannot understand. I can only guess it's a review of *Anatomy.* The book was also cited in the *Congressional Record* by Representative Pucinski, a Democrat from Illinois.

My article with John Welch about respiratory mechanics following major surgery is being published.[22] Thus everything is moving forward, and we are beginning to think about our next trip to Europe. The American Mathematical Society, of which I am a member, will participate in a major international Congress coming up in Moscow this Fall. For the first time information science will be admitted as one of the topics, and the Soviets have promised to unveil some of their computers. I have decided to submit an application to go there as part of the American delegation, in the hope this will give us an opportunity to try and find out discreetly what is happening to UFO research in the Soviet Union. Officially, of course, there is no such research.

Technological Institute. Thursday 24 February 1966.

It is 9:30 a.m. and I feel distressed. I reached my office, as I do every morning, after driving my little boy to the sitter's house and leaving Janine at the subway station. I parked the car in the large open area at the edge of the lake. It is icy cold here, the shore is frozen, the waves are carrying big white plates all the way to the horizon.

I sat down at my desk and started dreaming. I feel exhausted. Yet I am not sick, I have no valid reason to interrupt my work, to take a day off to rest. The truth is that for the last three nights I have slept very poorly. I keep tossing and turning. In the purple darkness of our room I feel unable to rest. My mind is confused and unsatisfied. Every day I get more letters from readers. I no longer answer them, as I used to do. I classify them and I forget them.

Bless you, Janine, you are all I have.

Chicago. Sunday 6 March 1966.

On Thursday a letter arrived from Hynek, who wasn't supposed to come back until next week. As he was on the campus of the New Mexico School of Mines in Socorro he suffered a sudden loss of consciousness and fell, breaking his jaw in two places. Now he can barely speak. The doctor says he must slow down. But I saw him the next day, that devil of a man, and I heard him complain between his teeth against this "stupid accident" which prevented him from jumping into yet another airplane to fly to Nebraska and interview the witnesses of a recent case. Perhaps this time he will be forced to stay in town for a little while: I took the opportunity to remind him we should be polishing up our research proposal. He gave me the green light. Don spent Saturday with Hynek and me, as we selected the strongest cases that we intend to append to the document.

It was a slow and amusing process: Hynek sat in his armchair in the corner of the living room, near the large hanging light, and I took the sofa, with my back to the lawn and to Ridge Avenue. We spent time watching television: Saturday morning cartoons are a major distraction for him, a sacred ritual that can only be interrupted under the most extreme circumstances. Then we listened to music. Hynek's preferences run more to Offenbach than to Mahler. In the middle of our discussions he occasionally picked up a pencil to direct an invisible orchestra through some fast movement.

Chicago. Monday 14 March 1966.

In spite of our progress with the day-to-day research I have reached a pessimistic conclusion: nothing significant is going to happen here. The Air Force has other fish to fry. Quintanilla even says that the recommendations made by the O'Brien panel will probably be rejected by the brass. The whole problem may then return to that obscurity from which mysterious people in Washington hope it will never emerge.

Part Three

PENTACLE

11

Chicago. Wednesday 23 March 1966.

After breakfast on Monday morning I sat at my desk at home to resume work on a computer program, and I turned the radio on. What I heard told me that nothing would ever be the same again. An information bulletin was flashing on every network, and it had to do with UFOs. Four objects were said to have flown over a farm eighteen miles northwest of Ann Arbor, Michigan. One of them landed near a swamp in Dexter. There were supposedly no less than sixty witnesses, including six policemen, and all the students in a dormitory at a college in Hillsdale.

At 9 a.m. I called Hynek to tell him about it. I also alerted Don Hanlon. Within the hour Hynek was calling me back: He had spoken to Quintanilla, who said he had no interest in the case, since it had not been reported to the Air Force!

"That's not very scientific," Hynek said scornfully.

"I don't give a damn," replied the Major.

Upset, Hynek just hung up on him. Half an hour later it was Quintanilla who called back:

"Can you jump into the first available plane and head out to Michigan?"

"I thought you didn't give a damn. I thought it hadn't been reported."

"Well, since I told you that, it has been reported."

"By whom?"

"By the Pentagon, doggone it! All three networks are talking about nothing else. The brass is having a fit!"

Hynek had classes he could not cancel, so the afternoon had arrived by the time we could hold a meeting in his office. He made plans to leave the next morning, on Tuesday, and said he would call Bill and me if he needed help in the investigation. However we quickly realized that we could not fly with him for lack of funds. His meager Air Force budget gives him no latitude.

So he left at 9 a.m. on Tuesday, in spite of his broken jaw. I had suggested to Bill and to Harry Rymer that we take Harry's plane to fly over the landing site, but nothing was decided. We went on talking and speculating instead of doing something; I got increasingly frustrated.

More sightings took place during the night. Again Bill spoke of going to Michigan. The radio did mention that Hynek was thinking of bringing other members of his "UFO Committee" to assist him, but he kept procrastinating. On Tuesday night I reached him in Ann Arbor:

"How do you want to proceed?"

"The most important thing you can do is to complete our proposal as soon as possible," he told me.

So I pushed aside my dissertation and the piles of computer listings to extract the draft of the proposal from my drawer.

Now a sudden blizzard has buried the entire Midwest under a foot of snow. The roads have become impassable. The temperature is dropping. Yet the flap continues to develop. Several people are said to have seen an oval-shaped craft in Normal, Illinois, on Tuesday night.

The papers, the radio, all the media are feverish. Hynek is obviously enjoying the VIP treatment he is receiving in Michigan. He has two Air Force sergeants at his disposal, a chauffeur and a Jeep. The military are worried. This is the first large series of sightings since Fuller's widely read article in *Look.* A radical change has taken place in public opinion, the situation is mature. Hynek should speak out.

Chicago. Saturday 26 March 1966.

Two thunderbolts yesterday. First the morning papers came out with what they claimed were "actual photographs" of the Michigan objects. I rushed out to buy some copies . . . and I dropped the paper in disgust. Anyone with elementary training in astronomy could see that the pictures simply showed Venus and the Moon. While the scientists snickered around us, the confusion was increasing among the public. There are new sightings all the way into Illinois and Wisconsin. All four states around Lake Michigan are now involved.

Second thunderbolt at 4 p.m. when Hynek gave his eagerly awaited press conference in Detroit. Every major newspaper, every television network in the United States was represented. He offered his final verdict about the two cases in Ann Arbor: *Marsh Gas!* He reserved judgment on the other cases. But the reporters were already out the door, rushing for the telephones. Marsh gas indeed!

Now the main witness is being ridiculed and harassed in his own community. Overnight his neighbors vandalized his house, broke his windows, threw stones on his car and phoned at all hours to call him "Head Martian!"

Later the same day.

Hynek has just called me: He felt the need to explain to me his Detroit statement, knowing that I was seriously disappointed, like everybody else.

"There was so much confusion, Jacques, you can't imagine what a pandemonium we had. This just wasn't a strong enough case to base a real offensive on it. I would have found myself in a very fragile position. . . ."

"What about the Hillsdale sightings?" I asked. "Sixty girls said they clearly saw the lights from the College."

"Well, I spoke to them too, but I couldn't get a straight story from them. Again, it was all muddled, and the lights they saw were very faint." He chuckled: "You might say that neither the lights nor the girls were very bright. . . ."

"At least you could have reserved judgment, you could have said, *This demonstrates the need for research once again, why shouldn't science undertake a serious study?* You could have shown there were clear patterns, even if no single sighting is going to prove anything. . . ."

He sighed. "Well, maybe I spoke too fast, but you understand everyone wanted a statement from me, some explanation right away. Too bad you weren't here with me. At least I have strengthened my reputation as a hard-nosed scientist."

In the meantime all the cartoonists are having a field day with the astronomer who thinks flying saucers are nothing but marsh gas.

Chicago. Sunday 27 March 1966.

A radical shift in public opinion is continuing right under our eyes. It is so obvious that the objects are real, and so many good cases have accumulated, that people have accepted the existence of the phenomenon. Hynek's statement in Detroit was such a caricature that it is violently rejected by the public—the same public that would have ridiculed and crucified him only a month ago if he had argued that UFOs deserved scientific study, and a significant expense of taxpayers' money.

It is hard to keep a cool head in this crisis. NICAP plans to give a press conference tomorrow in Washington; Congress may well start new hearings this time, a possibility that scares the Air Force out of its wits.

Hynek has just told us that if he were called before Congress he would hide nothing. He would expose the sloppy investigative methods Blue Book has been using all along. He even asked me to leak a copy of the full text of his Detroit statement to NICAP; it contained a recommendation for scientific study which the media have ignored:

> The Air Force has asked me to make a statement on my findings to date. This I am happy to do, provided it is clearly understood that my statement will refer to the two principal events as reported to me: the event near Dexter, Michigan, on March 20, and a similar one at Hillsdale, Michigan, on March 21. It does not cover the hundreds of unexplained reports.... I have recommended in my capacity as scientific consultant that competent scientists quietly study such cases.... There may be much of potential value to science in such events.

All the journalists, naturally, ignored this cautious preamble, and focused entirely on the misleading "Marsh Gas" explanation. But he deserved it; why did he have to say, "A dismal swamp is a most unlikely place for a visit from outer space!"

Chicago. Saturday 2 April 1966.

This week I finished writing our proposal. A media storm is now raging against the Air Force. Congressman Gerald Ford, the Republican from Michigan, has expressed outrage at the suggestion that his constituents couldn't tell marsh gas from spaceships, and he has written a formal letter to the Armed Services Committee of the House to demand immediate hearings. But who will get the job of studying the problem? Can the Air Force be trusted to be unbiased? What about NASA? The Space Committee of the Senate, which controls all the big NASA decisions, has already rejected any notion of considering the UFO question. Congressman Rumsfeld, whom Hynek knows fairly well, has explained to him that NASA didn't want to tarnish its image. So they threw the hot potato back to the Armed Services Committee. Perhaps they know that some dark and horrible secret may lurk in the basement, and they want no part of uncovering it.

On Friday morning Hynek left for Dayton, expecting to see Quintanilla in Wright Field, but the Major had already left for Washington, where he will stay for six days at the disposal of the Committee, ready to testify. Hynek just called me to say that he had to be in Washington himself next Tuesday. He wants to see me over the weekend to help him prepare his statement, which will be part of the record.

"You ought to see the Air Force people," he said with glee, "they're running around like chickens with their heads cut off!"

Chicago. Thursday 7 April 1966.

The hearings started with the committee in executive session, but the doors were soon opened to the press and to the public. Hynek made his statement without clearing it first or even showing it to Quintanilla. Things were so rushed that he didn't even have time to retype it, so what he pulled before the Congressmen were my four sheets of yellow paper covered with the wobbly type of my tired machine.

He was somber when I called him yesterday morning.

"We've gone too fast," he said. "Maybe we've blown our big chance."

The Secretary of the Air Force, Harold Brown, said in his testimony that he was in favor of having a new scientific committee study the 648 reports that are still carried as "Unidentified" in the files. Now Hynek is afraid the Air Force will think he is too independent and will push him aside.

Things were different at noon today. Bill and I had to wait half an hour for Hynek, who was in a phone conversation with an editor of *Science* magazine who had just read *Anatomy.* "Yes," said Hynek, "Vallee is a solid guy."

"Does his book describe the phenomenon accurately and fairly?"

"I think so."

"Then why did you wait so long, Doctor, to reveal to the scientific world that the problem was a serious one?"

The obvious answer is that the scientific world is as close-minded as an old pig and that his colleagues would have treated Hynek as a dangerous visionary if he had spoken out. He would have been replaced by some unscrupulous bureaucrat who would simply have swept the data under the rug. But can he say that to *Science* magazine?

Chicago. Friday 15 April 1966.

My thesis committee has met and I am ready to start work on the written part of my doctoral examination. Unfortunately I have another bad cold. The *Flying Saucer Review* has just published my "Ten-Point Research Proposal," but I have a strong feeling I am wasting my time with the believers. They don't need to do any research. They already know what the flying saucers are.

Every day there are more cartoons about Hynek, ridiculing his "Marsh Gas" explanation. Oddly enough this has the effect of placing him increasingly in demand as a lecturer. People want to see and touch him, and he enjoys every minute of this newfound celebrity.

Chicago. Sunday 17 April 1966.

It is raining on this city, which has turned to a silvery gray. Standing at the window I daydream, watching the shiny roofs of the cars and the puddles in the school yard beyond. The scene reminds me of rainy afternoons in Paris and in Pontoise: memories so powerful, so haunting, of passersby dashing out of bakeries with their bread under their arm; umbrellas opening everywhere like flowers.

I wonder what life would be like if we went back to France. I see a little house, and all my American books on a shelf; I see rain falling on a little garden. What is the feeling that comes with the rain, bypassing the conscious mind, when we have made love too long and too hard, when the brain has been dulled by great stunning waves of life? "I come from the great universe," says the raindrop perched for a moment at the edge of the roof. "I shall make a torrent that will sweep your silliness away into the gutter. I am here because men ignore their own greatness."

Chicago. Tuesday 19 April 1966.

Before starting work on my examination I took time to read a chapter from a book by Scholem recommended by Don, *Major Trends in Jewish Mysticism.* It describes the doctrine of the Merkhaba, the throne of God as mentioned in Genesis and Ezekiel. There used to be a wise rule that forbade the study of these subjects until the student was over thirty, says Scholem. Merkhaba mysticism is the oldest subject in the Jewish tradition. The dangers of letting consciousness "rise through the palaces of the Throne" are very great, say the cabalists, because Angels and Archangels attack the traveller to chase him away, and a great fire comes from inside his own body to devour him.

I think I know what that great internal fire is.

The relevance of these traditions to the study of the UFO phenomenon is obvious: people have always seen similar apparitions in the sky and have always given them religious interpretations. Thus in the Apocalypse of Abraham, the Patriarch is said to have heard "a voice speaking inside the celestial fire with the sound of many waters, the sound of the storming sea." The same loud humming or whirring noise is described elsewhere as "the hymn that the Throne of Glory sings to his King." And Enoch, like Ezekiel, speaks of being taken away from the earth "on the stormy wings of the Shekhinah." This secret is supposed to become universal knowledge "in the next Age."

Later the same day.

For the last week I have been dragging several ills: a persistent cold, the anguish of the examination and a sort of dizziness that may be due to a slight ear infection. I am also working too hard on my second book, trying to do too much. I am upset by our lack of progress with the Air Force. I had proposed that Quintanilla pay for the photocopies Don needs to make in order to preserve the airship clippings of 1897, a mere $200 or so, but he turned us down. And Hynek has received a letter from Senator Dirksen: "I support your idea of a scientific committee to study UFOs, but don't count on the Defense Department to fund it!" So far the Congressional Hearings in which the believers were putting so much hope have changed nothing: the Committee has simply given the Air Force the go-ahead to implement the recommendations of the O'Brien Panel. After all, it does appear logical to go on with an academic study.

Chicago. Sunday 24 April 1966.

Janine is now reviewing the first two chapters of *Challenge to Science* and I make changes based on her reactions. I am sending a package of UFO magazines to science writer Alexander Kazantsev in Moscow, in exchange for his articles about the Dogons of Japan and the archaeological mysteries of Baalbeck. He argues that the Dogon statues wear strange masks that closely resemble astronaut helmets and that the Baalbeck terrace is not only hard to explain in terms of weight, transportation, assembly and precision of the stones, but that it would form an ideal base for the launching or the arrival of spaceships.

The dizziness is gone now, and I feel much better although in moments of weakness I still dream of spending the rest of my life in a little house in France with an old wooden door and shutters on the windows and a good science-fiction library around me. Yet I used to have all that, and I left it behind for very good reasons.

Chicago. Monday 25 April 1966.

The presence of Death is what I suddenly feel this morning, very close to me. The steering wheel vibrates angrily under my burning hands. Something has come howling through my life, out of nowhere. It is like a high-speed train tearing through a herd of cattle lounging on the tracks, scattering bodies and blood. Whatever it was, it has left me inexplicably exhausted. It is barely 9 a.m. but I have had to turn around and come back home. I could not drive to the University in this state of emotional exhaustion. I felt like an old, drunken man.

Chicago. Tuesday 26 April 1966.

I was simply exhausted, physically and mentally. After a good rest things are going better. Tomorrow I will start on the operations research part of my long examination. Hynek has received a letter from Sagan, who is still working with O'Brien. Other investigators are trying to get official support and money for research.

The governor of Florida, together with four newsmen, has seen a UFO from his plane. It seemed to be under "intelligent control," but what does that expression really mean?

Chicago. Friday 6 May 1966.

What is happening is even more complex than anything we had imagined. We have received new, totally unexplained reports of sightings by policemen in Ohio and by an aircraft electronics instructor in Oklahoma. The latter involves an elongated object landing on a road and a normal human being next to it, an astonishing report. The cigar-shaped fuselage took off. It was without wings and without any visible propulsion system. Could that be a secret prototype?

Bill Powers and I have discussed the issue of Hynek's continuing work with Blue Book: should we advise him to resign? I don't have much time to dwell on the recent sightings. My own daily life is consumed by the theory of algorithms, Markov chains and dynamic programming.

Chicago. Saturday 7 May 1966.

The whole day was spent at the Technological Institute typing my answers. I am straining under this pressure. Hynek is different, too. He seems very tired and preoccupied, unable to relax as he did before whenever we talked. I should again point out to him that he is giving too many lectures, taking too many useless trips. Next week, for instance, he is going to cover most of the country again, giving three lectures, including one to the Northwestern Alumni.

"I know that you regard these lectures as a responsibility to inform the public," I already told him last week. "But the events that are taking place demand that you stay here, on the bridge, to steer this ship."

He knows all that, and he acknowledges the advice. Bill Powers is telling him the same thing. But he likes to fly around. He just loves to be on stage, in the spotlight, even if it is only in the lecture hall of some third-rate little college in Nebraska, or in the back room of some Holiday Inn.

At noon I ate a sandwich in an Evanston coffee shop, and I watched the nice students around me. They seemed harmless enough, clean, with no experience of life and passion. I admire many things about the United States, I have acquired huge respect for this country, but there is something deeply hypocritical in American life, and it is most obvious in this quiet, silly academic suburb. I can't blame Hynek for wanting to get away at every chance he gets. Evanston is puritanical and artificial. It clearly separates three communities, keeping the affluent students who drive the Jaguars their wealthy fathers have given them for their 18th birthday apart from the professors, a bunch of dreamers who go to the Campus on their bicycle or on foot, and the elder matrons who drive Lincoln Continentals, own entire neighborhoods and call their attorneys in a huff whenever anyone steps on the gardenias.

Hynek told me about the time when he had to pick up a distinguished lecturer at the airport.

"Would you like to borrow my car?" offered one of his wealthy students, trying to be helpful.

"Why would I do that?" asked Hynek, astonished.

"Well, for your image. You see, your car . . . you know, it looks a little beaten up."

"This man is coming to see me," said Hynek, "he isn't coming to see my car!"

By contrast with these rich, idle zombies, Don Hanlon is quite refreshing. He has a deep, solid intelligence. Yet sometimes I think he has already been hardened too much by life. He is a little too tough, too proud. He used to be a gang leader in the streets of Chicago, before he discovered Jung, the Kabala and mysticism. Is there enough love left in him, enough tolerance towards ordinary folks?

The other day he confided to me that he was somewhat disappointed the first time he met us:

"After reading *Anatomy* I thought such excellent scholarship could only come from an older man," he told me with his good sarcastic smile. "So I was sure the charming young woman with the sexy voice and the French accent who answered your telephone was your daughter. I came to your house all ready to seduce Janine, before I realized you were only a few years older than me, and she was your wife, you lucky guy!"

Chicago. Thursday 12 May 1966.

My examination has been going on for a whole month. I am trying to finish the section on compiler and automata theory. I am immersing myself in the literature of the topic, a mass of articles which contribute to the information explosion while claiming to manage it. Very few of these articles offer any real solutions to practical problems.

Two nights ago we watched a one-hour CBS documentary on UFOs sponsored by IBM, of all people! It is the documentary in which I had refused to participate, preferring to stay behind the scenes. I congratulated myself on this decision. Menzel could be seen pouring benzene over a bath of acetone to explain how mirages caused UFOs. Disgusted, I turned to Janine and said, "Don't forget to let me know next time it rains benzene, so we can go out and watch the saucers!"

Under the pretense of giving equal time to all sides CBS had prominently featured the ridiculous claims of old George Van Tassel and his

Venusian friends. This was followed by Sagan, who made speculative statements about the extremely improbable nature of extraterrestrial visits: he might be willing to accept one trip every ten centuries or so, but not more!

They also had the typical handsome officer with steel-blue eyes who came from NORAD to recite his well-rehearsed lesson: There are no unidentified echoes on the radar screens that protect America! I thought of Hynek's joke about the Air Force being secretly run by a little old lady in Nebraska who owned General Motors and was scared of alien invasions. After the handsome officer came a wise, reliable-looking astronomer from the Smithsonian who said there were no unidentified objects on astronomical photographs, either.

The last two interviews were carefully edited to completely isolate Hynek. Yet he was the only sincere man in the whole bunch, the only scientist genuinely searching for the truth. He could have proven that both the NORAD and the Smithsonian statements were outright lies: we have the pictures and the files to prove it. But the editing of the documentary made dialogue and rebuttal impossible, as usual. His most important statements have been cut out.

The Project Blue Book secretary in Dayton told him in confidence that Quintanilla had spent three days in Washington actually editing the documentary along with the CBS team. That is a very serious revelation. Yet the American public is naive enough to trust such documentaries, especially when they close on a picture of everyone's favorite uncle, Walter Cronkite himself, standing with his back to a blow-up of the magnificent Andromeda nebula.

All this does not make me as angry as it perhaps should, but it saddens me. After all, many of the facts are here, in my filing cabinets, so I don't need CBS or Uncle Walter to tell me who is lying and who is telling the truth. **But the American public is being taken for a ride**. I would like to pass on the real facts to the people who can use them, and I do not know how to do it, other than working as hard as I can on *Challenge to Science*. I hope that at least a few people will know how to read between the lines, and will do their own research once they know where to look.

It is still windy here. The cold and the rain feel good as I walk along the rocks of the shore where waves crash like cataracts.

Chicago. Sunday 15 May 1966.

Hynek called me this morning, with much bitterness in his voice. He was the guest speaker at the Midwest astronomers banquet in Madison last night, on the theme: "Flying Saucers I Have Known." He quoted the cases most susceptible to intrigue an audience of scientists. Yet Northwestern students who were in attendance later told him they had heard many negative comments among their colleagues. People were waiting for funny stories of naive witnesses and stupid farmers who confused Capella with a spacecraft and marsh gas with an extraterrestrial mother ship. Instead he described the Mount Stromlo case, in which all the witnesses were astronomers, and the MIT case, and the sighting at Monticello by the two anthropologists we had interviewed in Madison, too close for comfort. And the explanations for these sightings? There were none, he said courageously. The audience didn't like it.

Personally I am skeptical about the possibility of any genuine research in this country even if the Air Force does create a special commission. It seems obvious that such a group will be dominated by skeptical people. I can imagine it conducting very costly statistics like Battelle Memorial Institute did in the days of *Report 14*. They can easily bury the problem in technical jargon while pretending to study it.

Chicago. Monday 16 May 1966.

A Northwestern graduate student has overheard a conversation about Hynek among two astronomers in a Tucson restaurant: "With all the money he makes as Air Force consultant, it wouldn't be surprising if he turned out to be the one who has been starting all those silly rumors about flying saucers!"

Chicago. Tuesday 24 May 1966.

My written examination is finally over; I have submitted a book of answers 150 pages thick and my Committee has accepted it, clearing me for the oral part and my dissertation itself. Relieved, I wrote to Alexander Kazantsev to confirm I would travel to the Mathematics Congress and to give him the dates of our trip to Moscow, then I drove to the beach to look at Lake Michigan. I sat at the edge of the water and started writing a new article about the patterns behind the landings for the *Fly-*

ing Saucer Review. I have read the Valensole file again. I imagined what it would be like to stand with Maurice Masse in his field in Provence. Perhaps I will do it some day. The clouds rushing around the skyscrapers of Chicago and the howling airliners which seemed to hover on their way to O'Hare brought me back to reality. Lake Michigan was very quiet today. Miles away to the North I could see the two white domes of the new observatory.

Chicago. Wednesday 25 May 1966.

Harvey Plotnick has just returned from New York. He tells me that the recent success of *Anatomy* has opened up a new genre and triggered a chain reaction among the big publishers. The people who had written off the whole subject as utterly dead have revised their opinion now that they have seen the reviews. Every major New York house has a UFO book on its list of forthcoming titles, Harvey says, to capitalize on the market we have revealed. For example, G.P. Putnam is bringing out a book entitled *Incident at Exeter*, by John Fuller, which is the best of the bunch.

Chicago. Tuesday 31 May 1966.

The Memorial Day weekend was spent making more revisions to *Challenge*. I see things so much more clearly, now that I can refer to the French text which is already in print! I have paid a heavy price in time and hard work, for a few clear ideas. Now I start preparing for the oral part of my doctoral examination while listening to some appropriate music, Victoria's *Officium Defunctorum*, an extraordinary piece from the Spanish Renacimiento sung *a cappella* by the choral of Pamplona. Just what I need to take my mind far, far above the feuds and the recriminations of conflicting UFO theorists.

Chicago. Tuesday 7 June 1966.

Françoise Gauquelin, who is in town for a few days, came over for dinner tonight. She still works tirelessly on her statistics with Michel, and they make steady progress in their study of man's unexplained relationship to the cosmos. Are they rediscovering a forgotten science, which might have given rise to the astrological tradition among primitive peoples who kept the vision but lost the method?

Françoise told us that Aimé Michel had gone off permanently to his beloved eagle's nest, a farmhouse in his native village high in the Alps of Savoy. I don't blame him, but his Parisian friends are sad: he will lose contact with all the important trends, they predict. His enemies say: "We told you so, he's out of touch, he just wants to talk to himself." Few people understand his need for communion at a higher level, beyond the petty gossip of Paris, beyond the desire to explain his beautiful vision to the constipated rationalists who are the new arbiters of French thought.

Chicago. Wednesday 8 June 1966.

A major event has happened in the last few days. A friend of Brian O'Brien has launched a bold new campaign that is taking everybody by surprise. His name is James McDonald, forty-five years old, professor of atmospheric physics at the University of Arizona. Having suddenly become interested in the subject, he read many books, including *Anatomy*, and decided to do his own research. Through O'Brien he asked to be authorized to spend two days at Wright Field. He began by requesting to be shown all the cases of "globular lightning." He was amazed and horrified at what he saw: case after case that obviously had nothing at all to do with electrical discharges in the air. So he asked to see more and started reading the general files, getting increasingly upset as he kept on reading.

McDonald moved very fast once he realized, as he told us bluntly, "that the explanations were pure bullshit." So he bypassed the Major and went straight to the General who heads up the Base, to tell him exactly what he thought of Blue Book. After forty-five minutes, which is much longer than Hynek ever spent with the General, they were talking about the humanoid occupants! Then he flew back to Arizona and started contacting all the amateur investigators, one by one, from APRO to NICAP. He made an appointment to see Hynek.

We have just had lunch with McDonald today, and it is clear that an entire era has come to a crashing end. This man has many contacts, many ideas, and he is afraid of nothing.

He reached the campus about 11:30 and Hynek took him on a tour of the observatory. At noon I went to pick them up, and I drove them back to Hynek's office, where we all sat down. McDonald signed the Guest book, and I presented him with a copy of *Phénomènes Insolites*.

After that the serious business began, with a forceful attack against Hynek:

"How could you remain silent so long?"

I jumped in before a fight could erupt.

"If Allen had taken a strong position last year the Air Force would have dropped him as consultant and we wouldn't be here talking about the phenomenon."

McDonald brushed aside my comment.

"I'm not talking about last year. It's in 1953 that Allen should have spoken out! Public opinion was ready for a serious scientific study."

"In 1953 I was nothing, a negligible quantity for the Air Force," replied Hynek. "Ruppelt regarded me with considerable misgivings, as a first-class bother. He didn't like to have a scientist looking over his shoulder."

"Yet he says some nice things about you in his book."

"That didn't stop him from playing very close to the vest whenever I was around. He didn't let me see his cards."

The debate remained on that level, with McDonald insisting that Hynek had a duty to say something while Hynek would only concede that he had been "a little timid." Bill and I kept trying to explain to McDonald that any forceful statement by Hynek would have thrown him out of the inner circle. It could even have precipitated a decision by some General to put the files into the garbage.

Eventually we set aside our differences and the four of us went to lunch. At the restaurant the discussion became more constructive. Hynek retraced in detail the real history of Project Blue Book, truly an incredible tale. Thus he explained how, following Ruppelt's departure, he had seen a succession of unqualified, uninterested officers at the head of Blue Book. He was almost never invited to give an opinion. Hardin neglected his duties completely, he said. He spent all his time following the stock market while waiting for retirement—indeed, today he runs a brokerage office. McDonald was astonished, although he ought to have some experience of how the military runs. I can see how difficult it will be for the public to understand the situation, when the history of this incredible period finally gets written down.

Chicago. Thursday 9 June 1966.

Two o'clock, and my oral examination is over. Things had started very badly this morning. I woke up too early and I was seized with nausea. I managed to drive all the way to Evanston. When I called my adviser to tell him I might be late, he said, "If that can make you feel better, you should know your written examination was outstanding, we're all impressed...." Another professor told me my responses were among the best he had seen at Northwestern, and the oral part went well.

Afterwards Hynek bought me lunch. He assured me the road was clear for my thesis work. Naturally we compared notes about McDonald, and we discovered we had the same impression: extremely positive and enthusiastic at first, then a certain feeling of mistrust towards the man, an uneasy reaction that was hard to define.

Chicago. Sunday 12 June 1966.

Tomorrow Hynek goes to Wright Field to meet with the Base Commander, General Cruikshank. He wants to find out just how impressed he was with McDonald's arguments. He has also written to Harold Brown, telling him frankly how he felt about the whole issue. In his answer the Secretary of the Air Force says he has "carefully studied" his ideas: indeed the Air Force will go ahead with university-based investigations, which McDonald wanted to scratch as academic, worthless and irrelevant. The whole project is now in the hands of General James Ferguson in the Pentagon. Hynek and the Dean of the Northwestern Faculty of Sciences have an appointment with him. Other universities are being approached but it seems that Northwestern is the only one with enough guts to look seriously at the problem.

John Fuller is trying to set up a meeting with U Thant, the UN Secretary-General.

Chicago. Thursday 16 June 1966.

While Janine and I are beginning to bounce back from the recent turmoil, Hynek seems more preoccupied and tense than ever. The source of his worries is McDonald's abrasive, insulting ways, so diametrically opposed to his own gentle and witty personality. I pointed out to him that McDonald's radicalism would in fact make the way smoother for him. He is preparing a lecture before the American Optical Society in which he will argue that a serious, sober study is needed. In contrast, McDonald now advocates throwing everything overboard.

Our old black Roadmaster has finally died. The engine was still running but all the other systems were failing one by one, so we had to replace the monster. It is in a brand new Buick, a little blue convertible, that I drive Hynek to the O'Hare bus stop at the Orrington Hotel, and we have coffee together under white and pink parasols, discussing various hopeful scenarios. He and the Dean are scheduled to meet with Pentagon officials to propose the creation of a specialized research center at Northwestern. There is bright, hopeful anticipation in the air.

The Air Force has decided to get rid of the UFO problem once and for all by throwing it to the scientists, so the whole issue has become a political one. The Johnson Administration doesn't want any move that could embarrass its Secretary of Defense, Robert McNamara. The Air Force wants to be able to short-circuit all inquiries by referring any sharp question to some sedate university.

Chicago. Friday 17 June 1966.

I am trying to visualize what our research center could do, how it would be organized physically. But I am not sure I want to go on living in Chicago and raising our son here. We do have many acquaintances on Campus, and we are close to a few people with whom we have dinner occasionally, or go to a party or a movie. But apart from these few friends we see nothing but an emotional desert all around us.

Chicago. Sunday 19 June 1966.

Hynek has suddenly become fascinated with psychic surgery. A wealthy industrialist he knows has come back from the Philippines with a color film of "Dr. Tony," who is said to remove tumors and cure cancer by putting his bare hands into the bodies of his patients, only occasionally using a crude instrument like a pen knife or a spoon to scrape off some infected tissue.

Recently Allen announced that he was going to borrow the film to show it to his colleagues at the observatory. So I invited over two top surgeons with whom I am working downtown, namely Dr. Lewis (who invented the technique for blood refrigeration in open-heart operations) and one of his senior assistants.

After the movie, which was suitably gory, the astronomers were turning green but the two surgeons seemed delighted.

"Well, what did you see?" I asked them in Hynek's presence.

"We saw two kinds of things," said Dr. Lewis. "First of all we saw some absolutely fascinating primitive surgery, like the time when Tony removes a tumor by scraping it off the back of the eyeball, or when he breaks open a boil on the woman's skin. We are taught in Medical School that's where our own science comes from, but we rarely have a chance to see it done as it was done in the Middle Ages. By the way, we take the same steps Dr. Tony takes, only we perform the operation under anesthetics, with cleaner instruments and sharper knives. But make no mistakes about it, we do basically the same thing!"

"What about the psychic surgery itself?"

"When he was supposed to open the abdomen I think it was sheer sleight of hand. We couldn't recognize any internal organs."

I was impressed with that reaction. He never said, like so many rationalists, "I don't believe it because it's impossible." He simply said he did not recognize the internal organs. Here is the sign of a true scientist, a pragmatic man who deals with facts.

Chicago. Tuesday 21 June 1966.

Hynek returned to Evanston today with some important news. He had gone off to New York to see Fuller and U Thant, who was anxious to know what the United States is going to do about the UFO question, because some of the member nations had expressed concern. Hynek assured him that the Air Force had firmly decided to create an independent scientific commission, the only remaining question being to know where it would be located. This could be a step towards the UN setting up their own study, under the Space Committee.

"You know that I am a Buddhist," said U Thant. "We believe there is life throughout the universe."

"Most astronomers would agree with you," replied Hynek. "The question is to know how 'they' would ever come here, given the enormous distances involved."

"Perhaps their lifespan is measured in centuries rather than years. Coming here could be as simple for them as going around the block is simple for us!"

They went on to discuss the possibility of alien bases on Mars. U Thant also wanted to know about observations made by pilots. Hynek quoted *Anatomy,* adding in his usual cautious manner that such observations "had indeed been reported."

Finally they discussed possible action points. U Thant explained that the initiative could only come from member nations. It is their government which must bring the subject up before the General Assembly.

Hynek has requested my help in this. I will ask Aimé Michel if he can set up a meeting with a French government representative, and I will write to Fontes to get him to initiate similar action with Brazilian authorities.

While we were thus talking in Hynek's office a long letter arrived from the Secretary of the Air Force, and he read it to us. Brown said that Hynek's recommendation to turn the problem over to scientists was now getting the highest priority.

We also learned that Lyle Boyd, Menzel's co-author, would be happy to move to Evanston if we were to get the research contract. Finally, a group of unnamed scientists said to be from Wright Field has asked Hynek to supply his "twenty best cases." Who are these people? I advised him not to send them anything until we know for certain where these friendly strangers come from and what kind of hidden agenda they may have.

Chicago. Wednesday 22 June 1966.

What I heard today has left me very puzzled. Hynek and the Dean just came back from Washington, where they had conferred with General Giller, who serves on Ferguson's staff. Ferguson himself belongs to Air Force Research and Development. And Giller told them categorically that *under no circumstances would Northwestern get the contract.* The Air Force, he patiently explained, is looking for a university that has not had any previous involvement with the problem. Hynek says it is like opening a restaurant and looking for a chef who has not had any previous involvement with cooking!

The real reason is perfectly clear. They are only looking for a rubber stamp, and the last thing they want is the intellectual independence of Hynek's team. There may be a small silver lining here: perhaps the Air

Force will give us the task of organizing the historical data, while the major contract aims at future cases only. Another relatively positive aspect is that Hynek will now be free to speak out without an "Air Force" label. He will begin by writing a carefully worded preface for *Challenge.* I also recommended that he push harder towards the creation of a UN commission, where he would be the logical leader. And I added:

"Let's see how those who get the award will explain away the UFO Phenomenon. They don't know what they are getting into."

Chicago. Thursday 23 June 1966.

Hynek can't sleep any more, caught as he is between McDonald's vitriolic attacks and the Air Force's desertion. He turned a lot of ideas around in his head last night. Finally he picked up the phone and woke up the Lorenzens to share his distress with them. They told him that Jim McDonald had had a strong interest in UFOs for the last four years. So, why is he pretending to have suddenly "discovered" a scandal? Why has he picked Hynek as his primary target?

Yesterday, at the Pentagon, Hynek told the Air Force that he felt like Moses, who led his people to the Promised Land but was not permitted to enter it. One of Giller's assistants took him aside and told him that the contract would be renewed yearly, so that it might well come back to Northwestern. Hynek would like to believe it, but Bill and I told him he was naive. It is probable that those words were said simply to cushion the blow.

We are going to make a real effort to get the contract for the encoding and classifying of the files. If Northwestern is selected I will abandon my plans to leave the United States and return to France next year.

Under great pretense of secrecy, Winston Markey has "revealed" to Hynek that three men would supervise everything about the scientific program to be run at the university that will eventually be selected. And these three men are Hynek, Menzel and McDonald. We take that with a huge grain of salt, too. This man may be the Air Force chief scientist, but he seems to send Hynek into one blind alley after another.

"Why do you think the Air Force is so anxious to wash their hands of the whole problem?" I asked Allen.

"The Air Force is not a monolith," he answered. "I saw it clearly at the meeting with Giller. There are several groups, each with its own axe to grind. One of these groups is clearly thinking that UFOs are none of the Air Force's business, and that the problem belongs in an academic research setting. They are the minority. Most of those guys think the phenomenon is just a lot of hogwash, and they are looking for a convenient university to write a negative report and kill the whole business."

"Aimé Michel has written to me that he was still in contact with someone from the British Intelligence Service, who told him something similar: the Pentagon wants to wash its hands once and for all of the UFO question, so they will dump it on some university, preferably one that isn't going to cause any trouble."

"There's another scenario," said Don, who had been patiently listening to us. "Maybe all this is just a cover. Maybe Blue Book was a sham from the beginning, and Allen has been used. It is unthinkable that the four of us here should be the only group in the world seriously studying this whole question."

He had just voiced a thought that was in all our minds.

Hynek drew a few puffs of smoke from his pipe and said:

"I have often thought of that."

And yet, knowing how Government bureaucracies do things, it is quite possible that their scientists would be as puzzled as we are, even if the Intelligence services of the major powers (the US, the USSR and the British) arc sharing their UFO data, as another contact of Aimé Michel keeps telling him.

Chicago. Sunday 26 June 1966.

Jim McDonald called me yesterday from Tucson to get more data about power failure cases. We ended up spending an hour on the phone talking about the general situation of the field. He confessed to me that his radical campaign bore little fruit so far. It seems all he has accomplished is to antagonize the Wright Field people. He acknowledged he had not succeeded in convincing Kuiper either. Even his friend Brian O'Brien, with whom he had another meeting last Friday, remains skeptical. One would think he would learn something from this. Yet he continues to claim that the lack of interest in the subject among scientists is all Hynek's fault. He has clearly been indoctrinated by the folks at NICAP, especially Keyhoe and Hall. In a conversation with McDonald, Hall has even insinuated that Hynek didn't really know much about the UFO problem, and that he had only done research on "five or six cases," which is patently false. Hynek's only interest in the whole thing, Hall told McDonald, is the money he gets from the Air Force! It is not surprising that the field makes no progress, mired as it is in this kind of unfounded rumors, bitter infighting and deliberate calumnies.

Jim tried to recruit me for his camp.

"If it wasn't for your influence, and all the research you brought over from France, Hynek would still be arguing that ninety-nine percent of those reports are due to Venus or to marsh gas!" he said. "It's time for you to move on."

Yet I don't see what good McDonald's approach will do, if he keeps behaving like a bull in a china shop.

Chicago. Monday 27 June 1966.

For the last two weeks the weather has been very hot and muggy, and we feel exhausted, with no energy to work. Chicago bores me. I read a lot (right now, *The Revolt of the Angels*, by Anatole France). I just mailed to Allen the end of the *Challenge* manuscript and a copy of the dust jacket.

Chicago. Tuesday 28 June 1966.

While I was typing the appendices to *Challenge* the phone rang and I heard Allen's joyous voice. He was calling me from Colorado to report on a direct request by Quintanilla: he now wants to have our proposal for data reduction of all the Blue Book files as soon as possible. This would be a parallel effort to the main contract the Air Force has not yet awarded. The Major stated that Northwestern had the highest chances of getting the job, since both Allen and I were there. This would not be a sole source contract, however. At least one other university would be consulted to provide healthy competition.

Allen was elated at this news, and he seemed ready to jump back on the barricade and fight. He was reading the end of *Challenge*, he told me, and he congratulated me: "Reading it, I realize better than ever that the phenomenon embodies an authentic mystery. It shows up in your graphs with the hourly distribution of sightings," he said. We made grandiose plans for the Fall.

Chicago. Sunday 3 July 1966.

I have just finished typing the Air Force database proposal. Since Janine had to work this afternoon I went to her office with her to use the Xerox machine there. Everything else is closed because of the Fourth of July. We dropped off the package at the main Post Office, to ensure Allen would get it in spite of the holiday. I hope he will be pleased with my fast reaction. Clearly this project has his undivided attention now. We have had several years to refine our techniques and our arguments, so I am confident he will find few problems with the document I have drafted.

Chicago. Friday 8 July 1966.

On Wednesday morning I went to O'Hare with Janine and with the observatory administrator to greet Hynek as he flew in from Denver. We held a meeting in the TWA Ambassadors' Club to review the proposal in detail, far from the ringing telephones of the office or from his busy house in Evanston.

If the history of all this gets written some day, I take pity on those who will try to extract the deeper meaning and the actual motivations of the protagonists. The Air Force wants to get rid of the responsibility for a mystery it obviously can't cope with. If they lean too heavily towards the skeptics' side, turning all UFOs into marsh gas and globular lightning, the witnesses will have good reason to feel insulted, and they may complain to their Congressmen. But if they were to confess the truth, namely that many sightings go unexplained, the American public may simply get scared and feel unprotected. This is one battle the Air Force cannot win, especially when the scientific community looks down its august nose at the whole thing and refuses to get its hands dirty. The new concepts that started to evolve in our own minds five or six years ago have not spread among our colleagues yet, not even among the best informed of the ufologists. But sooner or later these new concepts will become the dominant framework. Historians will take them for granted. It will be very hard for those who will look back on this period to understand our hesitations, our painstaking maneuvers.

Hynek received the proposal on Monday. He had only a few changes to suggest, mainly in the budget section. I proposed that we add Lyle Boyd to the team. After all, Boyd wrote Menzel's book and would be an excellent source of knowledge to balance the staff of our project. Driving back to Evanston we rushed to a typewriter and we had the document polished up in time for Hynek to leave for Dayton. He promised to call as soon as he had anything new.

I was at home and asleep when he called last night about 11 p.m. The Air Force had already started to process our document. There was general confusion in Dayton, because they didn't know yet which university would get the main research contract: "The right hand of the Air Force doesn't know what the left hand is doing," he said, laughing. Our only competitor for the data analysis contract is Battelle Memorial Institute in Columbus, not far from Dayton. Many years ago Battelle compiled *Special Report #14*. It is said that they were paid the sum of $600,000 for it, a very large amount in 1953, showing that the Air Force was seriously looking for answers at that time.

Hynek did present our database proposal at a 10 a.m. meeting attended by Quintanilla, four Colonels and the Wright Field chief scientist, Dr. Cacciopa. The briefing was well-received, he said: "If it was up to the Dayton people we would already have the contract."

The Air Force is eager to get on with the job. They are getting a little anxious because another one of the universities they contacted has turned them down, and two leading scientists have told them they wouldn't touch the UFO research business with a ten-foot pole, no matter how much they were paid. Hynek keeps over-reacting to the criticisms voiced by McDonald. I have had to advise him against writing a strong UFO manifesto and sending it to *Science* magazine: "Let's at least wait until we have a signed contract," I recommended.

Jim Lorenzen called Hynek during his stay in Boulder. He told him that he and Coral were thinking of dissolving APRO. They are looking for someone to buy the files for fifteen thousand dollars.

Chicago. Saturday 9 July 1966.

I went to the edge of the lake alone, to think. What are we trying to accomplish? I came to the sad conclusion that our efforts are childish and probably hopeless, because this is a very shallow, desperate world that has no vision and takes no chances. *Anatomy* will soon be out in paperback. It will be in every airport and every drugstore in America, from Maine to Oregon, sandwiched between the latest report on teen-age sexuality and some Western novel of cowboys and Indians. What good will that do? People are closed to new possibilities, to their own mental potential. Even Bill Powers laughs in my face when I try to tell him about the achievements of authentic psychics like Croiset in Holland. The UFO groups are surpassing one another in the silliness of their statements, in the pettiness of their behavior, in an attempt to destroy or to undermine their competitors. This is a joyless, humorless world.

Chicago. Sunday 10 July 1966.

Aircraft mechanics have gone on strike, so Hynek is driving back to Chicago from Dayton in a rental car. Don, Bill and I are scheduled to meet with him at the observatory. When I think about the coming year it seems probable that the UFO scene will now be narrowed to two main groups, with our little independent team caught in the middle. On one side will be the university group that will be funded by the Air Force, and on the other side NICAP, which will find a strong supporter in McDonald. I think he has enough ambition to see the UFO problem as a springboard that can send him to the foremost echelon in American science.

The other day he told me on the phone that "Hynek's hesitations demonstrated he wasn't the man of the situation." The implication was that he, Jim McDonald, was the one who should lead ufology to its ultimate victory and that I should rally under his banner. He certainly is a true man of action, capable of organizing a vast campaign, leaving no detail uncovered. He does not have Hynek's subservient attitude towards power, his obsequiousness towards the military. For example, McDonald has clearly seen through Hynek's harmless pleasure at having a Jeep and a driver at his disposal in Michigan. What he fails to recognize in Hynek are the other important and subtle traits in his character: his profound sense of scientific adventure, his kindness, his ability to grasp the hidden side of situations. It is to those abilities that he owes his knowledge. He has understood this research domain deeply, with his heart, while everyone else was using only their brains, and laughing at the witnesses. Where was McDonald all those years?

Chicago. Saturday 23 July 1966.

Yesterday I brought back the galley proofs of *Challenge* to Betty McCurnin at Regnery, with some minor corrections. She has done a remarkable job on the book, with care, precision and intelligence. I also gave her Hynek's new foreword. He had written a first draft in which he was awkwardly apologetic, trying again to defend himself against McDonald. There is a French saying, "He who excuses himself accuses himself." I wrote to him in Canada, where he spends a couple of weeks every summer to get away from everything, advising that he take up the problem again from a more authoritative standpoint. Why should he be defensive before McDonald? It is to the public that he owes an explanation. McDonald, who is fast becoming the darling of the ufologists, is only another demagogue.

The observatory has received a letter from an aerospace engineer, the same man who was so secretive when he visited us last year. He is asking us to send him our "twenty best cases," a request already made recently by a mysterious group of scientists from Wright Field. Our files seem to be in high demand, but who are these people? Is this man working for some Intelligence agency? Or simply for the secret study McDonnell-Douglas is rumored to be starting? Does he think we're stupid? Bill has replied on our behalf in an evasive way.

In one week we will be back in France, for a short stay on the way to Russia. It will be good to see Aimé Michel again. Should I tell him everything? He advises me to contact Professor Rocard, a physics expert from Ecole Normale Supérieure who has access to De Gaulle. Should I speak to Rocard with complete frankness? Or, on the contrary, should I stay in the shadows, waiting for them to demonstrate that they truly understand the scope of the problem?

Chicago. Sunday 24 July 1966.

Bill Powers called me yesterday, relaying a discouraging message from Hynek. The Dean now refuses to sign our research proposal on behalf of Northwestern! After all this work, he is going to turn down the opportunity and kill the project. The Dean shakes in his boots, it seems, because a reporter for the *Saturday Review*, a fellow named Lear, has contacted him with the intention of writing an article about Hynek. He thinks Northwestern will then be exposed as a hotbed of false science. In fact, it is much more likely that the reporter is inspired by McDonald and will criticize Hynek for not doing enough to solve the problem.

The situation is made worse by the fact that no university has agreed to take the main Air Force project, which is still going begging, as McDonald told me on Friday. The academic community refuses to touch the problem, and Northwestern is now aligning itself with other campuses out of fear. Hynek is furious. So is Bill. Hynek talks about suing the *Saturday Review* if Lear's article costs us the contract. He also thinks of taking a one-year leave of absence, founding an independent research company, or even resigning altogether to take this work outside. Bill said courageously that he would follow.

Academic science is a circus. This is not what I want in life. I want to be part of an intelligent, organized effort with people I respect. I don't care to play politics, to convince the masses. I simply want to do things that have not been done before.

Later the same day.

I am finishing the study of the Air Force files for 1951. I find landings and large cigar-shaped objects, just as in Aimé Michel's classic work. I am beginning to think like McDonald: how could Hynek have missed this? The wave of late 1951 is astounding. Scientists will be amazed when they finally decide to look at all this.

Chicago. Tuesday 26 July 1966.

Hynek is still fuming. He talks of going to another university, to any college that would agree to let him work on this contract. His letter to Northwestern's Dean of Sciences, which was sent yesterday, is well worth preserving:

> I fail to understand the University's attitude on the UFO proposal. What wrong could there possibly be in reducing the Air Force data to machine-readable format? (. . .) They have finally taken my advice and have earmarked better than a quarter million for this—and now my University balks!
>
> I think the University is afraid of its shadow... You will recall that I brought up the whole UFO problem before the trustees at the

> Casino Club and asked permission to bring UFO work to Northwestern. Permission was tacitly given. Is Northwestern afraid to do a scientific job that needs doing because of a little possible criticism? Are we a University or an ostrich farm?

Feeling very hopeless in the middle of this debacle I have cleared my desk in anticipation of our trip to Europe. I have put the final touches on my cardiac simulation program. But the respect I once had for this University is gone.

Hynek just called from his cabin on the shore of a little lake near Blind River, Canada. He wanted to wish me a good trip to France and Russia and to finalize our plans. He isn't sure that his letter to the Dean will reverse Northwestern's decision. He is still worried about McDonald, who gives orders at Wright Field as if he owned the Base. Hynek has written an article for the British magazine *Discovery*. He asked me to take several copies of it to Europe, along with his foreword to *Challenge*, to make his position clear to all our friends abroad.

12

Paris. Friday 29 July 1966.

The joys of travelling with a group: even after the traditional pictures on the tarmac, where we huddled with other scientists, they still wouldn't allow us to go into the chartered plane. We were exhausted when we finally took our seats. Olivier slept in my arms while I watched the horizon, the arctic light over towering clouds, a fantastic picture of other worlds, like dawn rising on Saturn.

Annick, Janine's sister, was waiting for us at Orly. She took us to Maman's clear, airy new apartment on rue de la Mésange, behind those *Arènes* where I spent some of the best time of my student years. I had been awake for some twenty-six hours when I finally collapsed into bed. Aimé Michel came over and woke me up two hours later. We started talking, laughing and arguing. We had dinner at a restaurant near Place Monge. I told him about Hynek's meeting with U Thant, his recommendation for an international effort and our hope that French officials might put the UFO question before the General Assembly. He will try to reach Yves Rocard tomorrow. He is the only French scientist we know who has access to the proper government levels.

Paris. Saturday 30 July 1966.

Lunch with Aimé, on rue de la Huchette. I told him about our plans to visit the USSR as Hynek's unofficial ambassadors and to find out what was going on there, officially or privately. I have this urge to walk around Paris, to watch a thousand details and register them in my mind. Thus I strolled for three hours, loitering under the sunshine and the rain. And I came to realize what was wrong with Chicago, why I felt so dry and oppressed there, why the human dimension was missing from American streets.

Paris. Monday 1 August 1966.

Aimé has now read both of our research proposals and the foreword to *Challenge*. He is happily surprised to see how far Hynek's position has changed. A phone call came. Rocard was inviting us for lunch.

We found the professor in his office at Ecole Normale Supérieure, and all of a sudden I was thrown back ten years, even a hundred years. Dusty bookshelves, glass walls along endless dark corridors, creaking wooden floors.... Also Rocard himself, perhaps the most powerful physicist in France, cultivating the look of a schoolmaster in a nineteenth-century provincial college. He looks short and gray, with a sad little moustache. But he is a brilliant man under this unassuming appearance; his eyes burn with wit and with the sly caution of bureaucrats:

"In my position, there isn't much I can do, you realize that, don't you? Of course, I do have a few contacts...."

I had to keep reminding myself that I was speaking to someone who invented nuclear bombs for a living.

Craftily, he warned us that he suffered from very bad hearing. Aimé Michel took me aside and told me not to believe any of it: French journalists have often mocked "Professor Rocard's carefully simulated deafness," a feature which is convenient in politically difficult circumstances....

He called his chauffeur and asked to be driven to the Marly restaurant, where we made very ambitious plans over lunch: Could Aimé be given a semi-official role, so that he could be our counterpart if our research went ahead in the U.S.? Could we set up a meeting between Hynek and Peyrefitte, who is De Gaulle's current research minister? Nothing was resolved, but I decided to give Rocard the benefit of the doubt, to play with all my cards on the table. I gave him our proposals and our twenty best cases, none of which, of course, contain classified material. Hynek had advised complete openness. Now the ball is in the French camp. We will find out if they have the guts to do something with it.

Marigny. Tuesday 2 August 1966.

We arrived in Caen last night, and we drove on to Marigny. Normandy is cold, it rained the whole way. Fortunately the house is full of toys and laughter. In the attic we found a bull's head made of wicker; my nephew Eric wore it while I teased him with Janine's red raincoat for a wild and joyous corrida around the backyard, then I went back to work. I have to mail the galleys of *Challenge* back to Betty in Chicago this week.

Marigny. Saturday 6 August 1966.

A curious remark by Rocard comes back to my mind. He told us there were rumors in the administration that the French secret service was going to be reorganized by De Gaulle. He was hoping that would mean that our kind of research, all this forbidden science, "would not be discouraged as strongly in the future." What could he mean by that?

Rocard is in a good position to have experienced both sides of the equation. He directs the physics research for the French nuclear program, but he has also been viciously attacked as a charlatan by his rationalist colleagues because he dared to publish his private experiments on dowsing. This will surely cost him his election to the French Academy of Sciences, according to Aimé.

The professor has explained his experiments in a book entitled *Le Signal du Sourcier (The Dowser's Signal)* published two years ago by scientific publisher Dunod. It reviews his personal tests of the relationship between dowsers and the earth's magnetism. It also describes some formal physical tests he conducted by getting his students and colleagues to walk through a frame where he could unobtrusively create a current, hence a magnetic field. From these tests he concluded that the human organism could actually detect weak magnetic gradients, and he hypothesized that the detection mechanism had to do with the spin of the proton. Naturally the very fact that he would dare to set up such tests in his laboratory of Ecole Normale Supérieure, and that he would publish the results, is anathema for the narrow-minded French rationalists.

This morning another discouraging letter arrived from Hynek, who is still on vacation in Canada. The University's final answer to our database proposal is a definite NO. The Dean is afraid. We have against us many scared, gutless men who have painfully reached positions of apparent power as professors, as directors of various institutes. They don't know what's around them in this dark and scary universe, and they don't want to find out.

I keep taking notes in these diary pages, very religiously, because I know that only those who live through such crises of the mind can record them. In ten or twenty years, who will believe that Hynek's research, the work of his whole life, was blocked by the Dean of his own university?

Marigny. Sunday 7 August 1966.

Under the shadow of yesterday's bad news, I am afraid of what we will find when we go back to the U.S., and I dread the stale atmosphere around Evanston.

Is it the rain over Marigny, the dampness that permeates everything? Janine was exhausted when she woke up this morning. She works too hard. She places her hopes in me for our future together. And what can I do? Finish my dissertation, for one thing; drop out of any publicly visible effort on the UFO question; keep it as my private study. Later, next year if we can, let us come back to Europe.

Up in the attic is a grand old nineteenth-century clock that belonged to her grandparents. We will take it back with us to count the hours we still have to spend over there. Although we are in August the weather in Normandy is truly horrible, with rain pouring by the bucketful. This afternoon I will stay home and take apart the old clock. Tomorrow we plan to visit Janine's relatives in the little village of Yvetot-Bocage. And in less than one week we will be landing in Moscow.

Paris. Thursday 11 August 1966.

We are back in Paris, which looks sour and nauseous. The French landscape is changing with the new generation. Strange young men have appeared along the sidewalks of rue Saint Jacques. A kid with unkempt clothes and uncombed hair is typical of the new breed. Stoned to his ears with the fashionable new drugs, a jacket draped over his shoulders, he is sitting in the gutter. He pretends to be reading Kafka.

If I came back to France, I would do it my way, I thought as I ate a sandwich on Place Saint-Michel. I would follow my own course.

Another letter from Hynek, who is still fighting valiantly: our proposal has been submitted to the Air Force after all, but with a "restrictive notice" attached to it by Northwestern. I have never heard of a "restrictive notice." Everything will be decided in September.

Moscow. Saturday 13 August 1966.

It is five o'clock and we have just returned from the Kremlin. Janine has fallen asleep in the next room, exhausted. She is afraid that the bad cold she caught in the rains of Normandy will come back.

For the next three weeks we will live in two tiny rooms on the sixth floor of this huge university building that reminds us of various edifices we have known: the sitting rooms could well come out of an old Texas hotel, the cavernous empty hallways could be borrowed from some dusty museum. But the disgusting meals served on plates of dubious cleanliness have no equivalent anywhere in the world. We each have been allocated an old wooden desk, a narrow iron bed whose mattress has not been exposed to fresh air for years, a table with a top that falls off when you touch it. We are not even allowed the luxury of sleeping together in this place!

We do have a window, however. It looks over the gardens and the avenue which goes down Lenin Hill towards the Moskova River, while the building rises far above us. But the meals are revolting. There is no excuse for such food.

We left Paris yesterday from Le Bourget. As we flew above Brussels the fat Tupolev 114 entered thick clouds and stayed inside the grayness, so that we never had a chance to see much of the countryside. We went through customs as a group, a "Delegation." This gives us some power we would never enjoy as mere individuals. Everything here has been designed with groups in mind. Thus we went from the airport to the university as a unit, a formal *Delegation of Mathematicians.* Janine and I became friendly with a Seattle professor.

It was unfortunate that we arrived after dinner time. Very calmly, our Intourist group leader told us that everything was closed and that it would in fact be very healthy for us to stay a whole night without food.

This morning we rushed down to breakfast. It was a disappointing experience. After a cup of bad tea and a soggy toast we took the bus for Red Square, the Kremlin and the wonders of Saint Basil cathedral. We have seen nothing of Russian life so far, except for some very decrepit sections of the city. The buildings do not age gracefully as they do in Paris, they just crumble into ugly ruins, as in America. Except that the Americans usually blow up their own buildings with dynamite as soon as they have passed their prime, to make room for bigger ones. There is no such concept here.

The crowd is animated and carefree. One does get a European feeling about the people. They look like a bunch of kind bureaucrats. You almost expect them to be carrying stacks of forms under their arms.

Moscow. Sunday 14 August 1966.

This morning I registered for the Mathematics Congress, which is organized in a very professional manner. The main dining rooms are now open, which helps a little. We are close to Lomonosoff Prospect, where Kazantsev lives in an apartment building we can see from here, just beyond the computation center.

We have just walked around the grounds with our new guide, a delicate and sensitive Russian girl named Galia. She is a physics student who wants to become a teacher, and serves as a hostess at such international meetings during the holidays. Galia speaks English fluently. Unfortunately she does not know this university very well. She finds it ugly, too gigantic. She works in Moscow proper, and speaks eloquently of the beauty, the poetry, the proportions of the old city. I found my way to the observatory, the Chternberg Institute where Shklovskii works, but this is Sunday and we cannot visit the place.

Moscow. Monday 15 August 1966.

Noon. We spent the morning at the Chternberg Institute, where a friendly astronomer answered our questions through an interpreter. He showed us the time service, the solar department and the computing section where we saw our first Soviet electronic machines. They look absolutely identical to IBM hardware of ten years ago. We visited the small library and signed the guest book, where the name of Bertola and a few other Americans had already been recorded. Finally we saw the domes, which are inoperative, as in too many observatories I have known since Meudon. The building reminded me of some French provincial place, sleepy, solemn and dusty.

Later the same day.

We took the subway to Kropotkinskaya and walked around the old streets of Moscow with Galia. We visited the former aristocratic section which contains the houses of Pushkin and Tolstoy. We came back by subway again. Janine still feels very tired. Everywhere we find the same impression of decay and odd emptiness, but I may have understood the reason for it: this city is built in anticipation of the winter. These big green and yellow walls which seem so incongruous and inhospitable in the August sun must be admirable under the snow.

Moscow. Tuesday 16 August 1966.

The official opening of our Congress takes place this afternoon. Several groups of mathematicians have arrived at the last minute. Notable among them are the highly visible French. They came here in their noisy Citroëns, as if pushed by an evil Western wind. They brought wives and families, who yell at the crying kids, leave greasy papers on the marble stairs erected by the Proletariat, beer bottles at the feet of Karl Marx and candy wrappers around Lomonosoff. And just to make certain everybody will know which country is responsible for these improvements to the landscape of their hosts, they proudly display big badges that spell FRANCE on their jackets.

We arrived at the Palace of Congresses in time for the inauguration, and watched with amusement the curious spectacle of contemporary mathematics. The Russians fell into each other's arms with obvious glee. The French paraded around, posturing with strange rituals and mannerisms. The British seemed distant and disdainful. The Japanese were dignified and anonymous. In the grand amphitheater we heard a speech by the Soviet Minister of Automation, followed by the Fields medal ceremony, the mathematical equivalent of the Nobel Prize.

During the long intermission we walked inside the Kremlin. We were astonished at the elegance of the domes which seemed so heavy in the photographs we had seen. Up close they give a remarkable impression of clever use of space and volume. We went back to the conference center in time for the unavoidable *Swan Lake* ballet. When we came out again we saw the stars: The red stars atop the bulbous towers and also the little stars sparkling far away in the sky, the stars that are still beyond politics, beyond the reach of man.

Moscow. Wednesday 17 August 1966.

This morning I attended Atiyah's lecture on elliptical differential operators, which went over my head, and Meltzer's lecture on automated inference, which I found of special interest since it relates to my current work. Meltzer is from Edinburgh, where a major center of artificial intelligence research is starting to flourish.

In the afternoon I skipped the meetings to walk along the Moskova with Janine, right under the Kremlin walls. We sat in the park and we reviewed our problems: Northwestern, the Dean, my dissertation, Janine's career in business data processing, the future of my books. I found this assessment rather discouraging. It is true that the sight of Moscow does not inspire one to daring solutions. The heavy architecture, with the awkward columns endlessly stuck over dark greenish masses of masonry, seemed designed to enhance our melancholy.

It is hard to characterize the crowd. The way young people dress, they could blend into any group in Paris or Chicago. But the older generation look as if they had been stuck in a closet for twenty years: the men wear wrinkled suits, the women shawls and shapeless dresses. In the few contacts we have had people seemed to show a depressing lack of personal initiative.

The Marxists are always talking about the unresolved "contradictions" of the Western countries; but one of the many unresolved contradictions of socialist States has to do with their whole approach to human problems. Why do they feel they must make final decisions today on moral, ethical or esthetic issues that rightly belong to future generations? As soon as the "progressivist" is free to act on his ideas it seems that he must turn into a dictator and issue absolute edicts, unaware that in doing so he denies the progress of the very society he is trying to build, since he closes off its future choices. In the Soviet Union these progressivist bureaucrats have arbitrarily decreed what constitutes beauty, morality and intellectual worth. They have crippled the very soul of their own people.

Moscow. Thursday 18 August 1966.

At nine this morning there was a very interesting lecture by Lotfi Zadeh, a young mathematician from Berkeley. He was introduced by Kolmogoroff, who pointed out that this was the first instance of a lecture on control theory at an international mathematics congress, and warmly recommended this new domain to the attention of the more "traditional" specialists.

Zadeh presented himself humbly as a mere electrical engineer. He criticized current optimization theory as too narrow and simplistic; it fails to address the real problems, he said. For example, criteria functions are generally expressed as single values rather than vectors (multiple values). Too often, when we are asked to study a complete system we can only build a theory of a small sub-system of it.

He gave economic plans as an example. Anybody can define the ultimate objectives of a five-year plan, he said. The true challenge is to define realistic local goals on the way to the big result. To be more specific: it is easy to decide that tractor production shall double in five years. The real question is, how many tractor drivers should be trained each quarter? How many tires will be in use by the time 50% of the plan has been implemented? How much should be invested every month? This is far from being a classical problem.

Another real-world concept discussed by Zadeh is that of "fuzzy sets." In mathematics it has always been understood that sets had a rigorous definition, but in practice that is rarely the case. A physicist will often talk of quantities whose value is "close to ten" or "much larger than one." Such fuzzy definitions will be a key to the understanding of artificial intelligence in years to come.

In the afternoon we strolled along Lomonosoff Prospect and we found Kazantsev's apartment building. With Galia, who is witty and charming, we enjoyed a long visit to the old Moscow University. Later, leaving the campus, we followed an avenue that led to the Novodevitchi convent, a beautiful structure with elegant bulbs and lofty spires rising serenely above the trees. Fresh flower gardens led to a dark crypt where the floor was paved with sculpted tombs. In the large chapel tourists walked around while old women prayed in the flickering candle light. A wrinkled crone travelled along the walls like a fast-moving spider, crossing herself quickly and kissing each icon. Galia explained to us that a funeral had just been celebrated by the priest. This woman was praying for the soul of the departed. As a young Soviet girl she expressed her revulsion before such superstitious behavior.

A group of crying women suddenly rushed into the chapel, filling it with their lamentations. We retreated to a larger room where something like a lying statue was resting. It took us a while to realize this figure was in fact the dead man himself, his corpse embalmed, its skin shiny as wax. Galia's face turned purple as she saw it. She rushed out of the church as fast as she could. When we caught up with her she was still very shaken and we had to calm her down. Her encounter with the old religion had caught her unprepared.

Moscow. Friday 19 August 1966.

I attended a very exciting lecture this morning by Richard Bellman of the Rand Corporation. The room was filled to capacity, because his work is well-known here. He gave an overview of operations research, emphasizing dynamic programming techniques. He went on to talk about information processing, stressing the problems of delay and adaptation.

Later the same day.

As we were watching the night that fell over the meandering river a young Russian man approached us with an offer to buy foreign currency at twice the official rate. One of the ways for us to land quickly in a Soviet jail is to accept such offers, so we told him that we had nothing but rubles.

Another young man who had witnessed the scene then spoke to us. He said he was a graduate student in applied mathematics, who had few chances of practicing his English. (In fact, he spoke quite well. The level of language education here is remarkable. Galia speaks with very little accent, in a way which is full of precious little phrases. For example she will start every other sentence with "As a rule . . .") When I asked him if he used computers in his work he opened his briefcase and showed me a program he was writing. It was assembly code. He knew of high-level languages like Fortran but they were not yet in general use, he said.

The sun was setting, the air was turning colder; he was preparing to leave. Instead we invited him to have dinner with us, much to his surprise. During the meal we learned some new things about current conditions of scientific research in the Soviet Union. As in the West, he told us with some bitterness, fundamental researchers are paid far less than their industrial counterparts. It is only when they do something sensational that they are given a little money or an increased budget. This may explain some of Shklovskii's surprising statements about his beliefs that intelligence can already be detected in the universe. Generally speaking (or, as Galia would say, "As a rule"), the same bureaucratic rigidity is found here as in France. There is a young mathematician named Arnold who is so legendary among his peers that he has been proposed for membership in the Academy, but he was rejected because of his young age and independent spirit.

Our guest seems to be a good specialist who ignores completely what he calls "non-scientific questions." Thus he regards science-fiction literature as useless dreaming, and he says he leaves politics to the politicians. Also he doesn't see how there could ever be controversies in science since all the methods of science are exact by definition. He must have missed Lotfi Zadeh's lecture about fuzzy sets! I think such a dogmatic attitude reflects immaturity in Soviet science education.

We are struck by the lack of enthusiasm among the young. When we talk about the space program their apathy is obvious; it borders on contempt. Shklovskii seems to be practically unknown among the people we meet, while he is a big man in the Western media. There is a sharp division between the "literary" and the "scientific" minds, which I find disappointing.

Moscow. Saturday 20 August 1966.

We still have ten days to spend here, and time is becoming heavy. Today we continued to play tourists, visiting the Lenin mausoleum, the tombs of revolutionary figures, the throne of Peter the Great, and Catherine's opulent carriages. Once again we returned exhausted. I went to bed and read Hoffman's *Fantastic Tales.*

This was a book that strongly impressed my father, perhaps because he and my mother had enjoyed the opera drawn from it. He had owned a copy that burned with his whole library in 1940. When I was a teen-ager I saved some money in secret and bought another copy for him on his birthday. I had some difficulty with the bookstore owner, who told me "It isn't a book for young people." I never read the book, and I imagined it contained all sorts of delightful horrors. Now that I have a chance to read it, I find nothing but a series of good old nineteenth-century stories, a lot of dark hints about "magnetism" and some distant fumes of sorcery, diluted in well-behaved rationalism and wilted poetic notions. I fail to find anything to justify my father's excitement or the bookstore owner's trepidation. After all, when I bought it as a teen-ager I had already read Poe and more sulphurous writers, not to mention the truly grotesque and gory depravities described in Sophocles and Homer, which we were translating daily in Greek class, word for horrible word, dripping with blood.

Moscow. Sunday 21 August 1966.

A leisurely excursion took us to the Moskova-Volga canal today. The boat stopped at the edge of a forest. We ate bread and cheese, and we walked happily through the woods. This was a welcome change of pace. All day yesterday I had been in a dark and silent mood as I finished reading Ruppelt's book, *Report on Unidentified Flying Objects.*

When I compare Ruppelt's situation with ours I can imagine how broken he must have been when he left Project Blue Book, and I despair of ever seeing the problem posed in honest, realistic terms. Even after an expert committee recommended that the research budget be quadrupled, the Air Force did not follow through. On the contrary, they killed Ruppelt's excellent proposal to install cameras on radar antennas to establish once and for all if anomalous echoes corresponded to some real objects. He also wanted to study the correlation between saucer sightings and sudden jumps in atmospheric radiation levels, and he was never allowed to do this.

I had not suspected that so much had been done in the U.S. during the fifties. Ruppelt's ultimate failure is discouraging. Hynek is the only one who has not failed miserably, perhaps because he has remained in the shadows, "biding his time," as he often says, and observing events quietly as a spy inside Blue Book. But what will happen if he tries to go public with our ideas? His colleagues are already placing banana peels under his every step.

Here in Moscow the charm of a few initial discoveries, the friendliness of some contacts, has faded quickly. Given the global nature of the UFO phenomenon one would have to be completely stupid not to try to find out how it is manifested and perceived in the USSR, which is one of the largest countries of the earth. But our first feeling of discovery and adventure on this trip has been replaced by an impression of bitterness, of stifling oppressiveness. In spite of what little Galia claims, the city seems very dull to me: nothing happens. We sleep badly. Something is very wrong here, there is a sickness of life, an immense sorrow in the air.

Moscow. Wednesday 24 August 1966.

Another agitated night. I was worried with incoherent dreams on this camp bed where I can barely turn around. We have accomplished everything we could accomplish here, so we are now trying to arrange to leave the Soviet Union a week earlier than scheduled. The bureaucrats of Aeroflot are not helping us. The walls seem to be closing in on us. We feel trapped.

We succeeded in visiting the main computation center of Moscow University with a small group of interested mathematicians today. It wasn't easy. We experienced an initial rebuff when a stern-looking official dressed in gray declined our request with simple words which required no translation: "Machini? Niet!" Thanks to some energetic Russian students our group was able to get around the bureaucracy. We saw two machines, the M20 and its transistorized version the BSM4. Soviet engineers were willing to answer all our questions but as we began to realize how far behind they were the dialogue became awkward, even embarrassing.

The real surprise came for me when I pulled out a small deck of cards I had brought in my suitcase, a program for stellar motion calculation written in Algol. The Soviet specialists passed the cards among themselves in astonishment. I explained to them what the program did. They shook their heads, confused.

"Don't you have a compiler for the Algol language?" I asked.

"Yes, yes, it is a standard language, of course."

"Then we should be able to run my program on your machine."

"No, that is not possible."

"Why not?"

The man shrugged, a helpless look on his face.

"Because we do not punch the information in our cards the way you do."

"But your cards are exactly the same size and the same thickness as ours!"

"That does not matter, we do not write them in the same way."

They proceeded to explain how they had "improved" on the simple IBM concept of keypunching. They pack the information into the cards column by column, instead of dedicating one column to a single character. Thus they can put three times more data per card. Unfortunately this makes it impossible to read foreign cards. It also prevents them from printing out the contents of their own cards along the top edge as we do, so they never know what is on each card unless they make a special note of it by hand, which of course is very time-consuming! They were astonished by the pin printer pattern used by IBM. They have never built such a machine. So much for the international exchange of "standard" scientific programs!

The State has a monopoly on ideas here, and it channels carefully selected thoughts, one at a time, into popular consciousness. Open discussion of the UFO subject is clearly discouraged. Alexander Kazantsev, who is a recognized writer and engineer, is the exception that confirms the rule, because he has been smart enough to treat the question in a science-fiction vein, or in the context of archaeology. Shklovskii is an exception too, because he plays a role similar to Sagan, who keeps proposing gratuitous speculation without looking too closely at the data. If the saucers turn out to be significant, Sagan will take credit for having daringly theorized about cosmic visitors. If they are discredited he will claim he always saw clearly through their mythical character. Shklovskii has made sensational statements about the periodic signals received from source CTA-102 which he says may indicate an intelligent extraterrestrial source.

We need to grasp once and for all how alone we are in our research. This trip to the end of Europe has convinced me of it. Reading Ruppelt confirms my feeling that a deep and weird drama is unfolding around us. But what can I do with this knowledge? I am not a missionary at heart like the brash Jim McDonald. I do not intend to mount any crusade.

Later the same day.

Galia has insisted on showing us the Moskova by boat. She was right about the beauty of the trip. It is from the river that one must see the Kremlin, as it rises above the grayish haze. It is a great sight indeed, with a single red flag at the very top. This earth is a poor little planet, subjected to the whims of power-hungry men.

Moscow. Thursday 25 August 1966.

It is raining today and our charming guide has a cold. In spite of it Galia came over and insisted on taking us to the bookstores on Gorki street. I found two short stories by Kazantsev in an anthology of his works entitled *A Space Visitor*. One of these recounts the 1908 Siberian catastrophe when an unknown cosmic object impacted the taiga. I looked in vain for Menzel's book, which we were told was the only UFO book published in Russian. This led me to explain to Galia how we had become interested in this topic. In spite of the official censorship she was perfectly well aware of its reality. They even have something similar to the American contactee movement. Some members of the Academy have received letters from individuals claiming to be in touch with an outer space source, and even receiving psychic orders from it. They assume that they are mentally disturbed. However, according to Galia, a Soviet criminologist has analyzed these letters and came to the conclusion that a genuine new sociological pattern was emerging, based on some authentic, unexplained observations.

While the phenomenon is often tainted with religiosity in the U.S., here it takes a more extreme turn. Thus a crazy woman has killed a great Jewish Soviet plasma physicist, by throwing him under a subway car. She had heard a voice from space tell her that the Jews must die because their existence on earth is very miserable, and they will be much happier in the next world.

"And the Academician who has centralized these letters, was he acting officially?"

"No, people just wrote to him."

"Is there a government body that conducts investigations on these cases?"

She shook her head energetically: "Not to my knowledge."

"If something strange passed in the sky what would you do, Galia?"

"I would watch it pass in the sky."

"Would you tell the authorities?"

"Whom could I tell? As a rule there is no place to report something like this."

What she is really saying is that anyone reporting this would be suspect as a source of potentially harmful rumors. Spreading unfounded rumors is a crime in the Soviet Union, which explains why we have been very selective in our conversations on the subject, some of which I do not want to record here even in these private notes. There is a thin line between exchanging scientific information with people and "spreading rumors," and we are not always sure how the line will be interpreted.

This afternoon we went back to the Intourist office with Galia, who used her official clout and barked orders at them as if they were a bunch of ignorant, despicable moujiks. They snapped out of their lethargy and immediately found us two seats in the first Tupolev that leaves for Paris in the morning.

Sadly, we had tea with Galia for the last time. We told her we hoped that some day she would be able to come and visit us in the West.

Paris. Sunday 28 August 1966.

We thought our early return would be a private surprise for Maman, but instead we found a whole family reunion in her apartment. My brother Gabriel was there with his wife and kids. It was Eric's sixteenth birthday. A distant American cousin on her way back from Lebanon was also there with her two children.

I have been walking around Paris with almost mystical pleasure. I bought more old books, from Ovid's *Art of Love* to a history of religions dating from 1666. We are struck by one detail: nobody has asked us any questions about our trip to Moscow. The French people we know are intellectuals who already have all the answers about everything. They read newspapers like *Le Monde* that provide them with complete and definitive knowledge. What would they do with the testimony of first-hand observers like us? The actual reality of things is of no consequence to them. What counts is to joke sarcastically about current events. Elegant cynicism is the fashionable attitude, not informed opinion. The French wit attacks and contradicts, using brilliant paradox to demonstrate that the world's evils are grotesque and that mankind's sorrows are irrelevant, as long as the little cafés along Boulevard Saint-Germain don't run out of croissants.

I have made some progress in my research on the humanoids. A book of grimoires by Paracelsus (born 1491, fifty years after the invention of the printing press) contains precious definitions of what he calls Elemental Beings, who seem strangely related to our modern UFO operators. I want to dig deeper into this, and that means I need to find more texts by the Hermeticists, as these texts existed before they were adulterated by occult amateurs. Once I find these documents I will have to investigate the parallels with recent phenomena, a long-term effort.

Paracelsus writes that there are four orders of spiritual entities, namely Nymphs, Pygmies, Sylphs and Salamanders. He notes, however, that these names were attributed to them "by people who had not spoken to them." To be completely accurate one would have to include giants and other orders of beings, such as mermaids and sylphs. These beings are seldom seen, but their existence, he says elegantly, is beyond doubt:

> Myself, I have only seen them in a sort of dream (. . .) These beings appear, not to remain among us or to ally themselves with us, but in order for us to understand them. These apparitions are rare. But why shouldn't they be? Is it not enough for one of us to see one angel, in order for all of us to believe in angels?

Modern science, of course, would answer an emphatic "no" to this last question. Elsewhere he writes about the Gnomes, whom he also calls "Homunculi":

> They can appear at will small or tall, handsome or ugly. They know all the arts and use the light of nature. Some who saw them thought they were the souls of men who died shamefully or by suicide. Others thought they were empty spectres, some took them as manifestations of sorcery. **Think twice before becoming allied with them. As soon as you are linked to them you have to do their bidding. When they are angry they inflict heavy penalties. Sometimes they kill. There are proofs of it** (my emphasis).

Unfortunately he did not give the details of these cases. It is known, however, that he did extensive research among the miners of Germany, who often reported seeing dwarfish humanoids in the underground tunnels they dug in search of coal.

Paris. Monday 29 August 1966.

Two depressing letters have arrived, one from Hynek's secretary, the other from Hynek himself, mailed from Atlanta airport on his way to Dallas. The Pentagon has rejected our proposal once and for all. This solves our problem. Clearly the Air Force does not want to see any serious study of its files carried out. What are they afraid we would find?

Hynek's letter to *Science* magazine has also been rejected. In the words of Joanne Belk, editorial assistant:

> Dr. Abelson has given me your letter of 1 August about UFO's (sic). He found it very readable and would normally have considered it for publication had not the *Saturday Review* devoted a large part of its 6 August issue to the same subject. . . . You will agree with him, I know, that we could not follow the *Saturday Review* articles so closely as to run the risk of echoing their feature.

That is the most ridiculous turndown letter I have ever seen. Since when does a scientific magazine have to consider what others have published in the popular press before it decides what its readers should see? Furious, Hynek wrote directly to Abelson: "There has been a spate of books and articles, and there will be more. Vallee's *Challenge to Science* will soon be out, Fuller of the *Saturday Review* is shortly to have another . . . the existence of other articles might be construed as good reason to publish my letter now, rather than to wait," he pointed out, adding with considerable restraint:

> I had hoped, and still hope, that the pages of *Science* could carry a sober status report on the UFO phenomenon amidst the barrage of articles written by people who are journalists and who simply do not have access to the data, or if they do, could not possibly have had the time to do justice to it.

For good measure he has leaked his *Science* letter to a number of prominent people, including Dick Lewis, who is publishing extracts of it in the *Sun-Times* as "The letter *Science* magazine was afraid to print . . ."

Paris. Tuesday 30 August 1966.

We walk along the old streets, breathing the dusty haze that rises from ancient books. I immerse myself in humanity, in the million fragments of moving life, colors and melodies of Paris. When we travelled to the Soviet Union our first impression was one of strange peace, the delight of being liberated from the ever-present bombardment by advertising on radio and in the streets, from the intrusive billboards inciting you to buy various products. Yet when we came back the sensory stimulation of advertising was a welcome joy. I was surprised to find that after the first period of blissful peace I had actually missed it.

Janine has left for Normandy. I am without my tender, lovely brunette, and I cannot stand this separation even for a short time. I miss her conversation. All day I find myself talking to her about what I see, what I think. I miss her pretty face, her body. What would my life have been without her? She has brought me happiness, untold wealth.

Paris. Wednesday 31 August 1966.

This afternoon I paid a quiet fraternal visit to Serge Hutin in his mysterious house of Fontenay-aux-Roses. Even though I have long dropped out of the formal Rosicrucian Order (I feel that far better sources are available to me now) we are still very close in spirit. I had not had time to write, and he has no telephone, but I trusted that "the Cosmic" would take charge of our meeting, as he would say himself. Indeed, as soon as I rang at the iron gate his mother appeared in the alley and she asked me to come in. I saw the abandoned garden, the dark old workshop at the end of the yard, and the numerous cats they have collected.

The great scholar was there, to answer my questions about Hermeticists and Homunculi, Hildegard of Bingen and her visions, and about current trends in the occult underworld. He works in two small rooms, one of which serves as his Sanctum, where he is surrounded with cats, art objects, occult artifacts, engravings and paintings. In this environment I regress easily to former times, to my childhood itself, with the wind blowing in tree leaves, the ancient touch of the stones of ruined castles. It is the old house that speaks to me of forgotten eras, it is a dark passageway beckoning me to look back through the dust of memories.

I spent four hours in Hutin's Sanctum. Afterwards he walked with me to the bus stop and waved at me when I climbed aboard the 194, as if I had been a cosmic traveller leaving for another galaxy. He is a man with a great heart and profound knowledge. I could almost hear the old house groaning, lamenting my ungrateful departure.

It is midnight. There is an empty place next to me in the bed. Janine is not here and I miss her sweetness. I think of a thousand things I would like to tell her.

Paris. Thursday 1 September 1966.

My mother just informed me of U Thant's resignation from the United Nations. De Gaulle is in Cambodia, openly criticizing American policy in Vietnam.

This afternoon I indulged in another long walk through Paris, buying astronomy and folklore books, and Enel's treatise on the origins of Genesis and temple teachings in Egypt. I also read an article by Seguin stating that among 500 popular tracts printed between 1529 and 1631, no less than 180 related "marvellous phenomena" of which ninety-five were celestial events. All that begs to be dug up and studied again.

I am angry at Aimé Michel, who has just published a review of our book in *Planète* under his pen name Stéphane Arnaud. He used this opportunity to get back at Muller and at Kourganoff. Why is he hiding behind me to attack them? I have no quarrels with Kourganoff, a man I admire very much. And I have forgiven Muller, whom I do not expect to ever see again on this earth.

Janine comes back tomorrow with my little boy. On Saturday we will visit an art show at Saint-Germain-des-Prés where my mother displays four paintings. On Monday we fly back to Chicago.

13

Chicago. Wednesday 28 September 1966.

Autumn has returned, and I work full-time again at Dearborn Observatory. Nearly a month has passed. With the rejection of our UFO proposal I found a depressing situation when I returned. I haven't felt like writing. My current project is the reduction of photometric observations of stars recorded at Kitt Peak. Fortunately the computation center has a new director named Ben Mittman who has an open mind on frontier subjects. I have introduced him to Hynek.

There have been two other notable changes. Our legendary Sergeant Moody is no longer at Blue Book. A new man, Lieutenant Marley, has taken his place. He has just spent three days in Evanston. Hynek's letter to *Science* will finally be published. Small consolation indeed.

It is hard for me to shake the melancholy that took me when we came back here. Don Hanlon, whose humor and intelligence were so refreshing, and who would have made an excellent research assistant, enlisted in the Navy when it became clear that our proposal was dead. It will be hard to replace his wit, his knowledge and his friendship. And McDonald is still arousing the media, turning the same humorless crank, endlessly accusing Hynek of every crime. I am glad I have scrupulously kept this diary. It may preserve a record of the truth as we lived it.

A French newspaper someone had left in the plane that brought us home contained a very curious report: two young electronic technicians have died on a hilltop near Rio de Janeiro while waiting for a flying saucer they claimed to have contacted. The case is becoming known as "The Lead Mask Mystery" because the corpses had crude masks over their eyes. Inspector Jose Venancio Bittencourt, a veteran detective of the Rio police, is quoted as saying it is the strangest mystery in his career.

Chicago. Friday 30 September 1966.

My thesis committee met today to hear my defense of my dissertation project. They gave me the green light. Mittman said I had attacked a very tough problem but he felt my approach was correct.

When I drove downtown to bring a collection of UFO magazines to the binder, I found the city as I had left it: with a big hollow feeling.

Hynek and I have entered into a secret pact after I pointed out to him that we were doing more and more superficial things and less and less actual research. All our time is spent with journalists, curiosity seekers, tourists, believers armed with a thousand pet theories we have all heard before.

"Allen," I finally said, "this kind of activity will never lead us to a scientific solution, and you know it."

"All right, Jacques. Let's say that by Christmas, either we will have obtained a radical change, or we will go back underground. Let other people worry about the problem."

"Yes! Let's see what McDonald and NICAP will do to solve this mystery, since they are so damn smart!"

Chicago. Sunday 2 October 1966.

Now Aimé is urging me to return to France: "Get your doctorate, come back here to work in the computer business."

It turns out De Gaulle has finally discovered the importance of computers and has launched a great new project he calls the *Plan Calcul.* He is bitter at Johnson's refusal to sell him a CDC 6600 with which to compute his H bomb. He now wants to regroup the French computer industry around national objectives.[1] Economically that is not a bad idea. The problem is that the plan will surely be managed, once again, by mediocre and greedy men. Chances are it will fall under the weight of their petty machinations for control. Such grandiose schemes always do. The money will probably be frittered away in the form of subsidies to the big, inefficient electronics industry.

If I do go back to France I will simply be another pawn in this game. In my discussions with Rocard I understood this: he reached his position through a combination of scientific excellence and skills at dealing with bureaucrats and ministers. Yet that does not give him any real power. When France explodes an atom bomb in the South Pacific he watches the proceedings from the bridge of the flagship cruiser. But he is a great man in a little world, a prisoner of a mediocre system, like Costa de Beauregard who studies the physics of consciousness, and Rémi Chauvin, a zoologist and parapsychologist who teaches insect sociology at the Sorbonne. None of them is allowed to undertake the research they would really like to conduct.

Chicago. Thursday 6 October 1966.

Ben Mittman has offered me a full-time job in the systems group, and I have decided to accept it. I am not doing very much UFO research at the moment, other than discussing the subject with Bill Powers. We have lunch together almost every day. Actually we spend more time talking about psychology than about flying saucers. I am fascinated with Bill's novel ideas on control theory and on behavior as the control of perception, which he wants to expand into a book.[2] He reads me entire chapters of his manuscript at the cafeteria, and I give him my reactions and critiques.

Hynek has just told me that tomorrow the Air Force is supposed to announce the name of the university it has selected for the UFO contract. He expects it will be the University of Colorado in Boulder. Everybody else has turned down the job with contempt, including Harvard, Columbia, MIT, the University of North Carolina and the University of California.

Chicago. Friday 7 October 1966.

Allen came to open the door himself, in his robe and slippers. He was surprised to see me. I handed out to him the first copy of *Challenge to Science*.

"I brought you something to read," I said. "Have a nice weekend!"

"Come in for a minute," he proposed.

But dinner was already on the table, Janine and Olivier were waiting for me, so I left quickly.

Several memorable things happened today. Our joint article for the Astronomical Society of the Pacific has just appeared, linking Allen's name and mine in mainstream astronomy as they have been linked in UFO research. We had lunch at the Red Knight Inn, a funny old place which has become a sort of headquarters for us. There was a new member in our little group, a man named Fred Beckman who is an electroencephalography specialist at the University of Chicago Argonne Cancer Research Laboratory. He is also a great connoisseur of food and wine.

"I don't suppose you come here for the cuisine," he said as he looked disdainfully around the restaurant, observing the mediocre fare on neighboring tables and the stern matron who was our waitress.

"They do have some unusual dishes," I said. "Look at the menu. It even offers *Chicken Consumé*."

"They must mean *Consommé?* Chicken soup? What on earth is Chicken *Consumé?*"

"It literally means 'burned-out chicken.' Haven't you tried it?"

He laughed good-heartedly:

"I am afraid I have had it many times. Generally in places just like this!"

Fred Beckman is a witty man with an acerbic contempt for most of humanity, with an exception for orchestra conductors. He has a deep passion for complex, exquisite things: Oriental carpets, Chinese gongs, Mahler symphonies. He has read *Anatomy*, after finding a notice about it in *True* magazine, of all places! He told me it was the first book on the subject which he did not throw away after a few pages.

As Hynek expected, the Air Force has officially announced the award of a $300,000 research contract to a team at the University of Colorado led by Dr. Edward U. Condon, a nuclear physicist of established reputation.[3] Hynek and I will be invited to Boulder next month to help define their strategy. I suggested a structured seminar format rather than an informal discussion which would be a waste of time.

In the midst of all this Jim McDonald is trying hard to recapture the attention of the media, which is slipping away from him. He is now telling the press what he secretly believed all along: Flying saucers are extraterrestrial spacecraft. Big deal. He does not have any evidence to support his statement, so he has very little impact except among the small circle of the ufologists, who were already convinced. In the scientific world he carries far less weight now than he did last Spring. Besides, the gossip in Arizona academic circles is that his wife doesn't believe in UFOs.

The whole day has been magnificent. Dead leaves falling from the trees that line the avenues landed with crackling sounds inside my little convertible. The sky was full of stars and of life.

Chicago. Monday 17 October 1966.

Jim McDonald was going through O'Hare tonight. He took the time to call me, at the end of a gray cold day. He seemed much more calm than last time, perhaps because his personal position is now a matter of public record. He wanted to know what was happening behind the scene in France. I gave him a vague answer, without mentioning how high our contacts went. He told me he felt good about the Colorado team, given Condon's reputation as an intellectually independent man: he once refused to work at Oak Ridge because of the security requirements there, an action that led to his loyalty being questioned during the witch hunts. McDonald only regrets seeing no field-experienced scientist on the team. He is absolutely right on this point: all the members are laboratory men. How can they possibly relate to the human reality of the sightings?

I pointed out to him that it was a pity he didn't join forces with Allen. He did not reject the possibility of doing so in the future, noting that Hynek's statements had recently evolved in a more positive direction. I did not tell him that Allen had deliberately used the opportunity to write a foreword to *Challenge* as a springboard to answer his criticisms.

Chicago. Sunday 23 October 1966.

The whole weekend was spent gathering ideas and documents to complete my doctoral research.

Hynek called tonight. He had just returned from Colorado where he had had dinner with the Condon committee and with his friend Dr. Roach, the same man I heard in June 1963 discussing the scientific training of astronauts at the National Bureau of Standards.

"My first impression? They will approach the subject very seriously."

"Who was at the dinner?"

"Well, most of the committee, and the wives too. The wives thought the whole project was ridiculous, of course. Especially Mrs. Condon."

"What makes you think they are serious, then?"

"Roach and Condon feel they have a real scientific responsibility here. And the psychologist in the group, too, a fellow named Wertheimer. I read the Minot case to them, and you could have heard a pin drop."

"Is our trip still on?"

"Yes, they've invited us on November 11 and 12. And it will be a seminar, as you requested. I will speak for an hour, then a break, then I'm counting on you to take over for an hour. General discussion is planned over lunch and for the rest of the day."

Chicago. Thursday 27 October 1966.

My dissertation is written. It is entitled "Search Strategies and Retrieval Languages." It is really about the interrogation of databases in natural language. Janine and I spent a long, tedious evening making and binding six copies for the committee.

For the last three weeks I have felt exhausted. I saw a doctor today for an extensive series of tests.

If Europe is often sordid, the United States only offer two choices: the grandiose and the utterly slimy. For the last year I have not noticed very much that was grandiose around here. Chicago has become very ugly, with its rundown buildings, its drunks along Skid Row, its corrupt political machinations bordering on the criminal.

Chicago. Saturday 29 October 1966.

Contrary to what I feel, the doctor says I am in excellent shape. Back to work.

Raymond Veillith, the editor of *Lumières dans la Nuit*, has sent me a collection of the old French UFO magazine *Ouranos*. I am reading it with surprise and delight. The founders, Marc Thirouin and Jimmy Guieu, had grasped the nature of this problem well before 1954. Their magazine was outstanding. It also provided a useful service in recording the exaggerated confidence prevalent among the skeptics; witness the following statement by Le Lionnais, president of the Association of French Scientific Writers, and Great Inquisitor of Rationalism:

> The belief in flying saucers ... is one of the most regrettable hoaxes of our time. Naturally, no man of science places the slightest credit in it. Statements implying that the Pentagon considers it seriously are false. ... It would be highly desirable to eliminate this intellectual poison from our country: it can only weaken the mental health of the nation.

Chicago. Sunday 30 October 1966.

A review of the key sightings of the last five years occupied me all day. I have concentrated my efforts on analyzing three contact cases involving humanoids, namely the Betty and Barney Hill abduction, the Antonio Vilas Boas sexual episode and the Douglas case. I also compiled a summary of the Brazilian *Lead Mask* case in Niteroi near Rio, where two men who were known to be engaged in an effort to contact flying saucers were found dead on a hillside with no apparent sign of violence. Several witnesses had reported a strange luminous disk over the site the same evening when they died. This opens up a much wider enigma than I had suspected when I first read a notice of it in a French newspaper; it takes the subject beyond everything Blue Book and the Condon committee are doing.

Chicago. Friday 4 November 1966.

Mimi Hynek just called me to say we are scheduled to have dinner with the leading Brazilian researcher, Dr. Olavo Fontes, and his wife in a Clark Street restaurant. They have been in town for two days. From our first meeting it was clear that Fontes was very much up-to-date on the situation in the U.S., thanks to his APRO contacts. Also in town is Dr. Duggin, an Australian scientist who is doing research privately. We all attended an interview Hynek was doing for WGN and the three of us ended up on camera, lending an international flavor to the show.

In the evening everybody met at our Bryn Mawr Avenue flat. Bill and his wife brought an editor from the *Saturday Evening Post.* The Mittmans also attended.

Late that night I took Duggin and Fontes to their hotels and I drove back home along the lake at 2 o'clock in the morning, turning some of their statements over and over in my head. Fontes is convinced we are living the last few years before the revelation that a space invasion is in progress. I do not share his point of view, but I cannot prove that he is wrong.

Chicago. Saturday 5 November 1966.

Last night over dinner Hynek made an interesting remark: If the Condon committee were to reach the conclusion that UFOs did exist, responsibility for their study would immediately be taken away from him. A similar situation existed when the Atomic Energy Commission was created. Overnight, Fermi and Oppenheimer ceased to be free academic physicists. They became the employees of an enormous political machine.

"I have always been struck by that sudden abdication on the part of the atomic scientists," said Hynek. "After all, they were the ones with the knowledge and the power. The politicians could not do anything without them. But they were unable to defend this treasure, the impending energy revolution, against the greed and the thirst for power of the bureaucrats and the military."

The very nature of our problem has selected those individuals who choose to work on it, men with high ethical standards and personal integrity. Unfortunately these men are isolated, thousands of miles away from each other, in a divided world with poor communications. Their own work is not devoid of ego problems. On Monday Hynek received a confidential letter at the observatory. It came from Pierre Guérin, who was warning him against accepting a space science prize for which Aimé Michel had proposed his name. Naturally, Hynek asked me to review this "confidential" letter for him, after which he was disgusted with everything French.

I am not far from having the same feelings. Professor Rocard has not given any sign of life since I visited him. Yet I had taken the unusual step of bringing him our very best data, our most detailed cases. His interest in them is reflected in the fact that he never returned these documents, which are probably filed somewhere in the dark recesses of some French office. But he never bothered to give us an answer regarding France's possible interest in a debate at the United Nations. I feel I have been used, a feeling one often gets in working with the French higher circles. The next time an outstanding case comes my way, it is not to them that I will bring it.

Chicago. Sunday 13 November 1966.

Last night I came back from Colorado with Hynek, following two days in Boulder with the Condon team. All our hopes were distilled into those two days.

First it was Denver rising from the night horizon beyond the wing of the plane, Dvořak music playing in our earphones. At Stapleton Airport we rented a little white sports car. Hynek took the wheel.

"Well Jacques," he began as we drove on the freeway that connects the city to its academic suburb, "this is a truly historic occasion. To tell you the truth, I have been waiting for this all my life. Did I ever tell you how I became interested in science?"

"Wasn't your mother a schoolteacher? You told me she once gave you a book about astronomy that fascinated you."

"That's not what made me decide to take up science as a profession. So many people get into science looking for power, or for a chance to make some big discovery that will put their name into history books.... For me the challenge was to find out the very limitations of science, the places where it broke down, the phenomena it didn't explain."

"Had you studied the paranormal before you decided to become an astronomer?"

"I had spent a great deal of time reading about esoteric subjects. Of course I wouldn't say any of this to my colleagues, they would think I'm crazy. But as a student I read everything I could find about the Rosicrucians and the hermetic philosophers."

It was my turn to take a deep breath.

"I might as well confess to you that I have spent several years in the same studies," I finally told Hynek. "Until recently I even followed the Rosicrucian Order."

"Which one?" asked Hynek.

"AMORC, which is headquartered in San Jose."

"You know there are several movements that call themselves Rosicrucian, don't you? Among the hermetic writers I was very impressed with Max Heindel when I was younger, until I started reading the books by Manly Hall. Eventually that led me to Rudolf Steiner, who I believe is the deepest of the group." [4]

"I always admired the old traditions which state that there is no such thing as a physical Rosicrucian organization. The only valid Rosicrucian Order, they claim, is not on this level of existence. And they insist that the true initiation, the only illumination of the spirit that counts, cannot come from any human master, but only from nature herself. When I read this I dropped my membership to the San Jose group. I continue to wonder if there may be a genuine 'Rose+Croix' that remains invisible."

There was a silence as we both savored the realization that we had followed such a similar course.

"I have never stopped thinking about what must lie beyond all this," he said with a gesture that encompassed the dark shapes of the Rocky Mountains to the West and the vast plains to the East. "I never cease to be fascinated by the limitations of our science, as great and amazing and powerful as it has become. Now we are about to see how it handles this phenomenon of UFOs that has become so familiar to you and to me. Yes, we shall see . . ."

The next morning we were standing with our backs to the blackboard in a lecture room on the University of Colorado campus. In front of us, seated at a horseshoe table, fifteen members of the committee. Professor Edward U. Condon, internationally known atomic scientist and author of America's first book on quantum mechanics, was at the head of the table. To his left, project administrator Robert Low and physicist Dr. Franklin Roach. To his right, three psychologists including Dr. Wertheimer, the son of a Gestalt theory pioneer, and a cloud of assistants and graduate students. A secretary ran the tape recorder.

At several points during his presentation Hynek requested that the tape be halted so he could tell the whole story of the weaknesses of Blue Book, giving a blunt appraisal of the various officers who led the project. Ruppelt and Friend were the only ones for whom he felt some respect. The committee was fascinated.

In an effort to give an understated presentation I had entitled my own section of the briefing "Analysis techniques applicable to the UFO problem," keeping it suitably academic. Unfortunately the group did not include any information scientist, so some of my more technical recommendations missed their target. In spite of this the committee praised my summary, decided to follow the guidelines I had suggested for their own computer compilations, and adjourned for lunch.

Hynek gave a splendid press conference where he appeared as a man who was finally vindicated after years of unrecognized labor. He fielded some vicious questions quite well. After the reporters had gone we went back to work with the committee. I stressed the global nature of the phenomenon, with a history of the sightings as seen from Europe, an important element, I told them, for anyone trying to understand the whole problem. As I looked in Condon's direction during this presentation I was dismayed to observe that he had fallen asleep. Bob Low caught my eye and motioned for me to go on as if nothing was happening. I later learned it was quite common for him to snooze after lunch.

At six o'clock Hynek and I left the campus and we drove to the house of George Gamow, where we celebrated what we felt was a real victory with vodka and good cheer. Gamow's big guard dog was playing in the snow around the jumble of rocks we saw through the wide living room window. Hynek reclined happily in a big armchair and popped almonds into his mouth while Gamow told us stories of the world of physics before World War II. Here, I thought, was the man who first formulated the "Big Bang" theory, assuming that the universe cooled down as it expanded, a man who had worked in the old days of atomic theory with Madame Curie and with de Broglie. He reminisced happily about his visits to Paris.

"I was just a graduate student when I had a chance to go see de Broglie. I rang the bell at his mansion and a butler introduced me. Right there I realized we had a big problem, because I spoke no French, and de Broglie spoke no English, so that we had to conduct our exchange of ideas through equations hastily written on pieces of paper."

"A week later I attended a physics meeting in London where de Broglie was the keynote speaker..." Gamow chuckled: "The same man I had just visited in Paris went to the lectern with calm, measured steps and delivered a brilliant presentation, without any notes, and *in perfect English.* I was thunderstruck. Had he learned the language in one week? I enquired of an older American physicist who was acquainted with the European scene. "You fool, he told me, when you go see the Prince de Broglie in Paris, of course you're expected to speak French!"

I told Gamow how much pleasure I had had, as a young French student, devouring his "Mr. Tompkins" series, which was full of wit and gave me my first glimpse of relativity and quantum theory. He was so pleased to hear my confession that he rummaged through his library and came back with one of his books, which he inscribed to me.

We left Gamow and went up into the mountains for dinner with Bob Low and his wife. Low asked me bluntly how old I was. That question, it seems, had been the subject of an animated debate among the committee after my briefing. Having settled that point I brought up the international aspect again, a topic I had privately discussed with Condon at lunch: Did he know Professor Rocard? I had asked him. The French would welcome a discreet relationship with the committee, if only as a matter of mutual information. I also told Low the names of some French scientists of international standing who could help manage the long-term implications of this research if the University of Colorado ever recognized it as a significant problem for science.

"I had thought of Jean-Claude Pecker," said Low, lifting his cocktail glass. He seemed to reflect on that and changed his mind:

"Come to think of it, perhaps it would be better to keep the French out of this altogether. What we have here is a group of American scientists who are on a good track, they are really self-sufficient." Bob Low went on: "The French couldn't be expected to make any real contribution anyway. They would just muddy up the water, as they always do, with their personal quarrels and their big egos."

We went to the Lows' house afterwards to watch our press conference on the nightly news. The reporter commented that it was just as hard to pin down Dr. Hynek as it was to catch a flying saucer.

On Saturday morning we had a smaller meeting under the informal leadership of Franklin Roach. "What are the practical objectives we should set?" he asked. "And what tactics can we use to satisfy both the Air Force and the politicians up in Washington while following an impeccable methodology from the viewpoint of the National Academy of Sciences?"

"Clearly the report must answer the question, *is there or is there not a UFO problem?"* proposed Hynek.

"What about the computer statistics?" asked Roach.

"For heaven's sake," I insisted, "whatever you do, take the time to do it thoroughly. I know you only have a couple of years, but a rushed computer job on the basic data will bias your whole effort, you'll only get garbage."

We were free at noon. I bought the Denver papers to find out if our press conference had had any impact. Nothing. Clearly the media did not share our sense of a historic occasion. I did find a small article on an inside page, entitled "Two Giraffes are expected at the Denver Zoo." I showed it to Hynek:

"That could apply to me," I mused. "People are always commenting that I am too tall."

"You seem to forget that my name is High-Neck!" he replied as we both burst out laughing.

We flew back to Chicago with a feeling of elation and enthusiasm, certain that we had influenced the committee and the research which would follow.

"Do you realize that all those people at NICAP would have given anything to attend that meeting?" Hynek said as the plane took off. We went over the data we had presented, and he concluded: "I think we've done all we could."

Chicago. Monday 14 November 1966.

Yesterday's *Denver Post* did mention our visit to Boulder, under the headline "Respectability Given Saucers With CU Job." It quoted Allen:

> Hynek said it's highly likely there's life on other planets somewhere in the universe, but not very likely that any spacecraft have

visited earth. Flying saucers so far "exist only as reports," he said. Some of these reports have substance, "but there is no hardware," Hynek said—no tangible evidence of saucer visits.

Chicago. Wednesday 23 November 1966.

In the brand-new computation center at Northwestern I have a fine office looking out to Lake Michigan. I cleared another big hurdle today. My thesis committee reviewed my progress on the Altair system. After I fielded their questions, with Albert Grau, one of the architects of Algol, giving me a hard time on language formalism, they asked me to leave the room so they could deliberate on my fate. Finally Gilbert Krulee opened the door, stepped into the hall:

"Congratulations," he said, shaking my hand warmly.

"Make no mistake, Jacques, it's over," Allen said later as we sat in front of a couple of beers in Skokie. "You have finished, except for a little polishing up on the text of your dissertation. That was the last examination you'll ever take. You are a doctor, even if the formal graduation remains a few months away."

Is it really over? I am still trying to catch my breath.

Chicago. Saturday 26 November 1966.

Now I feel like a convalescent recovering from a dreadful disease. I am relaxing in bed, with the Air Force files spread out around me all over the blankets. The phone just rang: I did not bother to pick it up. This morning some television reporter called from New York. He wanted to do a show on UFOs. I turned down his offer to fly me to Manhattan. "That doesn't matter," he said, "we have a team right in Chicago to tape you there." Regnery usually filters these requests.

I want to be left alone to resume my research. I have nothing to say to the world, no message. I am leading no crusade. I have published my data and anyone who knows how to read can find it in my books. Now it is within myself I want to look for truth. I don't care if others approve, or even notice.

There is no true science without visionary thought.

Chicago. Sunday 27 November 1966.

I have just finished reviewing the Air Force files for the first eight months of 1965. In numerous cases I find myself in sharp disagreement with Hynek's conclusions. He rarely challenges the Air Force's rejection of the reports. When they write "Aircraft" he tempers it with "Possible Aircraft." Yet some of these so-called *possible aircraft* perform maneuvers that are well beyond the laws of aerodynamics. I realize he is still in a delicate position. He has been fighting like this for nineteen years. But he does present a broad target to critics like Jim McDonald. So far I am the only scientist outside the Air Force who has studied the cases one by one with enough attention to detail to see what was happening. A case in point is the Rapid City sighting of 2 August 1965, which the Air Force explained as "stars" with Hynek's blessing. Yet witnesses located in geographically separate sites saw the lights move to the East, a radar also tracked the objects and they were seen from an Air Force control tower and from several missile bases: How could they be stars? "You are much too timid," I told Allen again the other evening.

Chicago. Saturday 3 December 1966.

Long letters to be typed. They go to Marc Thirouin, to Rene Fouéré, to Raymond Veillith and Guy Tarade in France; to Galia in Moscow, with a copy of *Challenge* I am asking her to deliver to the Lenin Library; to Alexander Kazantsev; to Charles Bowen at the *Flying Saucer Review* in London.

Now it is two o'clock in the afternoon, everything is already dark, the electric lights are turned on in our apartment. Olivier is playing with the large parking garage I built out of plywood for his third birthday. Outside the snow is hard, dirty, crackling underfoot.

What a marvellous tribute, the description of Paracelsus by Strunz:

> He had a mind with powerful abilities whose rare maturity translated scientific problems into warm human terms.... His restless life never removed from him the enchantment that flooded the immortal impulses of his soul like a golden sun, the vision that belongs to a great poet of nature. Yet few men of his time recognized, as he did, the fantastic results to be gained through the

> empirical-deductive method.... Paracelsus felt like an artist and thought like a mathematician, exactly as he combined the laws of nature and those of the microcosm, that is, the laws of man and those of his consciousness, his feelings and his desires.

Chicago. Sunday 11 December 1966.

In a few minutes Janine will call me for dinner. I have spent two days of happiness, although an outside observer would say that all I have done is to retype the final text of my dissertation, and to redraw all the figures. My happiness comes from having Janine and Olivier here with me, making a paradise as I write. I cannot imagine a deeper sense of life than working hard, with both of them so close to me.

Last week I interviewed with a man from IBM to discuss work opportunities outside the United States. He said they would greet me with open arms if I wanted to join them either here or in France. They do have a facility near Nice. But why should we go and bury ourselves in sleepy Provence, when we could be changing the world?

Chicago. Wednesday 14 December 1966.

Marc Thirouin has sent me a very friendly and warm letter. I miss France and I am not looking forward to a third winter in Chicago. The wind fails to disperse the unhealthy haze that hovers above the city. The sunset is pink with chemicals and gray with dusty rain. Clark Street is greasy and fat.

Chicago. Saturday 17 December 1966.

Before I become too over-inflated with my new academic distinction, I should note what Paracelsus has to say about the title of Doctor:

> Doctors? What use is the name, the title and the robe, if one lacks the knowledge? To whom are due the honors, the ring and the mantle of the Doctor, if not to those who deserve them by their science? When the illness of our patients pushes our backs against the wall, then the splendor, the title and the ring become as useful to us as a horse's tail.

Last night we drove to Northbrook for Bill Powers' annual party. Most of the guests were local socialites full of conventional wisdom. We passed unnoticed, as we talked quietly in a corner with Sam Randlett.

We saw Allen between two planes. He was flying back from Dayton, where Quintanilla is still mad at him for his account of the Michigan fiasco in a recent *Post* article. Every time we are together the Marsh Gas case comes up in the conversation, and he tells me again: "Jacques, things would have been so different if only I had taken you to Michigan with me!" Yet I can't understand Hynek's recent actions. He is far too apologetic about the Marsh Gas case, and puts himself into a position of weakness that is unnecessary. And he travels more than ever. Over the last two weeks he has not spent five days in Chicago.

Allen dabbles in too many things. A few months ago he was fascinated with psychic surgery. Now he is publicly quoted as supporting an even more shaky affair, the alleged "psychic photography" of Ted Serios, a beer-drinking "psychic" who stares into a Polaroid camera and produces pictures of the leaning tower of Pisa and other monuments. It can be argued that such claims deserve to be investigated with an open mind. But why does Hynek need to make public statements about such matters when he has only met the man once, at an evening party, and has conducted no serious test? I told him things would be different if a special camera had been designed with detectors inside, to establish whether or not light penetrated through the lens when Ted Serios operated.

Chicago. Monday 19 December 1966.

Snow mixed with rain, darkness. Here is a cross section of newspaper headlines: More Army divisions sent to Vietnam. Who killed Kennedy? Is Queen Elizabeth really happy with her husband?

I will not join the Faculty at Northwestern. I have too little empathy for the students here: most of them are just rich brats who go through the motions of learning. And I certainly have no respect left for the administration.

This morning our Control Data mainframe refused my program: "Too many transfer addresses, too many transfer addresses . . ." Ah, my dear machine, if you only knew!

A griffin is roaring within me, a devil scratches at the waxing moon with his trident. I have a date on the hillside with nymphs and salamanders. Every tiny second is filled with pleasure, knowing that I am alive, and Janine and Olivier with me, and the countless galaxies above.

Chicago. Wednesday 21 December 1966.

I have to fight a lingering cold to drag myself to work. After lunch I get so tired that I drive home, listening to country music to try to boost my energy. (*Kansas City Star, That's what I are...*)

I have started to write a satirical novel called *Zorponna*, in which Parisian students attempt a revolution against the Establishment. The uprising fails utterly, of course, but not without some very Rabelaisian episodes. One of the characters is a giant named Major Syrtis who tells the students:

> Miserable dwarves, you would start a revolution? Ah, this is laughable! Already you are afraid that the guards might come here, and you speak of seizing power? No, a revolution with such servants would be little more than a joke.

Instead it is the giant who ends up destroying the pompous city of science when he realizes the pitiful state of academic knowledge. He starts crying rivers of tears which soon engulf the whole country and reduce it to little more than a salt island battered by the ocean waves.

I doubt if this will ever move a publisher to the point of printing the book.

Someone has sent me a packet of newspaper clippings from France. They show a distressing picture of errors and misinformation about the status of UFO research in the U.S. The Air Force is constantly misquoted. French journalists are so sloppy they even confuse NICAP with Blue Book! If somebody had paid them to cloud up the issue completely they could not have done it any better.

Chicago. Thursday 22 December 1966.

Hynek has returned from Dayton with pessimistic news again, having spent a long time with Bob Low. NICAP has made such an impression on the committee, and they have done such a great job bad-mouthing everyone else, that Condon has decided not to establish any ties with Coral and Jim Lorenzen, the leaders of rival APRO, because "Mrs. Lorenzen's ideas are too eccentric." Yet APRO probably has the best world-wide sighting files.

The committee has started to work on something they call "the

Wertheimer hypothesis." According to this newly hatched wonder theory the phenomenon does contain some authentic elements but they are beyond the scope of science. In other words, flying saucers are as foreign to scientific thought as the possible reality of God.

"I don't call that a hypothesis," fumes Hynek. "I call it an abdication."

These people are ignoring everything we have told them in our briefing, every practical fact. The tragic thing is that the rest of the scientific world is fully expecting an answer based on serious analysis.

Chicago. Tuesday 27 December 1966.

Two Northwestern professors are proposing the creation of an information science center of which they offer me the directorship. In the meantime the news from Paris is not encouraging. French scientists are simply hoping to stay well enough informed so they can take some of the credit if the wind should suddenly turn in America. But they are afraid of facing their responsibilities, of handling openly the observations that come from their own people.

Allen has just made a big decision: Stating that he "must act according to his national responsibilities in this field," he is going to transfer all his personal files to the observatory. He wants Bill and me to devote a lot of energy to casc analysis. He himself promises to devote all his spare time to it during the next quarter.

Chicago. Friday 6 January 1967.

We had another meeting at Bryn Mawr last night. Fred Beckman brought over University of Chicago mathematician John Thompson, of group theory fame, and two other senior scientists. Things did not go well, however. Instead of giving them his best arguments Hynek, who had just flown back from Los Angeles, practically fell asleep on our couch. Now he is gone again, flying off to yet another lecture.

The sad fact is that we no longer share the same viewpoint.

He loves so much to be in the spotlight that he squeezes the last drop of publicity his association with the Air Force brings him. Perhaps that is his way to seek revenge for the ridicule of the Marsh Gas season, or to compensate for the fact that the University of Colorado has been given responsibility, authority and money to conduct a study which, in all fairness, should have been entrusted to him. But his commitment to spend time with us on the analysis of the data is not being kept, and our personal motivations have started to diverge. Last September he had promised me that we would go back underground if a radical change had not happened by Christmas. He has forgotten that agreement. And he pays little attention to Fred's urging that our first step should be to create a support group of really top scientific talent.

Chicago. Saturday 7 January 1967.

Hynek seems to be repeating NICAP's old mistake, taking his case before the media, showing his face on every channel in every town in America. Now he is all excited about writing an article for *Playboy.* I told him rather bluntly that what he was doing now was show biz, not science.

Chicago. Sunday 29 January 1967.

I have been reading a copy of *Le Comte de Gabalis,* lent to me by Don Hanlon, a delightful story of occultists having intimate relationships with the fair creatures of the Elements.[5] Alas! The precious salamanders of my own reverie, if they ever look for me here, will come too late. Nothing can keep us in Chicago beyond this year. We may be going back to France as early as March.

In my last letter to Kazantsev I suggested that we collaborate on an article for *Priroda,* the Soviet equivalent of *Science* magazine. He answered me in Russian. It took me most of Friday to translate his letter, but we are snowed in, so I had nothing else to do. It was well worth the effort. He was asking me: "Why should we waste our time with a little publication which is only read by a few academic stuffed shirts? I have just spoken on the phone to the editor of *Young Technology,* which runs nearly two million copies a month. He is waiting for our article!"

Yesterday two of our astronauts died in the fire that destroyed an Apollo capsule. That means a six-month delay in the race to the moon. There is a tragic lack of logic in our rocket program. Between the politics of Vietnam and the management of our space technology, America is making very costly mistakes.

14

Chicago. Monday 30 January 1967.

"My poor children, I'll get you out of this mess yet!" I vowed as I sat at the kitchen table this evening, drinking my coffee alone, surrounded with shadows and half a dozen socks Janine had put out to dry.

I had just come back on foot, carrying Olivier through streets covered with three feet of snow and ice in the colorless twilight. Stalled cars are in the way and snowplows cannot move, even through the side streets. Janine has called. Her program must run on a computer that is not free until five o'clock. She will come back whenever she can.

In the Evanston train all the faces seem old and lost. The wealthy and the poor share the same look in this cold, as if they had dropped out of the sky into a place where city blocks follow other city blocks without reason.

Nothing is happening here. The very existence of the Condon committee is freezing all independent efforts. Hynek went to Dayton without being able to see Quintanilla, who is still mad at him. The Air Force seems to delight in the idea of shielding itself behind the University of Colorado.

Would our return to Europe be an abdication? Perhaps there are other solutions. My friend Jean Baudot, who is administrator of the computing center at the University of Montréal,[6] has written to inform me that Radio-Canada was flooded with reports from the Canadian public which no one has taken the trouble to analyze seriously. Perhaps we could try to open in Canada the "Second Front" which is proving so hard to start in Europe. As long as the U.S. Air Force has a world monopoly on UFO research, nothing will get done.

Chicago. Sunday 5 February 1967.

A journalist named Otto Binder has sent me a copy of John Keel's interview with Colonel Freeman, a spokesman for the Air Force in the

Pentagon. Freeman told Keel that he was currently trying to identify certain individuals who had approached and intimidated UFO witnesses. *They wore military uniforms*, he said, *but nobody knew who they were.*

When I asked Hynek about this, he told me that he had come across an instance of the same mystery. During his recent stay in Dayton he had heard Lieutenant Morley frantically calling up every Air Force facility in New York and New Jersey to find out who was responsible for certain military investigations reportedly under way at Wanaque, where a peculiar object has been seen over a large reservoir. It wasn't a Blue Book investigation, and no other group would admit to being involved, so the Air Force itself is forced to start looking under the bed.

Our relations with the Condon Committee are getting tense. Following a trip Bob Low and Bill Powers took together to Joplin, Missouri, the committee has formally requested that any future investigation be conducted separately rather than jointly in the future. Was Bob Low shocked to find out that Bill was a far more knowledgeable researcher than he was? Perhaps they don't like to admit that they prefer to study the witnesses rather than the phenomenon itself. Thus they have summoned Kenneth Arnold, the private pilot who coined the term "flying saucer" twenty years ago, to ask him *why he wanted to believe* in UFOs. They are preparing a psychological questionnaire aimed at witnesses. Physicist Franklin Roach has quietly left for Hawaii. This leaves the committee under the control of four psychologists around Condon, who doesn't seem to give a damn about the whole subject.

Chicago. Monday 6 February 1967.

The release of my Altair system is having a strange effect on the astronomers to whom it is dedicated, and there are some interesting social lessons to be drawn from it. This program is an artificial intelligence system that takes questions in English, questions such as "How many stars having a spectrum between G0 and K2 are triple?" which previously required the writing and debugging of complex special programs and a 24-hour delay in getting each answer by conventional means. Furthermore, astronomers who do this kind of work need fast answers and are not necessarily skilled in programming. For all those reasons Altair was a big sensation for about three weeks. Then, remarkably, people stopped using it and returned to programming things the hard way. This hasn't affected my work, since I have now moved beyond these experiments in natural language processing to develop generalized data-base management systems. But I am a little piqued by this lack of interest in a program that, after all, has delivered all the benefits it was designed to bring.

One of the astronomers who should be a primary user of Altair finally confided to me why he had returned to the old clumsy way of doing things: he has recently mastered a computer language, and he is so thrilled at this new skill that he is disheartened by the idea that the machine can solve his problem without the need to write a program at all, since Altair generates its own internal code and runs it without the user being aware of it. He prefers to run the tape through a different computer program every time he needs a new statistical answer.

This leads to an important observation about artificial intelligence. When the genuine power of the computer is unleashed, it is a threat so direct and so close to us that the intended users tend to develop strategies to ignore it.

In my work at the computation center I have now adopted a data-base language called Infol (Information-oriented language) developed by a man named Bill Olle and his team at Control Data. Infol is the first in a new generation of software called *non-procedural languages* because they don't require the user to specify how the computer should do something. They simply indicate what they want, and the system itself generates the appropriate procedure.

The Infol system is only available for a particular machine, and of course we have a different model for which it does not run. I have started to write a new Infol compiler in Fortran, a language which is widely used. This will make it possible to move it to any other machine when the need arises, and it will solve a large class of data-handling problems for the university community, from medical data-bases to geology or the management of student records.

This morning I returned a call from Jim Wadsworth, a research associate with the Condon Committee. He is working with psychometrician David Saunders, an expert in statistics who does most of the committee's computer work. Wadsworth wants me to turn over to them my whole computer catalogue of UFO cases, in order to integrate our work into their files. In the spirit of scientific cooperation I agreed to do it, perhaps foolishly, and I did not ask for any compensation for the many years of hard work Janine and I have put into this catalogue; not only into the data that was included, but into all the cases that we screened out.

My only fear is that they will drown the fish of our careful research into an ocean of unfiltered statistics and will come out with no pattern at all—hence no UFO phenomenon. And they will be able to claim that their data *supersedes* ours. This game has been played before, whenever statisticians wanted to kill an unpopular theory.

A terrible cold wave has come to Chicago. There is so much snow in the alley that I cannot get the little Buick out of the garage. Every morning I carry Olivier to the house of the sitters, who live seven blocks away.

A skeptical science writer named Philip Klass has added me to his list of favorite targets, and we have had one tense telephone conversation.

"Have you ever seen an airplane?" he began.

His style is that of the trial litigator who tricks you into admitting that you once made a wrong turn into a one-way street and deduces from that fact that you are capable of murdering your own grandmother.

Philip Klass explains all UFOs in terms of hoaxes, plasma phenomena and increasingly implausible atmospheric effects that leave far behind the most extreme miracles of Menzelian physics.

I declined to respond to his critiques in *Aviation Week.* Instead I sent him a short poem I wrote entitled *Klass dismissed* that rhymed in -ass from beginning to end. It was inspired by the story of Brazilian farmer Antonio Vilas Boas (another name rhyming with Klass) who once assured Dr. Fontes he had been abducted inside a flying saucer and was forced to have sex with its female occupant:

> Many nights ago I have seen, alas!
> A wanton plasma, a fiery mass.
> The glow of copper, the luster of brass
> Flashed across the sky with a tail of gas.
>
> It landed, my friend, better to harass
> My deluded mind, and a pretty lass
> Alighted, wearing a robe of canvas.
> "Come here," she said, "and show me with class
> How life goes on, and theories pass?"

It wasn't a very good poem, but it did the job. I have had no further problem with Philip Klass, who has a good sense of humor.

Chicago. Saturday 11 February 1967.

A Northwestern professor wants to organize a meeting of psychologists in Washington this September to debate UFOs. He is also inviting Wertheimer of the Colorado committee. What is the use of such academic games? Wertheimer is already convinced the subject has no scientific validity whatsoever. Fred Beckman tells me that the psychologists of the German school, and particularly the Gestalt theoreticians, have always been known as dogmatists and doctrinaires. The Gestalt collapsed about 1930 but some of its old branches show new signs of green life from time to time.

Chicago. Sunday 12 February 1967.

At the moment Fred is the most active member of our group. We are frustrated because there is a flood of new sightings coming from Southern Illinois and we cannot do anything about it. Even calling the local police using Hynek's name is hard to do: the Air Force won't pay for the calls, we don't have any money, and we can't get the University involved. Hynek himself is always away, giving lectures.

Fred keeps telling Allen he is wrong in not seeking serious contact with top-level scientists, and that he should stay away from publicity instead of courting it.

"All you will accomplish is to drag the whole subject down," he says whenever we can corner Allen for a few minutes.

"I have no illusion left about the integrity of my colleagues," Hynek replies. "I might as well put the problem before the public and hope that does some good. Just look at the episode of my letter to *Science* magazine: These people were acting in bad faith."

They both have a valid point, and they never solve their argument. Unfortunately Hynek ends up making ambiguous or contradictory statements that satisfy neither the academic world nor the public he is attempting to enlighten. As usual, none of the major U.S. newspapers has mentioned the current UFO wave. Yet the sightings have been sweeping Southern Illinois. The observations are concentrated around the point where the Ohio and Mississippi rivers meet, halfway to New Orleans from Chicago.

In preparation for our leaving Chicago I have been making duplicates of all the key sighting reports in my files, classifying them by type. If the Air Force ever disbands Project Blue Book or destroys the data I will be able to reconstruct the important cases in the history of the field from my own notes.

Chicago. Tuesday 14 February 1967.

Today I sent a copy of my entire computer catalogue to Jim Wadsworth and Dave Saunders at the University of Colorado. A little voice in my mind called me a poor naive fool, and told me I was stupid to comply with their request.

Chicago. Wednesday 22 February 1967.

Bob Low, who is emerging as the key man on the Condon committee, has spent most of the day with us in Chicago. Hynek had lunch with him and drove him to the campus. I then gave him a tour of the computation center, and Bill took him to the new observatory.

This visit has taught us a few new facts about the committee. Contrary to what we thought, the psychologists have lost a lot of ground. Bill Scott, one of the most radical, has left the group. The Wertheimer delusionary hypothesis they were considering last December has allegedly been dropped. "Now the real work begins," says Low. A couple of new physicists have joined the team. Condon, who will be sixty-five next week, feels tired and wants to prepare a revised edition of his complete works. He really has no interest in the UFO question, never did.

"After all," remarks Allen with irony mixed with some bitterness, "that's why he was selected to run the project!"

When we brought up Menzel's relationship to the committee we heard an unexpected diatribe. Condon did speak to Menzel on the phone for fifteen minutes, but when he hung up he turned to Bob Low and remarked: "You know, this man is insane."

Menzel has *demanded* to be allowed to spend four days with the committee and to speak to each member separately. We think this method will have the opposite effect to what he wants.

Low tried out a few interesting scenarios on Allen. What would he think of a final report recommending that Blue Book be closed down? His contacts in Washington have told him bluntly that the Air Force wants to get rid of the whole issue once and for all. Hynek doesn't know what to answer to these trial balloons.

It also turns out that the CIA is now collaborating with the committee in the analysis of photographic data, a fact which will be kept out of the final report. Low has been given an extensive tour of the classified facilities at the Photographic Interpretation Center and was very impressed.

We spoke briefly after Low had left the campus. The most important thing, according to Hynek, is that the committee now realizes that there is a real UFO problem: "They understand that the water was not an illusion after all. It's wet, and it's cold."

Chicago. Tuesday 28 February 1967.

Last Friday we left Olivier with the sitters for the weekend. He was delighted to sleep over at his friends' house for a change. Janine and I flew off to Montréal, where we were greeted in a warm and simple fashion by Jean Baudot and his wife Stella. I brought Jean up to date on the history of our various proposals.

The next morning Jean gave me a tour of the campus, notably the new linear accelerator. I presented him with a tape containing the same data I have given to the Colorado project. In the afternoon we visited the city, and the next day we drove off to the snowy Laurentians to spend time with some of Jean's friends who have a chalet there. It was a charming, nostalgic weekend. The little villages in the snow with their quaint churches and their enchanting names reminded us of the French countryside.

On Monday we discussed computer science again. Jean introduced me to the director of the center, who surprised me by offering me a position on the faculty. I promised to consider it very seriously. It is indeed a tempting offer because it would enable me to continue to collaborate with Hynek while gaining some healthy distance away from the U.S. and its complicated scientific politics.

Chicago. Saturday 4 March 1967.

The bookbinders of Chicago are staying in business thanks to me, it seems. Every week I bring them another batch of case data or another part of my archives to be bound. I am buying many books on all sorts of subjects, most recently the six volumes of Casanova's *Memoirs*, which I read with pleasure. It is a profound and honest work in spite of an unsavory reputation which is largely overstated. Certainly Casanova was a rake and an adventurer, but he was also one of the keenest observers of social and political life Europe has ever had. I also read Flammarion's *Pluralité des Mondes Habités*, Howard Hinton's *Fourth Dimension.* I have finished reading *Le Comte de Gabalis*, which gives me renewed interest in the old folklore and the knowledge that may be hidden in the grimoires of the Hermeticists, even if the evocation of amorous Salamanders seems a bit outlandish.

Chicago. Sunday 5 March 1967.

A current review of the history of the field cites *Anatomy* as the starting point of the reappraisal of the UFO problem. But what good is it, if no solution is in sight? At least we can better distinguish between what is significant and what is irrelevant.

Condon and the Air Force may well realize in two or three years that Hynek and I were right in 1965 to say that Aimé Michel and Ruppelt were right in 1956. While they are mulling over the problem in Boulder I can use those years more effectively, to do innovative work. I must go ahead into unexplored areas. I owe it to the poor idealistic student I was a few years ago, who was so thirsty for new challenges and uncharted journeys. I cannot betray him now by resting on a few laurels.

Chicago. Thursday 9 March 1967.

There is now an avalanche of garbage about UFOs in all the media, from television to the tabloids, the porno magazines, the occult press, and the respected national newspapers. This includes the most grotesque hoaxes and mediocre pieces about space visitors. I am utterly disgusted by this tendency of the professional press to take a serious subject and to turn it to ridicule, yet we should have predicted this would happen. By putting the UFO question before the public Allen has unwittingly created a market for every sensational hack. Since the tabloids have no standards of accuracy or reliability they can satisfy the public's demand for ultimate "answers" much better than any scientist.

We are also witnessing the beginning of a genuine wave of sightings. Many of the reports that reach us describe objects flying over the roads and projecting beams towards the ground.

My friends will be surprised when I turn down the Montréal offer and when I give up the leverage that my current expertise could give me at Northwestern. But I think it is precisely when we have mastered a subject that we should reject it. I have been observing my son, who will be four years old in December. I am amazed at the speed with which he learns new concepts. His secret is that he rushes without hesitation into anything that grabs his interest, and he works on it until the mystery is gone. Then he turns away from it and never looks at it again. If adults had the same intellectual stamina we could accomplish miracles every day.

Man never "conquers," "controls" or "discovers" anything about nature. He only conquers, controls, and perhaps, with luck, discovers himself.

Chicago. Thursday 16 March 1967.

Allen has been on a long trip to the Southwest and the West Coast.

"It would be unthinkable for Condon to reject the reality of the phenomenon," he told me when he returned. He visited Holloman Air Force Base with Bob Low and his secretary, Mary-Lou Armstrong. They had dinner together and they discussed Menzel's recent trip to Boulder. Mary-Lou laughed so hard as she recalled Menzel's speeches that she fell from her chair and landed flat on her back on the restaurant floor. Menzel's explanations for the cases were so ridiculous that only propriety and respect for a senior colleague prevented the members of the team, including Condon, from laughing openly in his face.

Bob Low also told Allen that if any additional computer work became necessary the committee wanted the contract to go to me.

It is likely that Condon will request a contract extension to produce his final report. One scenario would have him recommend the setting up of a permanent research group.

Costa de Beauregard would like to see me go back to France. He writes that the Blaise Pascal computing center is being reorganized. Why not join the National Center for Scientific Research (CNRS)? Another friend of Costa has spontaneously forwarded my résumé to Meudon Observatory. I burst out laughing when I read that. A very ironic turn of events indeed, if I were to go back to Meudon.... I wonder if the rats are still scurrying under the wooden floorboards at midnight, and if the Observatory still employs same nasty accountant.

The French Society for Operations Research has accepted an article I sent them on self-encoding systems.[7]

Chicago. Sunday 26 March 1967.

Last night Janine and I invited Allen, Fred and Ben Mittman to join us for dinner at a fancy suburban restaurant where Janine's hard-working brother Alain is now part owner, in Aurora. I wanted to get everybody together in a somewhat formal atmosphere to announce to them I was seriously planning to leave, and to tell them frankly and openly what I thought of our current research. I had conspired with Fred, who kept deploring the lack of an opportunity to talk seriously to Hynek. He had prepared his own statement stressing two major points: First, the UFO Phenomenon is available for scientific study; although it is not reproducible at will, it does reappear frequently. Second, the application of today's technology to the phenomenon could yield important new knowledge and should enable us to establish its nature, if not its origin.

Unfortunately, if Hynek listens politely to all this, he does so in distracted silence. He follows the discussion for a few minutes, then he seems to drift away, occasionally jumping back in with a few anecdotes that are superficially amusing but leave us frustrated.

In private Fred tells me that he suspects the Pentagon is eager to preserve the *status quo* at all costs. It should be relatively easy to get good measurements and good photographs of the phenomenon, the kind of information he calls second-generation data. Once enough of these cases are on record we have no doubt that the entire bulldozer of American science could start rolling over it, not just a few isolated scientists. Hynek is the one who should initiate this effort.

The tragedy is, I am losing my confidence in Allen. We have outlined a systematic scientific program but we are doing nothing to implement it.

Fortunately there are many other joys in my life. This evening the stereo was playing Katchaturian's *Saber Dance*. We thought Olivier had gone to sleep, but suddenly we saw him stand up on his bed, a blissful smile on his face, one finger raised up in the air. We played the record all over again for him.

Chicago. Wednesday 29 March 1967.

Lunch with Allen. He read to us the draft of a chapter he had written for a proposed collective textbook on UFOs. Although clear and lucid, it is purely a rehashing of things he has said before. Bill Powers, in the meantime, has asked me to comment on the new draft of his own book on behavior, which I read with attention a few pages at a time while my programs are running.[8]

This evening McDonald called me to say he was coming to Chicago in three weeks. He wanted to meet with me to discuss the results of my last trip to France. He is still trying to bring me into his team. I told him we should include Hynek, and he reluctantly agreed.

The man is as sure of himself as ever, he keeps telling everybody what to do. He has lost nothing of his arrogance.

Chicago. Saturday 1 April 1967.

When I mentioned to Hynek my likely departure from the United States his first reaction was to oppose it.

"The next two years are going to be vital, Jacques, they will decide everything," he told me.

"Come on, nothing will happen until Condon writes his final report anyway," I reminded him. He had to recognize that my arguments were solid.

"If you do go back to Europe, that will give me an excellent excuse to go and visit you there."

Chicago. Sunday 2 April 1967.

Janine makes me happier than I have ever been. Her beauty, her gentleness goes so deep inside me, it leaves me breathless.

Chicago. Thursday 6 April 1967.

The current time is precious, we should be dedicating it to serious research. But Hynek is drifting away from such research more and more. When I reminded him that Jean Baudot and the head of computation at Montréal were scheduled to meet us at O'Hare, he seemed to have forgotten all about it:

"I can't be there," he told me abruptly, "I have to speak to the students at a college in the suburbs."

"But they come here to facilitate the exchange of UFO data with Canada," I objected, "this is a high-level contact for us. . . ."

I have put a lot of effort into preparing this meeting. The Canadians have many UFO cases that are just begging to be analyzed, and which could have an important impact on U.S. scientists. But Hynek doesn't seem to be interested in these contacts. Nor does he remember what we had decided at our dinner in Aurora. In spite of all this he remains charming, captivating, charismatic, but the fact is that he does nothing of long-term viability. He has filed away and forgotten Fred's solemn lecture and our precise proposals for the application of science to the UFO problem. I do not think that he has committed in the past the grave mistakes of which McDonald accuses him, but there is no question in my mind he is committing them now. That is another reason for me to continue to think of returning to Europe.

Aimé Michel has exhausted his own contacts. He feels stumped in his efforts to get some serious attention in France:

> I don't see how I could go much higher than Rocard, other than seeing the Old Man himself (De Gaulle) but he doesn't give a damn. What I see is failure all along the line. For the first time I have no alternate plan. There is no hope for your idea of a *second front* of independent research here. Everybody is looking to America. . . . The new interest in the media is limited to reprinting extracts from Frank Edwards and John Fuller's books. The newspaper bosses tell me: "Maybe you are more knowledgeable than those guys, but our readers don't think so."

In response to my last letter, where I stressed that it was unthinkable that no one would be studying the problem seriously in France, he writes:

> There are some individuals who are doing their own study on a private basis, just as we do. Maybe there are some people who collect French UFO data as agents working on behalf of British, American or Soviet Intelligence. But for France itself, don't make me laugh! High Gendarmerie officials have told me what happens to the cases: the reports go up through the hierarchy depending on the level of notoriety of witnesses and the personal rank of the

> officer in charge who wrote the report. When they get to the appropriate level they are kept for ten years, after which time they are burned! Therefore all the reports of 1954 have now been destroyed. It is enough to make one want to bite off one's testicles in frustration, isn't it?

And he offers his personal conclusions:

> It is urgent to do nothing, at least openly. Once Condon publishes his report, which I already know by heart, we will just have to gauge the reactions.

Chicago. Saturday 8 April 1967.

Next week I am scheduled to attend a computer conference in Pittsburgh. Janine will take a few days off with Olivier, staying at Alain's place in Aurora.

Fred made a special trip to Evanston the other day to have the long-postponed serious discussion of the UFO problem he had been requesting for months. He found Hynek fumbling with his stereo, connecting this and that, adjusting bass and treble without once listening seriously to Fred's arguments.

"And all that to play some wretched *French can-can!*" remarked Fred with disgust.

The only thing that relieves the drabness of life in Chicago is the opportunity to have almost daily discussions with Beckman, whose expertise embraces areas as diverse as photography, classical music, electronic microscopy or neurology, and also with Bill Powers, whose book on behavior is making steady progress. Bill is another rare mind, who has analyzed the process of thought in a new way, looking at the brain as a system that handles cognitive information rather than as "an organ that secretes behavior," as he puts it.

Bill thinks that the brain must be the minimal structure that will enable man to compete and to survive, with all the flaws that implies. He has also speculated that the next step in the evolution of thought might be engineered rather than evolved.

"When I was born the triode had no practical application yet," he points out. "And look how far we have gone in the development of logic structures, first with transistors and then with integrated circuits. Do you know what that means? It's much more likely that we can *build* a better thinking structure before nature has a chance to *evolve* it."

Indeed, today's integrated circuits are on the way to a reduction in size that will approximate synapses, and it is possible, at least on paper, to imagine logical devices approaching the complexity of the brain, although it will be decades before we can build them.

Bill is thinking about automata that would have no preset task; he doesn't believe in the learning machines of Newell and Simon,[9] which start from a set of rules such as the rules of chess and gradually learn to play better and better, but do continue to play chess and nothing else. His idea, which is heretical but simple, is that intelligence should be regarded as an attribute of the environment. And it is foolish to try to simulate it. One should build an artificial thinking being, turn it loose, and let it learn.

"If you ever build it, I'll volunteer to program it," I tell Bill.

"Then we can get Fred to critique what we've done," he jokes.

The last news we have received from the Colorado committee comes from Bob Low's secretary and administrative assistant, Mary-Lou Armstrong. A week ago she sent along a stack of recent sightings.

"We haven't heard anything particularly interesting from our other sources: radio, TV, Associated Press, NICAP etc.," she says, adding:

> None of us has had the time to try and figure out possible explanations for these sightings—as we had, at one time, intended to do, so we'll certainly appreciate any comments you might have on them. The procedure we are following on the processing of sighting reports is still, at best, vague. It changes all the time. However, we are sending out questionnaires on the interesting ones, and all of them are getting punched for the computer. One of our biggest problems is that now there is only one person working on the actual coding and reading of the sightings as they come in. We had two people doing it, but one of them, Dan Culberson, was in a very bad motorcycle accident two weeks ago and is still in the hospital. Brain damage, the works, a real tragedy.

The Condon committee fascinates me. I feel that what I am observing here is the very heart of the scientific machine, as it gets corrupted by politics. Under the thin veneer of academic respectability, it is plain human baseness we see, with no sacred fire, no warmth, no lasting value.

Chicago. Sunday 9 April 1967.

Aimé Michel writes that he is afraid he has a tumor in the abdomen that may be malignant. I feel disheartened when I realize that a mind of his formidable caliber exists in France, and that he is left alone in a little village in the Alps.

Alexander Kazantsev writes to me that

> Interest in the UFO question keeps spreading among our scientists. I am currently a member of a group of physicists that I hope to put into direct contact with you. Like the Americans, they are primarily interested in discovering the secret of UFO propulsion. They are intrigued by the report of a measurement of radioactivity at Montville, Ohio, after a landing. They would like to know the soil composition there, to compare it with soil samples from the Tunguska explosion of 1908, in Siberia.[10]

I don't have any data or any physical samples from the Montville case, and I am a bit curious about this "group of physicists." If they are so interested, why don't they write to us directly?

Chicago. Tuesday 11 April 1967.

A fellow graduate student, an athletic beach bum from San Diego, is trying to convince me to move to California. Whether I go East or West, I must make a move, because I am increasingly discouraged by Hynek's distracted approach to our work. This morning an episode took place which was particularly depressing, although its comical aspects come out in the retelling. We had a business meeting scheduled with Harvey Plotnick and his editor Betty to discuss a contract for a proposed textbook. They came to the observatory punctually at ten o'clock, having driven all the way from the Loop. No sign of Allen. His secretary told us that he was at the new observatory having his picture taken, and that in half an hour he was expected in the Dean's office. Naturally he had completely forgotten our meeting. The three of us started talking to Bill about one thing or another.

Hynek finally arrived, returned a few urgent phone calls, uttered a few disconnected sentences, talked vaguely about the Dean and suddenly, having noticed a pile of old Ranger photographs on top of a filing cabinet, he grabbed them in a moment of desperate inspiration and dumped them in Harvey's lap:

"Here!" he told the astonished publisher. "I'll be back in an hour. Why don't you look at these nice pictures of the moon?"

Exit Allen, leaving us all embarrassed. I miss the days when he was not such a celebrity, when every discussion of "our subject" was a conspiracy, when every chance to advance it was treasured, every research opportunity eagerly seized. The topic has become fashionable entertainment, not serious science. Media men hire Allen as they would hire a guitar player. He rushes wherever he sees a spotlight, and if the spotlight moves he moves with it.

Pittsburgh. Thursday 13 April 1967.

The conference on Information Processing is taking place on the campus of Carnegie-Mellon. I skipped the banquet. I came back to my hotel on foot, stopping at the local Howard Johnson's for beef burgundy and coffee. I was briefly tempted to hire a cab and visit the city, then I thought better of it. What is the use of seeing another bunch of buildings made of steel and marble? We have all that in Chicago.

Three hundred and fifty experts are gathered here. They seem incapable of communicating with human beings. The masters of computer information are only exchanging a few superficial words when they are not reading aloud their own jargon-filled papers on retrieval statistics and balanced trees.

Chicago. Sunday 16 April 1967.

Kazantsev's secretary has sent me a copy of the Soviet magazine *Smena* with an article by Felix Zigel that quotes *Challenge* and reproduces two pictures from it.

Yesterday I went to Hynek's house. I had promised that I would help him reorganize his files, which overflow into shoeboxes and in a wicker basket in a little room on the first floor. He complains that he doesn't find anything any more, and I can see why. The reports of various years are mixed together. Pictures and letters get lost.

Having established a method to reclassify this mess, I began the real work early in the afternoon. By evening I had returned many of the documents to the places where they belonged, in neat folders and envelopes. But I had stumbled on something I felt was important.

I found it among the relics of Project Henry. It was a simple letter dated 1954.

It came from a cloud physicist at the University of Chicago who was studying for a doctorate at the time. Together with three other physicists he had seen a bright unidentified object in the sky over Arizona. The letter gave precise details and calculations.

It was signed James McDonald.

When I showed him the letter Hynek was dumbfounded. He examined the text, as well as the response from the Air Force, which was attached to it.

"I don't remember ever seeing this," he said.

"Evidently it went to Captain Hardin," I pointed out. "He's the one who replied to it."

"If I had remembered this I might well have contacted McDonald. We would have met much earlier."

"You could have had an influence on the policy of the Air Force, if the two of you had joined forces at the time. We could rewrite history here: when the 1954 wave swept Europe you could have contacted Aimé Michel...."

"The Condon Committee would never have happened," Hynek mused. But he shot the whole idea down: "I probably would have been labeled as a crackpot. Whipple wouldn't have asked me to work with him at the Smithsonian in 1957. Northwestern would never have hired me as the director of Dearborn.... You and I wouldn't be here talking."

This brought the discussion to the subject of McDonald's upcoming visit. I told Hynek he could not refuse to meet with him. He tried to get out of it:

"You could take care of it with Bill and Fred ... find out how he feels." He wanted to avoid another confrontation at all cost.

I tried to make him feel guilty: "You already forgot the meeting with the Canadians, and with our publisher. This time you've got to be there."

He finally agreed that we had to find out once and for all if he could work with McDonald. Mimi Hynek has a dinner party planned for Saturday, but it can be rescheduled.

We hear that McDonald is not having much success so far. An article he recently published in the *National Enquirer* has eroded whatever measure of respect he was beginning to gain among his peers.

On the Air Force side, Lieutenant Morley has just briefed Colonel Sleeper, who has responsibility for the Foreign Technology Division and reportedly would love to see them dump Project Blue Book altogether. In this kind of circuitous communication the facts are irrelevant. All the conclusions are already drawn.

My California friend has a job offer for me in San Diego. My first inclination is still to return to France. But Allen has vowed to keep me in this country no matter what happens.

Chicago. Tuesday 18 April 1967.

At lunch today I brought Bill Powers up to date. He told me he placed no trust in McDonald.

"All his views are negatively oriented," he observed. "He doesn't propose anything concrete, otherwise he would just do it and move forward. All he does is to complain and criticize."

Allen showed me an article by Walter Sullivan, science writer for the mighty *New York Times.* He uses his position to heap ridicule on this whole domain, ignoring thousands of credible cases to linger on a miserable joke in which Boulder students released plastic bags over the Colorado University campus. He delights in recounting a mistaken report of a "monster" in Pennsylvania where the witnesses may simply have come upon a stranger relieving himself in the bushes: an analysis of the soil only revealed traces of urine. Sullivan is predicting that Condon's study will point out some remarkable facts about the weakness and the fallacy of human testimony....

"That's a trial balloon," I said. "They are paving the way for a negative report. Just as Bob Low did the other day when he asked what you would do if they recommended closing down Blue Book."

"Yes, but don't you see? It's not enough to be right, Jacques, in a situation like this. You need heavy guns if you're going to fight against such people."

This evening I wrote to a few prospective employers in France; then I walked over to the drugstore to buy some stamps. The weather is still cold. I wear my winter coat with the collar turned up. In the black sky are

Mars, Jupiter, Venus and only a thin crescent moon.

Chicago. Saturday 22 April 1967.

Jim speaks today before the annual meeting of American newspaper editors. He has sent me the advance text of his remarks. This time all the cards are on the table. He repeats loudly what we have been saying in private about the Air Force, which he accuses of negligence, and about Menzel whom he practically calls incompetent. He also comes close to calling Hynek a coward. Yet his statement on the basic UFO problem is almost a carbon copy of Allen's own declarations, couched in a more sensational style. He accuses Menzel of not conducting a proper quantitative study, but he is guilty of the same thing. One minute he makes a big show of blowing up the doors that we had already forced open, the next he rushes forward crazily and hits his head against solid brick walls.

The fact is that Allen is not responding creatively. He does not follow Fred Beckman's advice on scientific methodology, nor does he listen to me or to friendly biologist Frank Salisbury. Instead he reacts to McDonald's attacks with superficial comments: "Did Jim get published in *Science*, as I did? Was he invited to speak before NASA?" And he doesn't pay attention to what we tell him. For instance, when I mentioned Zigel's article, he volunteered:

"I know somebody in Washington who has made some discreet inquiries about this fellow Zigel. It turns out he is a professor of aviation in Moscow."

"That's what I told you several months ago," I had to remind him. "We spoke about it over dinner in Aurora. Zigel is a friend of Kazantsev."

"Really? No, I don't remember hearing that."

Chicago. Sunday 23 April 1967.

When Fred came over to our apartment yesterday he announced that a new sighting had taken place at South Hill, Virginia, last Friday night. Two witnesses have seen an object on four legs in the middle of a road. It took off on a bed of flames, setting the asphalt on fire.

We had lunch with Allen. We pointed out that his audience was getting increasingly confused. Sometimes he talks about UFOs as if they didn't exist, other times as if they were extraterrestrial visitors, showing slide after slide of astronomical phenomena. But when he is pushed to express a theory he wanders off into semi-mystical statements about ghosts and astral travel. We stressed that he should adopt a firm position and keep it, instead of wavering constantly in his answers to reporters. Also he shouldn't continue to allow himself to be used by the Air Force. He looked at us and commented ironically:

"With two grand strategists like you I can only win!"

We had to laugh, because it is true that it's easy for us to criticize him. We don't have any more influence on the media or on the Pentagon than he does.

We drove back to his house in my car and I resumed the work on the files while he called up the witnesses in Virginia. The sighting turns out to be significant. It is confirmed by the police, and there are clear traces on the ground. I suggested that Bill Powers fly there as soon as he could.

In the evening Jim McDonald came over to Bryn Mawr. Olivier was ready for bed and Janine had the place in perfect shape. I hadn't seen McDonald in a year or so. I thought he looked much older, and he seemed to have lost some weight.

He told us frankly that his press conference, without being a complete fiasco, had not led to the major fight he had been hoping for. Menzel had skillfully avoided him. Klass had wasted his time. Quintanilla had not contradicted him. Reporters in the audience had asked questions that were too limited in scope.

Jim continued to view Allen as the prime culprit: Hynek should have opposed Robertson and Berkner in 1953, he claims. Hynek should have given his resignation to ATIC, slammed the door, alerted public opinion.

Allen, who was the target of all these accusations, arrived as Janine was passing the coffee cups around. McDonald, who had already vented some of his anger, was cordial with him even as he resumed his tirade.

"Jim, there was nothing I could have done at the time of the Robertson Panel," Hynek reminded him. "I was an unknown, just an assistant professor at Ohio State. I spent most of my time in the hallway, waiting to be called in by the panel when they needed some specific astronomical answer."

"What about 1957?" countered McDonald. "How could you stay idle when there were so many sightings going on?"

"I had other fish to fry in 1957. You forget *Sputnik*, the whole madness, the setting up of observation stations around the world. Also, I had no permanent contact with the Air Force at the time."

Gradually, Jim appeared to see the strength of these arguments. I watched him as he seemed to grasp the true reality of Blue Book. Janine offered more coffee and our traditional cake. We were all tired. McDonald and Hynek, seated a few feet away from each other in comfortable chairs, seemed almost ready to set aside their differences. I told them that in their personal arguments they were losing sight of the key scientific issues.

After midnight the ice was completely broken between them. The conversation came to Ed Condon and his mistakes, a subject on which they could naturally agree. McDonald said he had a source in Washington (which he would not reveal) who assured him Condon would soon be forced to take more money, to shake up Bob Low's loose management style, and to expand his operations, all of which gives him much hope.

They didn't leave until two o'clock this morning.

Same day, 3 p.m.

Hynek calls. Bill is now at the site of last Friday's landing at South Hill, Virginia. He has found traces one inch deep in the road surface. The dimensions of the landing gear, as deduced from the marks in the asphalt, are very close to those of Socorro. Bill had the local hardware store re-open so he could get some tools and plaster of Paris to make a cast. The object itself was dark and silent. It did not emit any flames, it turns out, yet the road burned underneath it.

When Allen called Bob Low to tell him what we were doing in South Hill, he found out that the Condon committee had not even heard of the case and had no intention to investigate it. Low is much more interested in directing his energy to the subject of McDonald. He is sending his Washington friends to interview all the important people Jim has visited there, the classic maneuver of a skilled bureaucrat doing effective damage control.

Chicago. Tuesday 2 May 1967.

Ben Mittman has spoken to me again about the information processing center Northwestern wants me to direct. I would have *carte blanche* on research. I would have assistants and a secretary. I could build the Infol compiler and continue my own parallel research. I am tempted to accept. On the professional level my work here is fascinating. But I am missing something more important than intellectual fulfillment here. I miss something of the heart and the soul.

Allen called me at 5 p.m. He was just back from a meeting with Condon, whom he had found so sick he became seriously worried about him. Condon has cancelled all his scheduled trips and was unable to attend Hynek's lecture. When Allen reproached the committee for not doing enough field research, Bob Low answered sharply: "We are not an operational committee." They had a fight about that, with most members taking Hynek's side.

In his private meeting with Condon, Allen stressed that the committee could not make the mistake, in their final report, of stating that all the sightings had conventional explanations. He also told Condon that Northwestern was ready to undertake an in-depth study.

Allen is quite optimistic. He still hopes that Condon might recommend that a follow-on contract be given to our group. If my information center was up and running by then, we could have an impact on the course of this research. There is only one thing that is forgotten in all these fine political maneuvers: the reality of the phenomenon itself.

Chicago. Friday 5 May 1967.

Every morning Janine leaves for work early and she doesn't get back until 6:30 in the evening. She's an excellent business programmer, and a perfectionist, so the computer consultants who employ her exploit every minute they can get from her.

This afternoon Allen told me Condon had suddenly become very curious to find out what the Russians were doing. He answered that he had no information whatsoever, but if Condon invited me to Boulder I could probably fill him in.

"I prefer not to be involved in that," he added, "it's your business."

What kind of comment is that? It is one thing to remain suspicious and vigilant when dealing with the Soviets, and it is another to pretend they do not exist. There is a large amount of provincialism among many Americans. But Allen, who has extensively travelled, should not suffer from this. To study UFOs is like studying meteorites, or an epidemic disease: the research has to cross national and cultural boundaries.

Aimé Michel writes that I might be able to get a job with Thomson, the French electronic giant. Aware of my secret priorities, he knows how to seduce me:

> Salaries are half of what they are in the States, but you would have a view of the Obelisk, the *Bibliothèque Nationale*, and the booksellers of the Latin Quarter.

We hesitate. Should we stay here after all? We keep reading the ads for houses and apartments in the North Side. We visit the high-rises, then we come back to our old street where loose papers fly around. The cold wind is swirling in the schoolyard. A kid on a bicycle goes from house to house, delivering newspapers. Little faces of children smile sadly at us. The corner hardware store is slowly being driven out of existence by the supermarket across the street. In desperation, the owner just stares out the window at passing cars.

If we stay in Chicago this is where I want to live, not in some fancy house in the suburbs with a three-car garage. The only real wealth of this country is in these modest, hard-working folks. Here, with enough energy, I could build a little empire within another empire. But it would just be another academic rut, and I would have to watch Janine coming back from her job later and later every evening, her mind filled with subroutines to be tested and flow-charts to be redrawn. This is not the life I want.

Chicago. Sunday 7 May 1967.

Will Spring ever come? The sky is clear and the sun stares at us from on high, but at ground level the cold wind is still dragging everybody down. Our research proposal for the Information Science center was well received by Rowena Swanson of the Air Force Office of Scientific Research, an influential woman who is a potential champion for our work in Washington.

Yesterday at the Midwest astronomers meeting Eric Carlson and I presented the results of our comparison of various mathematical techniques for the calculation of the radial velocity of the stars. At the computation center I make fast progress in the development of our new data-base software, and I now see clearly the solutions to problems that stopped me six months ago. Outside my job, however, I find no framework for our existence, no art of living, in spite of our expanding network of social friends. What people call their "way of life" is really just a series of mechanical actions designed to make a little more money.

Maman writes to me from Paris:

> The catastrophe that has befallen the Russian cosmonaut, after the death of the Americans, makes me very sad. I don't have any courage left in order to do my work. I was really hoping that this time "they were there." However Paris is sunny, the cafés are full of people. I went out to change my mood. At the Saint-Séverin cinema they showed a movie by Seban in collaboration with Marguerite Duras. It was called *La Musica.* Once again I came away disappointed: Nothing but dark philosophy, rolling by for two hours of monotonous conversation, without anything happening. I felt like screaming.

It is the spectacle of Paris that delights her:

> The real cinema for me is in the streets of Paris: La Huchette, La Harpe, Galande. They are full of characters whose origin I can't even guess, but what beautiful faces! You can't tell the girls from the boys any more. Same hair, same costumes. And those little food places that belong at the end of the world.... They let you dine for ten francs by candlelight in a room full of smoke. I don't dare go in there. I feel like a foreigner in my own city.

Charles Bowen writes from London:

> This morning, just as I was leaving the house for West Byfleet station I caught sight of the mailman going up the road. I slowed down and my daughter Pauline (who brings back the car) told me: "Don't stop, you'll miss your train." To which I replied: "There may be a letter from Jacques Vallee, I'll be glad to miss my train to get it." The mailman solved the question by leaning his bicycle against a garden wall and running onto the road waving a letter for me... your letter!

Dear old Europe. America still has a long way to go to become so touching, so human. From the other end of the world our little friend Galia writes:

> During my whole stay in Armenia I have regretted very much that you were unable to see the town with me, and that you could not go to the Echnyiadzin, the center of the Armenian church.... I must say Armenian priests are very handsome. In general Russian priests too are beautiful, but they don't have the big sparkling black eyes and the pale faces of the Armenians. The inside of the cathedral is very beautiful. Do you remember the Novodevichi convent, and the church near which I was trembling with fear because of the corpse of the old man? There were many icons hanging on the walls. The only difference between the Russian and the Armenian churches is the lack of icons in Armenia.

Chicago. Sunday 14 May 1967.

The news media report that the Pope went to Fátima yesterday. Fifty years ago three little shepherds saw a UFO and its occupant there, with the theological implications we all know. The Pope visiting a saucer landing site ... and American crowds watching the event through the wonders of the Early Bird artificial satellite. I find this mixture of mysticism and technology quite amazing.

On Friday I had lunch with Bill, Don (whose destroyer is about to leave for Vietnam) and an engineer named Bielek who works in Morton Grove. He says he is a close friend of naturalist and parapsychology researcher Ivan T. Sanderson. Bielek is a big man, about forty, wearing a brown suit and a very serious and rigid attitude. He behaves like a secret agent whose pockets are full of microfilms and who wants to make sure you know it: the exact opposite, naturally, of what a real secret agent would actually look like. Listening to him is like talking to someone who has just had breakfast with J. Edgar Hoover. He throws the most extraordinary hypotheses around, but never bothers to develop them or to resolve their contradictions.

"How closely do you think the Intelligence agencies are following the saucer question?" I asked.

"They are holding regularly scheduled meetings to which the Air Force is never invited," said Bielek provocatively.

On this point at least he may be right. They must have had a secret inquiry for years. It would not be logical for a low-level project like Blue Book to be the only game in town.

I have recently found myself mixed up in an old intrigue which has been another subject of fascination among American ufologists over the last ten years. I am referring to the story of M.K. Jessup[11] and Carlos Allende.

When the late Jessup published his book *The Case for the UFO* in 1954, someone anonymously sent a copy of it to the Office of Naval Research in Washington. The pages were covered with hand-written scribblings, apparently authored by three different people using color ink. The point of these notes was that the authors were extraterrestrials living on earth. At about the same time, M.K. Jessup received three letters from a man named Carlos Allende or Carl Allen, who claimed to have witnessed certain Navy experiments in 1943. Run in collaboration with Einstein, these experiments were said to have attempted to make a ship disappear.

Jessup seems to have become obsessed with these stories. He also became convinced of the existence of extraterrestrial bases in South America, where he made several trips. Did he feel he was going to go crazy? Was he afraid, as one rumor suggests, that people would think there was insanity in the family because of his strange beliefs, which disturbed the very conventional patterns of American life in the fifties? One evening he gave all his papers to Ivan Sanderson. The next day he got into his car, closed the garage door and let the engine run until he was asphyxiated and died.

"So, is Sanderson going to release Jessup's papers?" I enquired.

"He can't do that, because of National Security, of course," answered Bielek.

It turns out that last January I received a postcard from Mexico, addressed to me at the observatory. It was signed *Carl Allen* and offered me an opportunity to buy a very expensive copy of the famous annotated book, of which only 114 are said to exist. I replied politely at the time, declining the offer. But it is only later that I realized I had stumbled on the track of Carlos Allende. I have no interest in the annotated book, which I believe is sheer garbage, but I am fascinated with the way apparently rational, intelligent people fall victim to such cheap "mysteries."

Chicago. Monday 15 May 1967.

A new wave of sightings has begun. My early warning indicators have served their function: there was a Type II sighting (cloud cigar) in Akron,

Ohio, on April 8 at 22:00, and five days ago there was a Type I case (landing) near Dijon. I wonder if this combination always signals the beginning of genuine waves. I have resumed work on my master card index.

This morning I met with Hynek and I got his attention long enough to obtain his advice on the steps to take with the Russians and with the United Nations, which seem to have a continuing interest in all this. We also called up Mary-Lou Armstrong in Colorado. She told us that Bob Low had gone back to Washington, where the Air Force rapped his knuckles because he didn't dispatch enough investigators into the field. As McDonald had predicted, there is now talk of additional funds to train two teams. But Low, who is a great opportunist, is already planning to use the new money to go on a junket, taking a personal trip to Prague instead. An astronomical congress is scheduled there this summer.

Chicago. Saturday 20 May 1967.

Allen has learned that differences are deepening within the Condon Committee. Wertheimer is back in the group. Condon himself neglects his duties, either because of poor health or through lack of interest. Jim Wadsworth has just returned from the state of Washington, where he went to investigate some mysterious sounds. He has found an explanation for them—they were caused by an owl, he is telling the witnesses. Besides, a local farmer has now killed an owl and there are no more sounds.

Fred called me last night. In his opinion the "owl" explanation is a joke. A friend of his, who has done his own inquiry with the local civil defense authorities, has found out that high-quality recordings were made which show artificial signal patterns. The Air Force is not going to conduct an investigation because, as Quintanilla eloquently puts it, "sounds don't fly." Instead they are thinking of financing a group of physicists who think they can generate luminescent lens-shaped objects in a gas to test the plasma hypothesis.

Yesterday a Swedish journalist told us that last March, in the North of Sweden, two gray objects had hovered over a house, terrifying the four people who lived there. The Swedish Air Force explained the sighting by saying there had been an electrical discharge between two high-tension wires, a fact which certainly would not generate dark phenomena.

Something very grave has been going on for the last twenty years, and nobody knows what it is. I am going to go back into my quiet personal research, outside the established frameworks. I will cancel my appearance at the UFO panel of the American Psychological Association, and I will tell media people that I have nothing to add to what I have already said in my books. Talking to the press now is a waste of time. They are not looking for the truth in the phenomenon, but only for entertainment.

Chicago. Saturday 27 May 1967.

Our little group meets again tonight at Bryn Mawr. Allen now calls it "the Invisible College" in reference to the scientific movement that preceded the creation of the Royal Society in the early 1660s in England, at a time when it was very dangerous to be interested in natural philosophy. Hynek and Bill, Fred and physicist Jarel Haslett will be there. Yesterday Allen spoke to Mary-Lou again. Menzel and Condon were in the room with her, so she had to move to another phone to speak freely. She told him the plenary session of the committee last Thursday had been a joke:

"I have 150 cases I can't explain!" said Bob Low.

"Give them to me," answered Menzel without any hesitation, "I will explain them for you!"

In the meantime not only did the committee drop the study of the peculiar sounds in the state of Washington, but they missed an opportunity to study a recent first-class case, in Winnipeg (Canada). A prospector observed a landed metallic disk. He is now at the hospital with multiple burns.[12] Again, Hynek shrugs: *"Testis unus, testis nullus."* A single witness is no witness at all. I fought him on that point: the man has suffered serious enough burns to be hospitalized. "If we don't go study such cases we'll never find the other witnesses who may very well exist. And why should it be surprising for the landings to have few witnesses? If there an intelligence is involved that's exactly what we should expect."

Fred, who turns out to be a patient and realistic investigator, has contacted the APRO representative for Illinois, a man named Achzener. He has custody of two fragments of the famous "steel meteorite" which he claims to be pieces of a saucer and are frequently associated with Canadian engineer Wilbert Smith.[13] But there is more: Achzener himself has had numerous personal sightings, and he has built a simple magnetic detector, which has inspired Fred to build a similar one in his apartment here in Chicago.

Last night (Saturday, at 1:05 in the morning) Fred was quietly talking to a friend of his, a man who is in charge of aircraft maintenance on the carrier *Kitty Hawk*, when he heard the detector go off, after it had been connected for barely an hour. They rushed out into the street in time to see a small orange disk, larger than Jupiter, crossing their field of view from North to South in six seconds. After this they heard a strange sound following the same apparent path. It took fifteen minutes for the magnetic pendulum to return to normal.

Chicago. Sunday 28 May 1967.

Fred came to see us tonight looking very pleased with himself. He was carrying an instrument about four feet high. He set it down on the floor and pointed out a plunger hanging from a long wire into a loop connected to two batteries, with a red light and a bell to give off an alarm when a magnetic anomaly is detected.

Hynek came over wearing old yellow trousers and a khaki shirt, and he spent ten minutes crawling on the carpet to try and take a picture of Fred's device. While he was doing this he casually told us he had once seen an object similar to Fred's luminous disk, when he was working at Perkins observatory. It was reddish-orange and it flew too slowly to be a meteor. It crossed the field of view defined by the slit in the dome. When Allen rushed out he was on the wrong side of the building and could not see it again.

Fred told us some fascinating details about his meetings with Achzener, mischieviously adding, as a good connoisseur, that the man lived in a suburb called Brandy Wine. Achzener, who is a salesman with Motorola, claims to have seen no less than fifty unidentified flying objects since 1957, including a large metallic disk with a dome over Louisville airport. Another time he saw four saucers hovering above his house while a terrified cat, his hair standing up, jumped on him and held on to his clothes with all its claws. The disks floated slowly out of sight. Some of his other experiences involved telepathic communication.

The most interesting stories relate to Wilbert Smith's personality. Achzener assured Fred that he had witnessed some of Smith's experiments which produced "a five percent change in the force of gravity." Another time, during a meeting at Smith's house, three short whistles were heard on the FM radio. Smith's expression changed. He got up suddenly, excused himself and rushed out. He came back forty-five minutes later, his shoes covered with mud. The implication is that he had a secret UFO contact that night.

Another time Achzener had the thought that he'd like to smoke a cigarette. Smith, apparently reading his mind, suddenly told him, "let's go buy some cigarettes for you." They drove to a coffee shop and sat at the rear of the room, with Smith keeping his back to the wall. At one point Smith made a peculiar signal, which Achzener says could only have been seen by three other people: an old man, the waitress, and a short fellow about four feet tall who got up and drove off in an old car with California license plates.

"I wanted you to see this person," Smith said. "That's why we came here. He's an extraterrestrial."

Such are the stories that circulate among American ufologists. Achzener has given Fred two fragments of the "mysterious" metal found on the shore of the Saint Lawrence River and highly prized by Smith. They turn out to be made of steel with a high manganese content. The analyses done in Canada are so clear-cut that Haslett has not bothered to reproduce them. The metal is certainly of ordinary industrial origin.

The discussion among us then took a mystical turn. Allen said he was envious of those witnesses who had psychic abilities and whose vision could reach into the astral plane. Bill interrupted him rudely: "Every time I've met a mystic I've met an uneducated slob." Don Hanlon, who could have nailed Bill down with a few words on this topic, wisely refrained from saying anything.

I like this group very much. Fred is refined and clever, a walking scientific encyclopedia. Jarel Haslett is a clear-thinking physicist. Hynek is both secret and transparent, subtle and irritating, an adventurer of the mind but not a man of action, a scientist patiently looking for the limitations of science itself.

"It's hard to give a definition of UFOs that doesn't also embrace ghosts, or the Virgin Mary," I pointed out in response to Bill's comments. He didn't believe me, so I pulled down from the shelf the book by Delaunay, *A Woman Clothed with the Sun*, and I read to them the record of the 1868 Knock apparition in Ireland.

"How do we know all this happened?" asked Bill. "Do the witnesses even exist?"

Many scientists would react as he did. They misunderstand the scope of the mystery.

Fred has changed since he has seen "his" UFO. He is getting more assertive in his opinions. He even raises his voice occasionally when Bill threatens to cut him off.

Chicago. Wednesday 31 May 1967.

Yesterday was Memorial Day. I spent the afternoon at Allen's house classifying the files, and I proposed that we meet with my computation center colleagues to draft a joint proposal. They have agreed to let me bring all the files to my office on campus this summer for an in-depth reorganization in preparation for the next step in our research, which we hope could be funded by the Air Force. Such a project, if it took off, could make me stay in the United States. We're not going to give up easily.

This evening Allen asked me to fly to Concord, New Hampshire, as an observer in a contact experiment Betty Hill wants to conduct with John Fuller, Dr. Simon and a man named Hohmann, who is an IBM engineer. Betty believes she has become a "transducer." She thinks she can make a flying saucer appear in the sky and land.

Chicago. Tuesday 6 June 1967.

Lightning war in the Middle East. The Egyptians want to "throw the Israeli rascals into the sea, exterminating them to the last man." The Israeli respond in kind. Everywhere soldiers are in control of this wretched planet. At the moment there is no danger of a world war. Nonetheless the feeling of helplessness is inescapable. How can man have a chance in the long run? How can rational thought be a significant factor in such a world? The kind of wisdom generated by the best minds is nothing but a moldy deposit growing on the hull of the great ship of State: something only fit to be scraped off in the next port of call. And scientists are pawns to be moved around by governments, only useful for inventing atom bombs.

I still must believe we can survive. Yet if the battle had turned against Israel yesterday I probably would not have the luxury to think about all this because the war could already have escalated and the greatest conflict in the history of the planet might be in progress. Will men of the future try to understand our time, our folly? I imagine a subtle being who looks over my shoulder as I write, an unseen presence. Together we explore the flow of time. He assures me that there will be a future, whose distant echoes fascinate me.

The room is warm and humid. We are working hard, thinking of progress. And we are at the mercy of some Colonel who sits on the hatch of a tank, raising columns of dust in a forgotten desert between two dry seas. Life is in our throats, bitter, full to the point of vomit. Who is talking about progress? We have not even begun to glimpse the horrors of which we are capable. In a few years Israel and the Arabs, India and the Pakistanis, will have nuclear bombs. Then what?

Every minute a jet flies over to land at O'Hare. In the heavy atmosphere its engines leave a long and strident scream. There is a little bird down there in a hole in the wall. I wonder why he stays in town.

At the library this afternoon I searched for references on apparitions and demonology. I unearthed a microfilm of a book dated 1577. The original is kept at the British Museum:

> A Strange and Terrible Wunder wrought very late in the parish church of Bongay, a Town of no great diftance from the Citie of Norwich, namely the Four of this August, in ye yeere of our lord 1577, in a great tempest of violent raine, lightning, and thunder, the like whereof hath been feldome feene.
>
> With the appearance of an horrible fhaped thing, fenfibly perceived of the people then and there affembled. Drawen into a plain method according to the written copye, by Abraham Fleming.

A big black crazy dog rushed into the church in the midst of a horrible storm and broke the necks of several parishioners. As he jumped around in the nave among the flashes and thunder, lightning destroyed the bell tower. The author appealed to Divine Providence, to the wonderful anger of God. Poor Abraham Fleming, how I understand you! I would have done the same thing.

The sky of the astronomers is dead. I am not expecting that saucers will come down with the answers for us, as Betty Hill does.

The streets of Chicago are dead, too. And I have no illusions about what I will find in France if I return there. It is of the fresh and intimate streets of unknown lands that I dream right now, clear skies after the rain . . . but research is my world, I belong where science is made. I feel no indulgence for the sellers of abstraction, for the philosophers who do not build things, real things, with their own two hands.

Chicago. Friday 9 June 1967.

Bob O'Keefe, my Psychology Department friend at Northwestern, is now using his students as subjects in an interesting case study on *assimilation and contrast.* He requested that I supply him with all the reviews of *Anatomy* and he has now asked his class to rate each piece according to two parameters: first, the position of the reviewer on a scale of no belief to total belief in UFOs; second, the position that the reviewer *believed I occupied* on the same scale after reading my book.

The results show a remarkable bias among the reviewers, with evidence for the classic "*either with me or against me*" delusion: the strongly pro or con reviewers evaluate my neutral or even moderate positions as far closer or far more in contrast to their own viewpoints than they really are, ascribing to me some extreme ideas that are nowhere in my book!

Based on no less than 125 reviews, Bob and his students have found almost a perfect correspondence between the bias of the reviewer and the position he or she *ascribes to me.* This psychological bias in judging was known to exist among cretins and political extremists. O'Keefe has now found it among the most prestigious reviewers of the major papers, who claim to be impartial intellects bent only on informing the public.

This morning Allen told me he had two reasons to be happy: his son Scott has just passed his Ph.D. preliminary examination at MIT, and his research proposal on supernovae has been funded. But he also has one subject of bitterness. When he went to Dayton on Wednesday there was an immediate scene between him and Quintanilla. He asked to see the Project Grudge report. There was only one copy at Wright Field, and it was in very poor condition. He was told he could have a copy made but the Air Force would charge him a dollar a page: there are some 500 pages. Angry that he could not even have a copy of a report he had largely written himself, after serving Blue Book for twenty years, poor Dr. Hynek actually sat down and started transcribing the introduction by hand.

Later Hynek happened to see an interesting sighting file lying on a desk. He asked a secretary if he could take some notes from it. Quintanilla jumped up: although the file was neither secret nor even confidential, it came from OSI (the Office of Special Investigations) and bore the words "Official Use Only." Quintanilla insisted that nothing could be copied because "information channels could not be compromised." Hynek got mad and mentioned all the rumors we had relayed to him about Blue Book being nothing but a cover, a smokescreen. Are the real UFO investigations run secretly under OSI, whose records are classified? This made Quintanilla even more furious. Hynek almost resigned on the spot.

Quintanilla finally gave in, taking the first step to improve the relationship; he invited Hynek for a drink in town, promising that he would extract the pertinent information from the case file. Later, when Hynek was alone with Morley, the lieutenant told him it was a fact that *he just didn't see all the reports.* Quintanilla frequently tells him: "We won't send this one to Hynek, it's better if he doesn't know about this particular sighting."

I am very interested in this turn of events. Hynek is more puzzled than ever. We already knew that Blue Book was far from getting all the official reports. For instance, if some unknown object is seen by a pilot or by a radar operator the data generally goes to NORAD, not to the Air Force. And when a case does come to an Air Force base it is only forwarded to Dayton if no "local explanation" is found. This obviously leaves the door open to any manipulation.

At lunch I stressed to Hynek that he was in a precarious position indeed. In fact McDonald is now on solid ground when he accuses him of being a willing strawman for the Air Force. Allen has accepted the job of scientific watchdog with Blue Book. ***If it turns out that the information he sees is deliberately biased, censored or distorted by the military, it is the entire research community of the United States which is being betrayed on a potentially crucial topic.*** The fact that the academic community may *want* to be misled doesn't matter: if Hynek knows about the manipulation and says nothing he is automatically an accomplice.

Later the same day.

A melodramatic letter has arrived from Bob Hohmann in preparation for our weekend experiment with the Hills. It talks about my role in what he calls *Phase Two:* that's when I am supposed to take command of the interaction with the Aliens when they land. Betty Hill has written to him:

> I follow your suggestions regarding the message and I hope that the next manifestation will take place closer than that of Saturday evening in Kingston. Very close, very clear and just above the trees.

Although the letter is signed "Betty and Barney" it is entirely written in the first person. Clearly Betty is the dominant mind here. It is this interaction we want to better understand by going to New England.

Janine is carefully reading Fuller's book *The Interrupted Journey.*[14] She finds in it some important facts that I had missed. The whole story of the abduction of the Hills by saucer people didn't originate with Dr. Simon at all but with this man Hohmann himself! This is the way it happened: when they came home to Portsmouth after their sighting of 19 September 1961 Betty first mentioned the incident to her sister Janet, who in turn spoke to the former chief of police of Newton. They were told to contact Pease Air Force Base, where practically nothing was done.

Two days later Betty found one of Keyhoe's books in the library. She wrote to him and the case was forwarded to Webb, a NICAP consultant, by Richard Hall, about one month after the sighting. In the meantime Robert Hohmann and C.D. Jackson had come to Washington where they had lunch with Keyhoe. The case was brought up. On November 3rd they wrote to the Hills and they met with them on November 25th, interviewing them from noon to midnight. They even asked them if they used fertilizers for their trees, because they thought saucer pilots might want to steal the fertilizer. Hohmann and Jackson also told the Hills there might well be life on Alpha Centauri. It is this team of self-styled scientific detectives who discovered that some sixty kilometers and some two hours were "missing" from the couple's itinerary. A friend of the Hills who attended that meeting suggested hypnosis.

On 25 March 1962 Betty and Barney consulted Dr. Quirke, of Georgetown. In the Summer of 1962 they saw Dr. Stephens in Exeter and he sent them to Dr. Benjamin Simon late in 1963. In other words, Simon didn't actually start hypnotizing the Hills until two years after the sighting.

Chicago. Monday 12 June 1967.

We have returned from New Hampshire without having met any extraterrestrials. When we stepped off the plane in Boston we found a heat wave in full swing and John Fuller waiting for us. He was jovial, ebullient, relaxed. We liked him immediately. We climbed into his Land Rover and we drove to Arlington, joking half in French and half in English.

We spent a long time stuck in the dreary Boston traffic, looking at run-down wooden houses and narrow dusty streets where sweating humans sought every chance to rest in the shade or to just lie there, waiting for the sun to go away.

The mansions of the Boston upper crust are located on the hill of Arlington. Dr. Simon's two big, mean African dogs made an effort to come after us, panted, and decided they were too tired to attack after all. The doctor was in the dining room in the company of two older women who immediately offered to feed us. We declined. Simon wore no shirt, and the only thing I really remember about him is a huge bulging stomach that bounced around the room ahead of him. He introduced us to various people and to his son the doctor. I explained why Hynek couldn't make the trip, and we just waited around for a signal that it was time to leave for Kingston.

Eventually John Fuller looked at his watch and moved towards his car. We drove through Woburn and Haverhill but we could still feel the pulse of Boston. The very notion of flying saucers ever landing in Massachusetts seemed preposterous.

A little road twisted among trees and we found the house. We turned and drove down to the pasture where Hohmann and his entourage had already pitched their tents. The sweltering heat moistened everything.

"There will not be any relief tonight," said John.

We changed to lighter clothes, but the mosquitoes attacked as soon as they saw our short-sleeve shirts. Ignoring them we walked over to introduce ourselves to Betty, Barney and their dog Delsey. We found Hohmann in the midst of a grand speech about the telepathic powers of Australian Aborigines.

Dr. Simon's Cadillac arrived. I took Hohmann aside: "What are we doing here?" I asked. "What the hell do you mean by *Phase Two*? Or *Phase One* for that matter?"

"Betty has become a transducer when she was in the saucer," he told me flatly.

"What's a transducer?" I asked. Surely he didn't mean those little hardware boxes we put on telephone circuits.

"You know, she can communicate with Them, she has acquired that capability. The Humanoids."

He firmly believed *They* needed something from us. What exactly? That wasn't clear. But we did need something from *Them*: the revelation of some new truths.

"For the last ten days Betty has been sending messages asking Them to come back. A numeric code has been established to indicate the contact date: June 10th."

All that was perfectly childish, like the folding table in the middle of the field and the large chalk circle where we were supposed to gather to demonstrate our goodwill to the Aliens. Hohmann wanted Betty to broadcast another date when *They* would come back and show their craft to millions of people. It was like a bad movie. But what if something did happen? We came this far, let's keep an open mind, I told Janine.

I had brought a Geiger counter and a good little telescope. Hohmann's circle was drawn at the bottom of the field, where we had the least exposure to the sky. He was willing to help me move the table uphill. I oriented myself and I began a temperature log. I went to a phone and I called Allen, promising to alert him if anything did happen, and we went back down to the camp for dinner.

Darkness came at last. Group conversation returned to lofty cosmic truths while Delsey licked noisily what was left of the food. I set up the Questar. Time passed, bringing Mars into view, followed by Arcturus and Vega. The mosquitoes danced and everything got wet. Simon went back to Boston. Fuller couldn't stand still for five minutes, he had to drive to Exeter "to see if there was anything new there." What could possibly be new in Exeter?

I sat in the grass with Janine, feeling thankful for her presence but guilty for having dragged her along. The sky remained clear. Every hour I noted the temperature. The Geiger counter clicked away in the darkness. Artificial satellites crossed the sky, then a meteor and lots of lightning bugs: at every such incident Betty jumped up excitedly. But if I had met a group of little aliens on a lonely road, and they had dragged me inside their craft, perhaps I would be as inclined as she was to see flying saucers everywhere.

At three o'clock in the morning everything was covered with dew. I folded up the telescope and we went to bed in a trailer behind the house. Three hours of uneasy sleep. Instant coffee. The people in Hohmann's group left. At last we had a chance to talk quietly to Betty and to Barney. They were nice and warm people once they didn't have to play transducers in front of a bunch of crazy scientists any more. They took us to the lake shore and later they let us listen to their recordings made under hypnosis by Dr. Simon, certainly the highlight of the whole trip. Why the fear, the sheer terror in their voices, unless the abduction was genuine? But at what level? I remember Simon's remark yesterday when I had asked him:

"Was their experience real?"

"No question. It was real to them."

"Doctor, let me ask it another way. If you and I had been driving with them when this thing happened, would we have seen a flying saucer by the side of the road and five little Aliens opening the car doors?"

"I have no way to tell you that. All I can say is, the abduction is part of their reality."

Jackson drove us back to Boston and we flew home, very tired. It was just as hot and muggy in Chicago as it had been in Boston. Olivier was sweating and flew into an uncontrollable rage as soon as we came into the apartment. I had to grab him and force him under the shower to calm him down. We took a shower after him, and I went to bed, my head aching.

This evening I am writing my report on the Hill case for Hynek. We have had a good night, a slow day. Olivier is babbling away, telling us long nice stories. We have learned a few things in Kingston but we know nothing more about the alternate world which Barney and Betty may have briefly glimpsed. They are sincere people facing a problem that is beyond their grasp. Barney is an intelligent, balanced man who understands the unique nature of his situation. But Hohmann has clearly planted many strange ideas into the couple's mind.

Simon is a sly old practitioner, an empiricist of vast experience. In my opinion he doesn't really care whether or not they have been abducted: that is not relevant to his psychiatric assessment. He places no value in the time loss experience: "When you travel you don't pay attention to the time it is," he told us. "Also, remember these people were lost."

The whole business about Betty being a transducer began when she said as a joke: "The Aliens ought to pay a little visit to Mrs. So-and-so," and a few weeks later, sure enough, a strange light was seen over her house. Everybody became convinced that the extraterrestrials were picking up Betty's thoughts and acting on them. She claims to have seen a saucer at treetop level again, just the other day.

Janine, who has not forgotten the lessons she took from Piaget on child psychology, took the time to talk to the kids. She asked them what they thought the light in question was, above Mrs. So-and-so's house. They answered it was just a light, a tiny pinpoint source, "it could have been anything."

From Woburn to Portsmouth every bush, every telephone pole has become a saucer sanctuary since John Fuller's best-seller has appeared. Here is where Betty stood when she sent her first message. Here she saw "one star too many in the Big Dipper." It's like visiting the Holy Land. And there is a new social pecking order in Kingston. It's not how big your house is or the color of your Cadillac any more, it's how many saucers you have contacted. Betty and her sister Janet are the supreme arbiters.

Not enough has been said about Janet. She saw her first UFO in 1956 and she has been seeing them ever since. A few weeks ago a mysterious light crossed the road above her and she felt "the ray."

"How did that feel?" I asked her.

"Like when they take an X ray at the hospital."

"You can't feel an X ray at the hospital," protested Barney.

"Well, maybe you can't but I sure can," was the peremptory answer.

Even when we recognize that Betty has been influenced by the believers and that Janet and others now recognize a UFO in every starry light, the original 1961 sighting does remain unexplained. The hypnosis tapes took us back to that night in the White Mountains, on 19 September. Their speech found a deep, tragic, Wagnerian tone:

"It's big, GOD! It's BIG! What is it? WHAT IS IT?"

A scream of sheer horror, and Barney's voice:

"I've got to get my gun!"

From the slanting craft with its double row of illuminated windows a light filters towards his eyes, he perceives a voice:

"Don't move. Stay where you are. Don't move."

There are about ten occupants who work on wall panels. A single one stands at the window and stares at Barney. He manages to break away, running to the car where Betty is crying and where the dog is cowering under the seat.

Later they turn into a secondary road and *it happens:* the craft is blocking the way. A series of fast beeps vibrate the whole car. At that point, as John Fuller's book describes it, Betty reproaches herself for being so ignorant:

"How stupid I am! What if it was Morse code . . ."

Her voice abruptly changes. The Bostonian accent vanishes, replaced by a flat, deep, grave voice. No hesitation. Short sentences. Factual.

"They come towards the car. There are three of them. Two others behind them. They open the door."

During one of the last hypnosis sessions Betty states that the one who walked ahead of the group held something pointed towards her.

"How big was it?" I asked.

"It could have been a gun," she replied. "But it could just as easily have been a cigar."

That last detail reminds me of Masse, paralyzed at Valensole two years ago by a similar device a little man had taken from his belt.

I think something did happen to the Hills. But what was it? Should we believe the story of interaction with the operators inside the saucer, the ludicrous (or symbolic) medical examination that comes straight out of a horror film? Have the true facts of their abduction been overlaid by a false memory? Their adventure suggests a grave and terrible mystery.

Chicago. Tuesday 13 June 1967.

The director of the computation center at the University of California in San Diego has called me again, and again I have declined his offer to join his research staff. I would lie if I didn't confess I am tempted. The beaches, the mountains, a happy active life for Janine and our children: yes, I want to take part fully in the novel, exciting frontiers of our creative time. But I also long for old experiences: good ancient books, rain falling over narrow cobblestone-paved streets, although I fear real cataclysms are coming to Europe, to sweep away the sand castles people have erected to protect their privileges.

Setting up a double refractor in Pontoise, about 1957.

Janine in Texas, with Antoinette de Vaucouleurs, 1963.

Dearborn Observatory in Evanston, December 1963.

My mother in Pontoise, April 1964.

Photo © 1964, *Denver Post*

Dr. J. Allen Hynek, mid-1964.

Working on the research for *Challenge to Science,* mid-1964.

Discussing the capabilities of *Altair* early in 1966.

The new observatory at Northwestern.

In New Hampshire with Betty and Barney Hill, June 1967.

With Olivier at Saint-Germain-en-Laye, April 1968.

American delegates at a computer standards meeting, Washington, 1969.

With Aimé Michel at Paris Observatory, October 1969.

With my children in California, 1971.

With Allen Hynek in New York, 1978.

I read over the early pages of this Journal and I wonder how on earth I can think of returning to Paris: I was so unspeakably happy to leave in 1962, escaping from all the idea-killers, from the Paris intellectual slaughterhouse where the pundits of *Le Monde* were in control, mocking anything new, anything out of the ordinary. . . . My generation should have shaken all that a long time ago.

There is a genuine deep current that carries the world towards its future, from the convulsions of Mao's China to the secret California where everything is crashing through the old barriers. I feel this current carrying me forward, even if the heat weighs too heavily over Chicago tonight, even if these hasty words are too partial, and the dust that blows from the big empty Middle West too stifling.

15

Chicago. Sunday 18 June 1967.

Step by slow, painstaking step, I am uncovering the real story. I made an important discovery yesterday. Hynek is away in Canada but he has given me a key to his house. I went there during the week to take the Air Force files away so I could go on sorting them. And I found something that dramatically changes the whole landscape.

Since the Marsh Gas episode Allen has been unable to keep up with the volume of new data. Copies of the files Quintanilla sends him end up in the small room where he works on the second floor of his pink house on Ridge Avenue. Papers are piling up everywhere: in the drawers, on the chairs, on top of the wardrobe. For the last few months he has been trying to put some order into this mess, but it is a hopeless task, so we finally agreed that I would transfer everything into my office during his absence to straighten out the files over the summer. Thus we hope we can sift out the garbage, extract the most credible cases and prepare a strong new research proposal in the Fall.

What a challenge! I sat alone in Allen's room and I contemplated the disaster. "If I don't do this, nobody will. . . ." I thought as I started putting all the files dated before 1964 into cardboard boxes. I carried them to the Buick. I drove over to the computation center and I began sorting. I have been sorting ever since. First I set aside the statistical summaries which cluttered the files; I eliminated numerous duplicates; I replaced the old stained folders with clean new ones, neatly labeled with places and dates.

The worst section of the files concerned the history of the Air Force projects themselves, from Sign and Grudge to Blue Book. Hynek had misplaced many of these documents. And it is in that section that I found a letter which is especially remarkable because of the new light it throws on the key period of the Robertson Panel and of *Report #14*. It is stamped in red ink "SECRET—Security Information." It is dated 9 January 1953. It is signed by a man I will call Pentacle.[15] It is addressed to Miles E. Coll at Wright Patterson Air Force Base for transmittal to Captain Ruppelt. It begins with the statement that the document contains a recommendation to ATIC regarding future methods of handling the UFO problem, based on experience in analyzing *several thousands of reports.*

This opening paragraph clearly establishes the fact that **prior to the top-level 1953 Robertson Panel meeting somebody had actually analyzed thousands of UFO cases on behalf of the United States Government.**

After pointing out that many of the reports contain insufficient information, and that it would be highly desirable to obtain "reliable physical data," the letter goes on with the blunt advice to cancel or at least postpone the meetings of the Robertson Panel until the full results of the analysis could be made available, a very natural recommendation. Failing this, they wanted to have a formal agreement between ATIC and the staff of Project Stork as to "what can and what cannot be discussed" at the Washington meeting with the five leading scientists.

In other words, the representatives of this top-level research group were against convening the Robertson Panel! But what was Project Stork? Whatever it was, its members were doing some excellent scientific thinking: They had noticed that the distribution of cases over the United States was not uniform, and this led them to identify areas of high reporting probability, which they proposed to set up as experimental areas.

In such areas they wanted to place observation posts with complete visual skywatch, radar and photographic coverage, and all other instruments "necessary or helpful in obtaining positive and reliable data on everything in the air." They added, even more ominously, that many different types of aerial activity should be secretly and purposefully scheduled within the area.

What these people were recommending was nothing less than a carefully calibrated and monitored simulation of an entire UFO wave.

For whom did Pentacle work? Did the proposed experiment take place? Who were these people who calmly sat around the table with the CIA and the Air Force and who, many years before us, understood the need to acquire second-generation data?

Their plan for getting such data covertly makes perfect sense. But what this document shows is that the scientific community has been led down a primrose path, beginning with the Robertson Panel and its group of prestigious physicists.

Another team of analysts had already reached the point where they could form some scientific hypotheses about the phenomenon and they were ready to test them. Yet the high-level advisers who reviewed the data were not briefed on that part of the Air Force's research efforts, even though their own recommendations were also classified at the time. What was going on? What kind of a game was being played?

Hynek once assured me that if it ever turned out that a secret study had been conducted, the American public would raise an unbelievable stink against the military and Intelligence community. It would be an outrage, he said, an insult to the whole country, not to mention a violation of the most cherished American principles of democracy. There would be an uproar in Congress, editorials in the major scientific magazines, immediate demands for sanctions. This memorandum does not prove that it has gone that far. And yet...

Let us go back to 1953: the Intelligence agencies have determined that unknown objects are flying over the United States. If these are controlled machines they are far beyond anything we have. Public opinion demands some action. What could be simpler than assembling a panel of scientists? Perhaps not the best informed, but the most prestigious. They are shown a sample of the reports, pre-selected by the Air Force. They find no reason, of course, to revise the current edifice of science on the basis of what little they are shown. And once the panel has been disbanded and public opinion quieted, what a wonderful opportunity for the military to resume its research in secret, with its own scientists, its own laboratories. . . . If such a research project exists it certainly does not need the Blue Book data. It could operate independently, at a much higher level.

I cannot discuss any of this with Aimé Michel or anyone else in France, and certainly not with Rocard, since communications with him have proven to be completely one-sided. I do not even dare copy the whole memo into this Journal. Yet this document, if it were published, would cause an even bigger uproar among foreign scientists than among Americans: it would prove the devious nature of the statements made by the Pentagon all these years about the non-existence of UFOs. These official statements have been taken as Gospel truth by most foreign nations, discouraging their own efforts to start independent appraisals of the situation.

Olivier just came into the room to kiss me good night. I see our future as a fine, quiet beach at low tide, about to be covered by new waves, new insights, and a lot of turbulent waves.

Chicago. Wednesday 21 June 1967.

Aimé Michel recently proposed a new idea to Clérouin, who is now a Colonel: Why not bring me back to France under some new structure so I could work seriously on the UFO problem with a small group of assistants? Clérouin appeared excited with the idea and promised to bring it up with his superiors. He gave no sign of life for a whole month, then he had dinner again with Michel and with his friend Latappie, and he dropped the following harangue on them:

"We in the French Air Force, we are sick and tired of hearing all those unjustified attacks against the USAF experts who are impeccable scientists. Their investigations are top-notch. . . ."

"Wait a minute," said Aimé Michel, "Do you think we're stupid?"

The Colonel would not listen.

"There is another point which is even more important. Why should we embark on trying to do a job nobody is asking us to do? The Pentagon is perfectly right to downplay the whole thing. **Society is working just fine the way it is, why should we encourage all this saucer business?** It might stir up social repercussions we couldn't control. So the longer we wait the better it is for everybody. Let the scientists do something about it if they want to."

Aimé Michel believes that Clérouin didn't go to his French superiors at all but directly to his American counterparts. This is an area where the USAF continues to have control over what the French services do, in spite of all the lofty Gaullist talk about independence and *grandeur*. And oddly enough, the leftist and Marxist "rationalists" who are a majority among French scientists are eager to believe whatever the Pentagon says about the futility of the phenomenon: it matches their expectations. There is little we can do against such a powerful coalition.

McDonald has just sent me a statement he read before the United Nations committee concerned with outer space affairs. It remains to be seen whether his highly visible approach will serve the goals he is hoping to reach. It may simply backfire in his face.

This morning I went back to Allen's house. At the top of the stairs there is a little room which remains locked, and the key is at a certain nail. It is in that room that he prepares his lectures and works on his files. I returned all the documents on Blue Book history, which I have now separated from the sighting files. They are too sensitive to remain in my office on campus. I picked up the loose files I had been unable to carry away earlier. I spent most of the afternoon sorting through those papers until I was sick of them.

Chicago. Sunday 25 June 1967.

When I met Hynek for the first time four years ago we discussed Aimé Michel's book, *Flying Saucers and the Straight-Line Mystery*.

"I can't believe that if such events had really taken place in France your scientists would have done nothing," he said. "Surely a special committee would have been appointed, investigations would have been conducted. . . ."

Well, just look at us. For the last several months a wave has been in progress here. And there happens to be in the U.S. a fully funded scientific committee with the highest authority, created specially to study the problem. What does that committee do? It argues, it dreams, it theorizes. Its top administrator Bob Low is getting ready to play the tourist in Prague. Hynek will meet him there: *Le Congrès s'amuse*. The scientists are having fun while the poor witnesses cry for help.

Fred is equally disgusted at the lack of response to the sightings. Yet at Northwestern, when Ben Mittman approached the Dean again with the idea of a new research proposal on UFOs, he got the same old answer: "It's all bullshit."

In Moscow Kazantsev and his secretary Tikhonov have made new contacts among Soviet scientists. The Russians love organization above everything else. They have turned bureaucracy into an art form. So I am not surprised when Kazantsev announces:

> You should know that Zigel, one of my close friends, has taken the initiative of forming a committee for the study of UFOs within the Central House for Aviation and Cosmonautics of the USSR. Air Force General Stoliarov was elected President of the committee. Tikhonov is secretary. Zigel is vice-president.

Kazantsev says that the excuse for the creation of this group was my suggestion for an international meeting under the aegis of the U.N. Clearly a committee is required in order to elect Soviet representatives, should the international meeting actually materialize! Hynek has given me a copy of a letter he has received from Abdel Ghani, the head of the space affairs group at the U.N., to forward it to Kazantsev. But I have little hope that anything concrete will happen in New York.

Chicago. Friday 30 June 1967.

This has been a strange, agonizing week. I keep thinking about the Pentacle letter. Fred is the only person I have taken into my confidence.

"There cannot be any doubt any more," he said after seeing the document. "When they talk about *what can and what cannot be discussed in Washington* they are clearly referring to a manipulation of the Robertson Panel."

Since I found this letter I have been going back through the Air Force files with a fine-tooth comb. I don't want to jump to unsupported conclusions. The first step is to build a statistical index of all the data, so I now spend several hours a day punching cards: dates, places, names of witnesses for the 11,000 cases we have assembled. A trained operator could do this in a much shorter time than I can, but I have no money to hire anybody. Besides I want to fix these names in my memory. Statistics are fine, but the real work is what goes on inside the brain. And I'm not bad with a keypunch machine.

In all this it is becoming very apparent that Allen has been much too timid. In many cases he even believed in the explanations he was producing. For eighteen years he was an amused, open-minded spectator, but he kept his thoughts to himself. "I'm not a fighter," he often tells me by way of apology. Still, he could have taken a more active role while remaining behind the scenes. As scientific consultant he could have assembled the most solid cases and he could have presented them privately to selected colleagues. But he also tells me:

"All those years.... All right, so I didn't speak up. But I kept expecting that a really good case would show up, don't you see? Dammit, if this phenomenon represents a genuine new physical effect, wouldn't you think that some day it would manifest in an absolutely undeniable fashion, leaving the kind of evidence a scientist could take back into the lab and study? Yet such a case never happened."

He tried to remain friendly both with the military goat and with the academic cabbage, antagonizing no one. He is too clever to allow himself to be chewed up along with the cabbage, but he certainly will not grab the goat by the horns.

To his defense it appears more and more, as I read these documents, that he is right in saying he never had a chance against the big guns of the Air Force and the Robertson Panel, that any effort to put the phenomenon before the academic community would have been ignored: the topic was not mature, the timing was wrong. My guess is that Jim McDonald will not get very far either.

Hynek should have taken the time to re-analyze his files. If he had, he would have noticed some sinister incongruities. The New Guinea photographs taken by Drury in 1953 are a case in point. I had come across a clipping indicating that the Australian military had forwarded the film to the U.S. Air Force.

"That's impossible," said Hynek. "The pictures are not at Wright Field in Dayton, I have never heard of the case being reported to Project Blue Book."

Yet I have just punched a card about the Drury film, based on the Air Force's own index. They even have a formal conclusion about the case: "Insufficient Data."

But it is preceded by an asterisk, which means that it is classified.

According to the Australian researchers who have seen it, the film shows an object emerging from a cloud and making a perfect double right-angle maneuver at high speed. This is exactly the kind of photographic data which could prove convincing to the scientific community. The altitude of the cloud can be determined with good approximation, hence the size, distance and luminosity of the object can be computed. The accelerations can be calculated too. Is it for that reason that the Air Force has classified the film? Has someone been working on this data in secret since 1953? Where would this research be going on today? What would be its goals? Could the whole problem have been solved by now?

Three years before the Drury film, on 15 August 1950 to be exact, Mr. Nicholas Mariana shot a movie showing two UFOs in flight. He submitted it to the Air Force which returned it, he claims, after clipping out the first thirty frames, where a dark rotating line was said to be visible on the objects. The Robertson Panel thought they were only airplanes, but they never saw the beginning of the film where greater detail was noticeable.

Now I am reading a letter to ATIC signed by A.H. Rochlen, public relations vice-president, summarizing an independent analysis that was done by the Douglas Aircraft Corporation "within the framework of a study . . . having to do with unconventional propulsion methods." The study was headed up by Dr. W. B. Klemperer, chief engineer of their missile division. The report was distributed to R. M. Baker and A. M. Chop, the same man who was in touch with Keyhoe in the early years and eventually became the spokesman for Project Blue Book. What was he doing on the staff of Douglas and what was this study of "unconventional propulsion?"

In April 1964, on the very day when Hynek and I were visiting Wright Field, Lonnie Zamora saw the now-famous egg-shaped craft landing on four legs in the New Mexico desert. It bore an insignia which closely resembled the logo of Astropower, a company founded about 1961 as a subsidiary of Douglas, under the presidency of propulsion expert Y. C. Lee. I have now found a full-page, four-color advertisement for Astropower in the *Proceedings of the Institute of Radio Engineers*, special issue on computers, dated January 1961. Astropower "offers its services in research and development of advanced space propulsion systems to government contractors and agencies." Its program includes nuclear, chemical and electrical propulsion systems, solid-state elements and energy conversion. Mr. Lee seeks to fill key positions at Astropower with "scientists having advanced degrees in plasma physics, nuclear physics, and thermodynamics."

Later I found another ad in the monthly *Notices* of the American Mathematical Society, where the same company was seeking to recruit statisticians and probability experts wishing to go into artificial intelligence development.

Was the Socorro object an advanced prototype of human technology rather than a genuine UFO? Have the government and its contractors been conducting secret research to study and emulate the characteristics of these objects since 1953, while it used Hynek and Blue Book as a screen to keep everyone else distracted?

All we have to do if we want to create a fine uproar is to show the Pentacle letter to McDonald and to NICAP. This would stir up a lot of trouble but it would solve nothing, of course. It would only generate more confusion. I even wonder if I should tell Allen himself about this document or simply put it back in his files where I found it, and forget that I ever saw it.

Chicago. Saturday 1 July 1967.

A correspondent of mine in Missouri announces to me the death of Frank Edwards[16] due to a heart attack. Allen's secretary tells me that he already knows about it: a lecture agency has contacted him to fill out Edwards' public appearance schedule. Hynek immediately jumped up and agreed to do it. Yet a few days before going off to Canada he was telling Fred and me:

"It's finally over, all those lectures: only two more and I give all that up. I'm wasting too much time, the real research isn't getting done."

He was going to rest at his cabin at Blind River in Canada and start the study of the problem again from the beginning. Alas! The music starts playing, the spotlights are on, the drums roll, and there is Allen rushing onstage again.

Chicago. Sunday 2 July 1967.

We have spent another weekend working on the Archives. Fred came to the computation center to familiarize himself with the system while Janine and I were punching cards. I told him about the Frank Edwards lecture tour.

"Allen is crazy to do it," he said. "He has become obsessed with publicity."

The time has come to seriously prepare our return to Europe. Yesterday I received my Doctor's diploma in the mail. I have not bothered to attend the graduation ceremonies. I have emptied all the drawers and cleaned house. I threw away most of the UFO magazines I had been amassing. I destroyed most of the obsolete orthoteny calculations, the computer programs, the charts. There are only two interesting research directions now: first, the long-term history of the phenomenon, which will tell us how and under what conditions it recurs through time. And second, the study of the landings and the humanoids, possibly leading to hard physical data. The kind of statistical analysis we have been doing can only be useful in providing a framework, in testing broad hypotheses. It is time to move into more specific directions.

The Robertson-Alvarez-Page Panel was manipulated by the American secret services. Nobel Prize winners can be fooled like high school kids. Blue Book is a smokescreen. I should use this knowledge. Or hide it.

Chicago. Monday 3 July 1967.

I have decided to go through Blue Book's archives again with a microscope. This afternoon I went back to Hynek's house and I read the Minutes of the Air Force Advisory Board, the report on the incredible spirit seance of 1959, and I tried to understand Allen's own role in all this. Will posterity take the side of McDonald against him? Will people understand Hynek's patience, his exceptional integrity? It is important to read his letters, his exchanges with Ruppelt and Gregory, his tense correspondence with Friend and Quintanilla, to understand how sharp was the knife edge on which he had to stay balanced.

In spite of my disagreement with his recent behavior I remain struck by one thing: of all the scientists who have considered this problem, Hynek is the only one who has understood the simple fact that the witnesses were not lying, that they did live through an extraordinary experience. Never does he fall into the skeptics' pedantic arrogance, or into Menzel's stubborn dogmatism.

Chicago. Saturday 8 July 1967.

On Thursday Janine's mother, her sister Annick and her son (our nephew Eric) joined us here. When we are not driving around Chicago and Evanston with them, I go on stubbornly punching statistics.

Chicago. Monday 10 July 1967.

Today I had lunch with Allen and I was finally able to ask him some hard questions. How is it possible that he never saw the Drury film? I showed him the Blue Book index where the entry is clearly marked. He didn't have an answer, except that the Air Force decided which cases its "scientific consultant" should be consulted on. His position didn't automatically give him access to all incoming UFO data.

"Jacques, that brings up an important point: the Air Force has sent me a new contract draft and I don't know, frankly, whether to sign it or not. I want you to read it and tell me what you think."

We ordered our meal and I read the document over while we were waiting for the food. The contract, I was surprised to read, was not really with the Air Force but with the Dodge Corporation, a subsidiary of McGraw-Hill.

"What's McGraw-Hill doing in the middle of all this?" I asked without trying to hide my bafflement. "Is that some sort of cut-out?"

"Oh, they are just contractors to the Foreign Technology Division," Hynek replied. "By working through companies like McGraw-Hill, which is a textbook publisher, it's easier for them to hire professors and scholars to conduct some Intelligence activities, keeping up with Soviet technology for example. Many academics would be nervous about saying they were working for the Foreign Technology Division."

The contract puts Hynek under the administrative supervision of a man named Sweeney, who is not a scientist. And it clearly specifies Hynek's task as *evaluating* the sightings of unknown objects to determine if they represent a danger for the security of the United States.

We find here the same contradiction that plagued the Robertson Panel: Scientists are not asked to make a scientific study but to give an answer in terms of a specific policy issue. Oranges are mixed up with apples and cherries. Such an approach is absurd. Only a deep analysis of the sightings, conducted along scientific lines by people filled with the sacred fire of research, could lead to a determination of the phenomenon's implications and potential threats or advantages to the nation.

The contract also shows that the project under which Hynek works with McGraw-Hill is not called Blue Book but Golden Eagle. We were going from discovery to discovery now.

"What was it called before?" I asked.

"White Stork."

A light bulb went on. The Pentacle letter mentioned Project Stork.

"Let's go all the way back to the days of the Robertson Panel," I asked Allen, who was wondering what I was driving at. "What was the project called then?"

"Let's see, at the time I was a consultant under Battelle Memorial Institute, you know, the large scientific think tank based in Ohio. They were responsible for Stork. Later the contract went to McGraw-Hill."

So Pentacle must indeed have worked at Battelle.

"Those details may be important. Why did the project change in structure?"

"The Battelle people were terribly anxious not to get their reputation mixed up with Blue Book. Ruppelt was always talking about 'our consultants back East', although Columbus, where Battelle is located, is only sixty miles by car East of Dayton!"

We both laughed.

"Who was Miles E. Coll at ATIC?"

"I don't remember hearing the name."

"What about the following people?" I named Pentacle and seven individuals whose initials appeared on his secret memo.

He replied without hesitation:

"They were all administrators or staffers with Project Stork, including the man who sent me on a clandestine survey of astronomers in 1952, to find out discreetly what my colleagues thought of UFOs.[17] Pentacle himself was a leader of Stork."

"What were their relations with the Robertson Panel?"

"Practically nil. Battelle wanted to remain outside all that."

"Could it be that Battelle had certain elements about which Robertson, Alvarez and the other luminaries were not briefed?"

"I don't know. I never saw any indication of it."

Therefore he didn't know he had a carbon of the secret Pentacle letter in his own files, presumably since 1953. He may have forgotten it. Fourteen years is a long time. And as he said, he "wasn't looking under the bed at the time." He added pensively: "Perhaps I should have."

Chicago. Tuesday 11 July 1967.

Last evening Fred Beckman, Bill Powers, Jerry Haslett, Janine and I met at the computation center. Mary-Lou Armstrong and Dr. Levine, a young physicist from the University of Colorado, came to town, returning from an investigation on the East Coast.

"A luminous craft was reported to have landed in a field," Levine told us. He laughed: "It turns out the witnesses had only seen spark discharges between the poles of a transformer during a foggy night."

"How do you go about evaluating such a sighting?" asked Bill Powers.

"The interesting question, of course, is to what extent it supports a given hypothesis."

"What do you mean?"

"Well, for example, take the ETH, the Extra-Terrestrial Hypothesis," Levine continued. "We investigate to what extent does this observation support the ETH rather than the hypothesis that we are dealing with a natural effect or a purely psychological phenomenon."

"Since when does the scientific method call for fitting observational data into thc slots of prc-conceived hypotheses?" I asked. Levine quickly realized he didn't have a leg to stand on, so he brought the question to a different level:

"The reality of the problem for us is that our committee has to give Congress an answer about the validity of the ETH. Is this hypothesis (a) proven, (b) acceptable in principle, (c) improbable or (d) rejected?"

"In other words, you're playing multiple-choice questions with Congress?" I remarked sarcastically. He shrugged:

"Politically, that's what the American public wants to know. All people care about is whether we're being visited by aliens or not."

The discussion turned to the coding system used in Boulder. Dr. Saunders has responsibility for it. They encode every report with a series of keywords like "Bright blinking green light," and an estimate of the value of the testimony and the strangeness of the sighting. This covers both the explained and the unexplained cases and it is an unholy mess. Someone has already cranked the computer through some statistics that prove that the unexplained cases are blue more often than the explained ones, at a statistically significant level! This is rank amateur work.

"What's the status of the Winnipeg injury case?" I asked Levine and Mary-Lou.

"Someone was sent there by Professor Condon. He has brought back some data."

"Did he visit the site in person?"

"No."

"What about the sounds recorded in the state of Washington?" asked Fred.

Mary-Lou expressed some surprise at the fact that he knew about Jim Wadsworth's investigation there and the small wave of other sightings in the area.

"Nobody said anything to us about a wave," she stated.

Fred told her the whole story. A local investigator had wanted to take Wadsworth to meet the other witnesses but all Wadsworth was thinking of was catching the next flight back to Boulder, so he missed most of the data.

Mary-Lou told us in so many words that the committee was simply not interested in field investigations. They are still pursuing the theory proposed by Wertheimer, one of their psychologists, who has "discovered" that if witnesses in a car report that their headlights fail when they are close to a UFO this is generally due to a perception phenomenon which is related to Gestalt psychology: these people have simply been blinded by globular lightning so that they don't realize that their headlights are still on.... We all laughed at this, although Jerry and Fred felt more anger than amusement.

"How come the light bulbs in the headlights have to be replaced in so many such incidents?" I asked Levine sarcastically. "Is there such a thing as a Gestalt voltage surge?"

He retreated hastily, speaking highly of "the great experience" of our group, and saying nice things like "it would be desirable for the committee to benefit from your knowledge.... We will certainly recommend that more detailed studies be undertaken.... More money is needed."

It was my turn to interrupt him.

"Why not start now? With the money you already have?"

"Our analysis is not over."

"But you have just admitted that your study is going nowhere. Why continue to waste your time and taxpayers' resources?"

Levine didn't answer. Mary-Lou seemed rather amused. I went on, getting angry:

"I hate to tell you this, but your computer coding system has no meaning. Saunders' catalogue is worthless. You've drawn arbitrary lines through your data space, without any calibration. Selection biases are overwhelming the signal you are trying to extract."

They backed down. They had no idea of the real scope of the problem when they began the job, they said. Condon didn't give a damn whether or not the phenomenon was real. Bob Low was an opportunist. But none of that excuses botching up the job so badly.

Fred, Jerry, Janine and I went out to have something to eat. We discussed the Hill abduction case and we went home very tired.

Chicago. Wednesday 12 July 1967.

A newsman from San Francisco has called me three times. He finally managed to get me on the line this morning.

"Please make a statement for Jim Dunbar's program."

"I've already said everything I had to say. You'll find it in my books."

"But new sightings have happened since then."

"The Condon group has responsibility for studying them."

"What's their objective?"

"To find out if the problem can be approached scientifically."

"Why should the American taxpayer have to shell out half a million bucks to the University of Colorado when you've already answered that question in *Challenge to Science* which costs $7.95?"

"Beats me. Ask the Air Force. They are the ones spending the money."

They would like me to comment on the air. "That would only take a minute." Sure. Then they'll hit me with a question like "Dr. Vallee, how long have you believed in little green men?" all in the name of entertainment.

Later the same day.

At noon today Allen and I went back to the Red Knight Inn for lunch. He often comments on our first meeting there four years ago in this same dining room, with Janine.

"A lot of water has passed under the bridge, Jacques," he said with a smile and a puff of his pipe. Yet the atmosphere has remained the same. The folded umbrellas and the hats at the rack on the wall, the advertisements that give the place an old-fashioned air, and the rental car office across the street, the toy store, the insurance company building.

Hynek told me he understood my eagerness to move faster:

"I should have asked more questions over my years with the Air Force," he said.

"I suspect there was a turning point in the whole project at the time of the Robertson Panel," I told him without mentioning the Pentacle letter.

"You may be right, Jacques. You're forcing me to open my eyes."

"Let's go over the early years again."

"Well, when UFOs first appeared, the Air Force did not even exist yet, as you know. Forrestal set up the initial project, which is referred to simply as Project Saucer. This was followed by Sign which was created in September 1947. When they realized they needed the assistance of civilian scientists, Project Stork was created under the responsibility of Battelle. I served as consultant."

"Did you visit their facilities in Columbus?"

"From time to time."

"Did you follow what they were doing at the time of the Robertson Panel?"

"I didn't go there during that period. To tell you the truth they ran the operation in such a cloak and dagger fashion I thought it was laughable. Perhaps I was quite wrong."

He drew on his pipe, watched the smoke rising over us, then he went on:

"The Robertson Panel put an end to Project Stork. The Battelle Memorial Institute wrote their famous *Report #14* at the end of 1953 but it was only released by Project Blue Book in 1955. The project was then called White Stork. It is much later, under Colonel Friend, that Blue Book became the responsibility of the Foreign Technology Division. Two years later the consulting project became Golden Eagle and the contract went to McGraw-Hill."

"How many other scientists are working for Project Blue Book within the framework of Golden Eagle?"

Hynek seemed taken aback by the question.

"To my knowledge I'm the only one. The other people are civilian experts on plasma, propulsion and aeronautics, who are analyzing material gathered by the Foreign Technology Division, primarily data about Russian aerospace technology."

"Then let me ask you this: If we were to discover that a secret study had been done apart from Blue Book, should we reveal it?"

"Why the hell not? A real crime would have been committed against science, against everything the Constitution holds sacred."

"Do scientists deserve to know the truth? For example, should they and the public be told if it turned out that some evidence had been withheld from the Robertson Panel, and we could prove it?"

"Why shouldn't we tell them?"

"I'm not sure they've earned the right to know. Look at the dogmatic attitude of someone like Condon. . . ."

The mention of Condon's name made him react:

"Did I tell you the latest about him?" he asked. "Levine and Mary-Lou Armstrong spent the night at my house and they told me some new things. The cold, hard fact is that Condon has no interest in UFOs whatsoever. Never did. From the beginning he thought it was all a big joke. Recently he entertained himself by attending a meeting of so-called 'scientific ufologists' in New York. That episode reinforced his impression that all believers were crackpots, jokers or fast-buck artists. He talks about nothing else for hours, repeating how ridiculous these folks are."

Evidently there are now some serious rifts within the committee, and growing opposition between Low and Saunders. The latter fancies himself as a great expert on computers, and indeed he is knowledgeable in psychometric statistics and variance analysis, but that has little to do with data-base management. Mary-Lou has suggested several times that he ought to come to Chicago and see what I was doing. He promised to do it but each time found some pretext to cancel the trip. I have written the committee to let them know I was going to build a complete index for the Air Force files: they never answered. Mary-Lou passed my letter on to Saunders and he put it aside: "I don't believe he can do it," he told her flatly.

Chicago. Sunday 16 July 1967.

Saunders must have changed his mind: when we met with Dr. Roach on Friday night he gave me a list of coded sightings "at the request of Dr. Saunders" and he suggested that I go back to Colorado as a consultant. In particular they would like me to write a computer program that would convert my whole catalogue into their new format.

In spite of that small personal vindication, the meeting with Roach left me sad about the state of the research. Fred's impression was even sharper:

"That's a scandal," he said calmly. "The public is expecting serious answers from a committee of experts. Yet the truth is, these people won't even have taken the trouble to become superficially familiar with the problem by the time they write their report."

Roach asked Hynek how the Colorado committee should handle Washington when their final report was ready to be presented:

"Allen, you're used to testifying before Congress. What kinds of questions do they ask? What is it they want to know? Do they expect the truth, or do they just want to show off before their constituents?"

Hynek shrugged and gave one of his direct, unambiguous answers:

"They strictly want to show off. It's all politics and posturing. Truth is the last thing they care about. Your presentation before Congress, if there is one, will be a circus. They will ask Condon, 'Well, Doctor, do you believe in extraterrestrials now?' And they will quiz him superficially about the Michigan case, Marsh Gas, highly visible stuff like that."

"In other words, they won't really be testing the scientific value of the report?"

"No, their only motivation is political. When I testified before them for the first time I was Associate Director of the Smithsonian Observatory. Our tracking teams had lost one of the satellites, just like that, lost it! The Congressmen were trying to make me say that it was the fault of the Air Force, who were supposed to provide us with the orbital elements. They didn't give a hoot where that satellite was. It was only an excuse, a political vendetta against the Pentagon."

Yesterday Roach and Hynek have had a long private talk. Hynek tried to convince Roach to stay with the committee, because he is the only real scientist there. Roach is not stupid. He can see that the final report which is already taking shape in Boulder will be a pile of garbage. He would prefer not to have his name associated with it.

Exhausted with our long emotional talks on Friday, Hynek fainted while he was arguing with Roach. Fortunately he was sitting down. He didn't break his jaw this time. But that's the second time such an incident has happened. He assured Fred it was only low blood pressure, and he will be more careful with his health in the future. How fragile all this is!

Chicago. Monday 17 July 1967.

Bob Low must be very well informed, because he is now trying to get rid of Roach. He has "not been able to find an office" for him on campus, so Roach has to work from home. McDonald continues to act like a bull in a china shop. Not only is he telling every journalist he knows that the United Nations want to undertake a "discreet" (!) study on UFOs, he is instructing everyone to send massive amounts of UFO magazines and books to the "Moscow committee for UFO studies." I have just received an alarmed letter from Tikhonov. The Russians are concerned at this bumbling attempt to help them: their committee has only been approved on a preliminary basis. Any premature publicity in the West will kill their fledgling efforts, he warns.

Things are also closing down in France. Aimé Michel has been told that a memorandum on UFOs was going to be circulated among military bases, but this idea has not been mentioned again.

Ignoring all that, I go right on punching masses of cards, hour after hour, day after day. I want my general index of the Air Force files to be finished at the end of the Summer, whether Saunders believes I can do it or not.

Chicago. Wednesday 19 July 1967.

Yesterday morning I spent two hours alone in Allen's house again. I sat in that little room where he works and I immersed myself in his universe. His books are there, his tapes (they do not contain as many UFO interviews as German and Italian operas) and his file folders. Many of them are labeled but empty, because he grabbed the contents as he rushed on the way to some lecture and never put them back.. There are at least six folders entitled "Follow-up" with only a few scanty notes in them. Everything is confused.

Through the narrow window I looked over the small garden and the garage. I imagined him at work here: his secret mind, amused and curious, looking for adventure, for novelty. His good heart, his simple kindness, all that is contained within these four walls; the precious naiveté of a very great man.

The chaos of his files reflects the trouble in his mind, a trouble to which I contributed four years ago when I rushed in here, brandishing the French cases, stressing the international scope, speaking of global patterns. Then came Socorro, Michigan, the media storm. Hynek had to rush ahead, testifying before Congress, explaining his unfortunate Marsh Gas comments. At the same time he was building a new observatory at Northwestern, giving lectures everywhere, running the Department of Astronomy. His class is the most popular on campus.

Now he seems lost, disoriented. While the Condon committee was wasting time and money on irrelevant issues the real phenomenon appears to have changed again. If journalist John Keel[18] is right, there are thousands of "silent contactees" who have seen flying saucers at close range and wouldn't tell their story to NICAP or the University of Colorado for anything in the world.

Chicago. Thursday 20 July 1967.

A week ago Hynek was interviewed by John Wilhelm, a journalist from *Time* magazine. Aggressive, inquisitive, precise, Wilhelm pushed him around, squeezing him into narrow corners. Hynek answered frankly, but the questions kept coming back about the handling of cases by Project Blue Book, about the false astronomical explanations, about the statements made to Congress by the Air Force, about the Condon study. When Wilhelm halted the interview to go get a sandwich Hynek called me at the computation center:

"It's not going well, Jacques. Not well at all. This guy wants to know everything. He is looking for blood. He is going back all the way to Project Sign. He keeps asking about the possible role of the CIA. Obviously it's McDonald who is sending him here. You know *Time* magazine: they need a juicy story. They need a simple answer and a scapegoat they can throw to the public. But here there is no simple answer. If they could put everything on my back, that would make a lot of people happy. I'm trying to be patient but there are limits. . . ."

I could tell in his voice that he was genuinely afraid. He has given himself over to this crowd with his many public appearances, his lectures. But the public and the UFO believers are clamoring for more. McDonald is giving them what they seek, dramatic accusations and simplistic conclusions. On one side is Hynek, an older man who is troubled, who constantly questions his own life in search of deeper truths. On the other side is a fiery champion who thinks he has all the answers: UFOs are extraterrestrial, we're being watched, even invaded. Perhaps he is right. But the way he is handling the problem stinks.

"Give the McDonald sighting to this guy Wilhelm," I suddenly said. "It's the obvious solution. Why is Jim hiding his own sighting, pretending he has just discovered the problem, if he is so sure of the answers?"

At six o'clock I called Susan, Hynek's secretary. Allen was still in conference with Wilhelm, behind closed doors. It was only much later in the evening that I was able to reach him.

"I followed your advice. Wilhelm was visibly shocked when I confronted him with McDonald's own sighting."

With a very tired voice he added: "After a day like this, Jacques, I wish I had never heard of UFOs." I feel the same way.

The pressure and ridicule the scientific establishment places on anyone who dares raise this forbidden topic is incredible in its vicious character, in its unfairness. Following Hynek's letter to *Science*[19] a biologist wrote in to record a personal sighting. Hynek requested more details. He received the following answer:

> I could simply ignore your letter and drop the whole story of my UFO observation. But that would be impolite and I don't want to be. Therefore I am answering you but I do so with regret.
>
> I have been subjected to fearful trouble since my short letter in *Science*. In reality I am not at all a sensitive person. I have spent thirty years in Federal and State biological research and I think I have withstood all kinds of criticism. But I must admit I no longer want to receive repulsive remarks from my friends (are they still my friends?), from my associates, from crackpots and others. The recent letter by Stibitz (*Science*, 27 January 1967) irritates me and causes more confusion that I intend to stand. So, let us forget this whole story, please.

No wonder scientists do not find the evidence convincing: the very best cases, the reports from their own peers, don't reach them because people are too embarrassed to describe phenomena which contradict what they think science is.

Chicago. Friday 21 July 1967.

It is in the evening that I feel worst: morose, stranded in the emotional desert of Chicago, and eager to leave. We are not without friends. I have tried to describe my anguish to them, but can they understand? We go to movies with them, we attend dinners and parties, but we form no lasting close relationships. Exceptions involve a few Europeans like Hans Rasmussen, a Scandinavian doctor, and his wife Inge, who have the sense of humor and the joy of life that is missing from most of our acquaintances here. They feel somewhat stranded in the Midwest, as we do.

A letter has arrived from the Shell Petroleum Company. They are recruiting foreign nationals, particularly computer scientists educated in the United States who are willing to return to Europe.

And the UFO phenomenon keeps evolving. Don Hanlon's research on the 1897 wave keeps turning up some extraordinary facts. When I compare some recent landing cases, or even the Hill abduction, with some of his old airship sightings I cannot help but feel disturbed at the similarities. And a new wave may be starting in France. Last Wednesday a young girl who was walking in the countryside with three children saw three little men running away from her, talking to one another in bird-like sounds.

Fred argues I should level with Allen and put the Pentacle letter in front of him before he goes off to the Prague Congress. But I don't think he is ready. His current confusion is obvious: back in Canada he is now writing an article for *Playboy*, of all things. That's certainly not the right way to place the subject on the map of respectable science.

Chicago. Saturday 22 July 1967.

My Rosicrucian friend Serge Hutin would say that "the Cosmic has spoken." Everything suddenly appears in a new light. I am ready to leave America, and I am at peace with myself again. I find new beauty in the whole world: sky and creatures, winds and tides, *vents et marées*. . . . Is it possible to love to the point of madness without knowing what it is we love? An invisible presence passes through my life and caresses me this morning with indescribable tenderness. I submit joyfully.

Chicago. Tuesday 25 July 1967.

John Keel has written a strange letter to me. The first part is rational, serious and well thought out. Then he suddenly slips into a disturbing, prophetic style:

> From all over the country people are calling me for help. Next week will witness a chain of events that will shake the planet. Religious fervor will engulf the world. A prolonged black-out will take place in the next few days, probably on the 26th or 27th....

After Jessup and Smith, could it be that John Keel will become obsessed with the phenomenon to the point of prophesying disasters?

Hynek writes to me from Canada, sending me an action plan he calls the "Blind River Manifesto." Unfortunately I disagree with almost every point it contains.

A telegram has arrived from Shell. They invited me to call them in New York. I found the conversation quite refreshing after the stream of bureaucratic letters I have been receiving from various scientific and technical institutions in France. We do want to return, but not until I have a job that makes sense, with people I can respect.

Chicago. Friday 28 July 1967.

Allen called me from O'Hare this afternoon. He was on the way from Canada to New Mexico. He told me that Keyhoe had just fired Richard Hall from NICAP. Hall, who is divorced, recently married a NICAP secretary.

"I don't see what's wrong with that!" I told Hynek.

"Me neither. But upon his return from their honeymoon in Las Vegas they found they no longer had a job. Hall tried to get hired by Condon who turned him down. Things are boiling over in Boulder, too. Mary-Lou Armstrong had a big fight with Bob Low, and now she only works half-time. Must be the Summer."

Chicago. Tuesday 1 August 1967.

Ben Mittman came back on campus this morning and showed me an encouraging letter from Rowena Swanson at the Air Force Office of Scientific Research. She says of my proposal that it is "one of the most sensible statements I have had the pleasure of reading about the work that can and should be done with data acquisition systems."

She is especially interested in the stellar data-base and the follow-up to the Altair work, so she would like to see us go ahead with the setting up of an information science center.

Hynek is back on campus, too. I met him in the hallway of the computation center. He seemed tired, ill at ease, disoriented. When I showed him how his files had been reorganized into the three shiny new cabinets, he was amazed at the progress made in a short time.

Chicago. Wednesday 2 August 1967.

Sticky heat, crying kids in the street. Things are dragging along. Another Chicago summer. Fred and I had lunch with Allen. When I brought up Battelle he changed the topic of conversation. So I stressed the old sightings, 1947, 1948. I have been studying them under the microscope for the last two weeks. He patiently repeated what he had told me before:

"Back then I was certain that all those phenomena could be identified. Of course when you read the reports now you see things differently."

He made an apt comparison, drawing a parallel with astronomy:

"If you know what you're looking for, you can take an old plate of the night sky and pick out Pluto. Of course it was there all the time before Tombaugh discovered it. People just didn't know it was there, so they didn't see it. Well, it's the same thing with UFOs."

We turned to the subject of hypotheses. Hynek would like to write a set of axioms from which he would derive the whole study of UFOs. A laudable goal. His first axiom would have to do with communication among the worlds, a very old hermetic concept.

"I simply cannot conceive of a universe where communication among remote planets is made impossible by a theoretical limitation. Everything man can imagine, he can realize sooner or later. Well, dammit, I really want to know what's going on in Andromeda. Nothing will convince me that's impossible."

I agreed wholeheartedly. It may take a hundred or five hundred years for us to learn how to do it, but eventually we will.

Second axiom: Mystical experiences demonstrate the possibility of mental projection. Hynek quoted the tests done with Ted Serios[20] and some recent LSD experiments.

"The limits of the mind are poorly known, we underestimate the human psyche."

Here I cautiously agreed with him, although Fred and I always grit our teeth when Ted Serios is mentioned, because the experiments seem inconclusive. If we want to bring in psychic abilities there are better examples. The Dutch research with Croiset the paragnost, which is being conducted at the University of Utrecht, is much more solid.

Third axiom: The physical plane is not the only level of existence. The hermeticists speak of an "astral plane." The spiritists talk about materializations, *apports*, ghosts that go through walls. Who knows if UFOs are not observational devices that are materialized into our world by the denizens of another?

I brought us back to more immediate matters.

"Since our 1952 files are incomplete," I asked Hynek, "why not contact Battelle to request a copy of their punched cards, or a listing of the cases used in *Report #14?*"

He seemed embarrassed.

"Well, that might not be easy to do. They are more and more involved in military contract work. In fact the Pentagon has prevailed upon *Time* magazine to eliminate any reference to *Report #14* from Wilhelm's article. You're not even supposed to have heard the name Battelle in reference to UFOs. If they sign my new contract I'll go straight to the historical files to check up on all this."

Fred and I exchanged a quick look.

"What was the attitude of Robertson and his people when they got that panel together? What questions did they ask? Did they demand to know where the sightings came from, who selected them?"

"I can only speak about the sessions I attended," said Hynek. "I had the distinct impression they didn't care."

"Allen, let's be serious," interjected Fred. "If somebody came over to the University of Chicago this afternoon and told me he had a bachelor's degree from West Texas College in animal husbandry and he was going to teach me how to take apart my electron microscope I would throw him out. How could Robertson, Alvarez, Page and other scientists of that caliber listen to some Air Force officers as if they were hearing the Gospel? They were in control. Their reputation was on the line. Why didn't they demand to be briefed exhaustively about the Battelle work?"

"It's very likely they just never knew about it," said Hynek, pensively drawing on his pipe. "The CIA and the Air Force may not have told them."

Incredible. Fred and I spent the rest of the afternoon rummaging through old papers, and we spoke sadly about the future. Fred told me about his experiments with the brain, and Aldous Huxley's observations under mescaline. The human mind relies on a delicate chemical balance. When that balance is altered, other worlds open up. Is there a single solution to the equation of thought?

We also discussed my imminent departure from the United States. Fred told me I shouldn't do anything hasty. But if there is a secret study in Washington, what good is it for me to continue all this hard work? I might as well go on with my own life, my own career. After all, I am a rather good computer scientist. I can make an honest living. I don't need all this political crap.

"Do you really think there's a secret study?" Fred asked.

"Yes," I said, "although it may be nothing more than a bunch of space cadets engaged in a misguided, technocratic effort at duplicating the UFO propulsion system. It may be coupled with an in-depth analysis of traces and material fragments."

I thought of Astropower, and of the McDonnell-Douglas company, who is rumored to have a secret team, employing a physicist named Stanton Friedman to collect physical data in a hush-hush manner.

"All those people are glorified rocket engineers, they can hardly put their arms around the real problem."

"Don't underestimate them," says Fred. "It's impossible to suppose that they would study UFOs without thinking about their origin, and about the beings that pilot them. . . ."

Fine. But I would like to know the answers to a different set of questions. For instance, to what extent are these strange humanoids related to the beings so often reported in our own history? If the same phenomena have been around us for centuries, and if it is only a cultural accident that makes us interpret them in terms of spacecraft and extraterrestrials, then it's a whole new ballgame, isn't it? How would we prove it? That research can only be carried out in Europe, where the best sources for historical material are located.

Will I have the courage to leave? Evanston itself is a very silly town, but every time I set foot on campus it is as if I stepped into a magic circle.

Hynek told us about his visit to Colonel Freeman when the Air Force had just signed the contract with the University of Colorado:

"He was laughing, rubbing his hands together. Obviously he regarded me as one of the boys, someone who was 'in' on the big joke. He acted very freely in front of me. He's sure Condon will write a report that will conclude that UFOs don't exist and he'll be rid of the whole thing at last. There was no question in his mind that the Colorado people would reach a negative conclusion."

Chicago. Thursday 3 August 1967.

This morning I finished punching 1963 and I spent the afternoon screening 1948, where I found yet another series of remarkable, forgotten cases. Now I can say without fear of contradiction that I know the Air Force files better than Quintanilla. I do have the information at my fingertips. And more significantly, *I know what isn't there.*

This afternoon Hynek called from Wright Field to ask if we had a copy of the August 1952 Bellefontaine, Ohio, case. The answer is negative. But the information is not in Dayton either, which is embarrassing because the file has been formally requested by the University of Colorado. The Air Force will be forced to admit that some important pieces of data are getting lost, an accusation I had already made in Boulder last November.

I also spoke briefly to Quintanilla, who is always very urbane with me. I described my ongoing work on the Blue Book index and asked his permission to go on with these statistics. There is no problem, he said, everything has been declassified.

We have just learned that one of Allen's colleagues, Northwestern astronomer Carl Henize, had been officially selected as a scientist-astronaut by NASA.

Chicago. Tuesday 8 August 1967. 19:00

Today Fred came to my office to read the reorganized files. I am getting to the end of the card punching: two long metal drawers full of cards. I have written a validation program to detect duplicates and make sure the index was clean. It also generates some basic statistical breakdowns on a weekly basis.

We no longer have a sitter for our son, so this morning with a heavy heart I left Olivier at a day care center, two dark dusty rooms on Clark Street. I am eager to move away from here. Rowena Swanson has told us that federal funding for new research such as our proposed information center was blocked because of the expanding Vietnam war.

Allen leaves for Prague tonight. He will be gone until September 5th. I saw him in Evanston last Saturday and we had coffee together. At my suggestion he had gone to Sweeney to request a copy of Battelle's card catalogue from *Report #14* days. He was coldly told that they were no longer in existence. Sweeney pretended to be outraged: "It's a crime, it's unthinkable. . . ." But Pentacle is still with Battelle, and he has told Quintanilla that the cards had actually been thrown away.

Chicago. Thursday 10 August 1967.

Ben Mittman took the news with concern today when I told him I was leaving. I am presenting him with a complete Fortran-based Infol compiler that is a few steps ahead of the Control Data version developed by Bill Olle and his team.[21]

The new French Institute for Information Science has expressed interest in hiring me . . . whenever they officially become operational. And Dave Saunders has called me with some new ideas.

"Can you come to Boulder as our consultant?"

"When?"

"As soon as possible."

"I'm working full-time on a proposal for our computation center here."

I am beginning to like Dave Saunders, but I am still suspicious of anything the Condon committee does. If I became connected with their work I might provide some implied measure of support for their conclusions in a project which has been biased from the start.

"Dave, if you don't mind, I would prefer to have Allen Hynek included in our discussions. Can the meeting be delayed until he returns from Prague?"

"Well, perhaps you could both come to Boulder when your schedule clears up, say around September 15th?"

We agreed tentatively on that date. But I have a feeling that I will be a long, long way away from Boulder, a long way from the United States by September 15th. He went on: "Anyway, we'll see each other in Washington when you give your talk before the American Psychological Association."

"I won't be going to Washington."

He seemed shocked. "Why not?"

"Well, don't you think it's going to be another one of those meetings where everybody talks and nothing concrete gets done?"

The committee has started to do some statistics, but they work from a very heterogeneous sample which is mostly garbage. They want me to write a program that would convert our 3,000 screened and documented cases into their code, and naturally they would also like to get their hands on my more recent catalogue. But simply mixing up a lot of data from various sources is not going to provide a workable data-base.

Chicago. Tuesday 15 August 1967.

The first seven volumes of the *Archives of the Invisible College*, each about three hundred pages thick, have come back from the bindery. They contain some unique documents, original letters, analyses of confidential sightings, and the most interesting Air Force reports, after all the trash has been screened out: as close to the real UFO phenomenon as we can come.

My friends think they can keep me here, that at the last moment I will decide to stay. They are wrong. I have lost interest in what happens here, in the outcome of the Condon committee study.

I draw my strength from Janine. She is more ravishing than ever, but it is her inner beauty that fascinates me. Today we took a trip downtown together to see the city for the last time. We watched the splendor of Chicago. Little blue workmen perched on steel beams at the thirtieth floor of a new building, sparks flying from a welder's torch, a back alley filled with broken glass, an orange truck in the shade, jackhammers on granite, Janine's needle heels on marble tiles, the new Picasso sculpture, the waves along the Lakeshore, red concrete trucks rumbling along, green steamrollers on a new expressway, the setting sun splashing over our windshield. Picasso is not appropriate here. Chicago is a Dali, a giraffe with drawers, a soft machine, a burning rhinoceros eating the wind, spitting fire like a mad dog.

Chicago. Saturday 19 August 1967.

We are packing up all our books into boxes to be shipped across the Atlantic. With the end of the Summer, the Europeans are coming back to their offices after their long vacations: the job offers I receive are becoming more precise. Thomson wants me to work in their Bagneux research center, South of Paris. But everything in their letter reeks of bureaucracy, of rancid procedures. Shell, on the other hand, has a concrete proposal. They will pay my way to Paris for an interview: Can I fly there early next month? I feel Chicago receding away from us, already becoming a foreign city again in my mind. Rain falls on dirty sidewalks. I drive along dilapidated Clark Street. Over the car radio, Johnson makes a contemptible speech to explain why he has to bomb the length and width of Indochina. The pale, haggard silhouettes of drunkards stagger in front of my car.

I will take Olivier to France with me on my way to the Shell interview, and I will come back alone to help Janine pack. My short-term priorities are to find a secure place for the general card index, for this Journal, and for the Archives.

Chicago. Saturday 26 August 1967.

It is ironic that all the civilian research groups in American ufology are collapsing precisely at the time when the University of Colorado is affording them their first glimmer of respectability. In Tucson Coral Lorenzen has become so obsessed with flying saucers that she locks herself inside her house, according to people who have recently visited APRO. In Washington, since he has fired Dick Hall, Major Keyhoe is going broke. He has launched another one of his heroic appeals to the pocketbooks of his members, assuring them that the solution to the problem is in sight at last. Is it?

I am still struggling with the ethical issue of what I should tell Hynek. Should I show him the Pentacle letter? After much debate, Fred and I have reached the conclusion that Allen alone could make use of it. Right now he is in Prague, trying to get information on UFO sightings in Communist countries so that he can mention them in *Playboy*. Somehow I don't think this is the best scientific way to approach the problem.

All the people we have consulted say it is simply "inconceivable" that the military hierarchy would rely on Blue Book alone to investigate a phenomenon which can interrupt radio communications and interfere with missile base radar signals, as happened recently in North Dakota.

We keep hearing of old secret government projects that are not documented anywhere. A former engineer with U.S. Intelligence in Germany has told us that the National Bureau of Standards had conducted an investigation of UFOs under Professor Dryden as early as 1943, along with an investigation of German research on jet aircraft. They were already aware that UFOs interfered with internal combustion engines at a distance. They suspected electrostatic effects.

In 1963 a Polaris nuclear submarine suddenly came up to the surface, interrupting its stated mission, which called for a long submerged cruise in the Atlantic. All personnel were told to remain below. A few superior officers went up to the tower. They are said to have come back down with three humanoid bodies in clear plastic bags. The sub dived again and rallied to the East Coast at top speed. The vessel had accomplished none of its stated objectives, which included the test firing of several missiles. As for the beings, they looked like shaved monkeys. Perhaps they were indeed monkeys, recovered from a classified space experiment?

There is also an employee at an Alabama military base who told an extraordinary story: he had unloaded several gray-blue humanoid corpses from a helicopter to place them on a waiting plane whose propellers were already turning. He was told the plane was going to Colorado. Naturally, it has not escaped our notice that all this could well relate to a super-secret space project involving primates as test subjects, and that the "crashed saucer" tale could be the actual cover story rather than the underlying truth. As Fred puts it so well: "Government is an obscenity."

Chicago. Sunday 27 August 1967.

There is no proof that any of the above stories actually relate to UFO pilots. In that respect the Pentacle letter is very valuable because of what it does not say: it mentions no crashed disk, no hint that humanoid pilots have ever been recovered.

Chicago. Monday 4 September 1967.

Deep inside, I still believe strongly in the value of science. Yet the kind of scientific process that will crack the UFO mystery open has not been developed. Contrary to Hynek's hopes, I believe research will continue to stagnate, and that the United Nations will do nothing.

Our American friends, even the most enlightened, look at me with some suspicion when I tell them that a war like Vietnam will do more damage to this country than it will to Indochina, no matter who wins in the end. This does sound like an absurd statement, since there is no combat on American land. Yet the harm the war does to American society is deep and irreversible. You can see it everywhere.

Chicago. Tuesday 5 September 1967.

Allen has called me, proposing lunch. He has returned from Prague with the strong feeling that astronomical opinion is changing. De Vaucouleurs and Guérin have told him they were impressed with the reports of landings. Kuiper has told Hynek he was going to submit an article to *Science* on the subject of extraterrestrial life.

Hynek gathered a rather heterogenous group of colleagues at his Prague hotel. He even gave them a nickname: *Commission Zero.* It was composed of most of the attendees who had an interest in UFOs: Bob Low and Frank Roach from the University of Colorado, Morrison from MIT, Sagan and Menzel. But why did he neglect to ask Guérin and De Vaucouleurs to join? Sagan invited Shklovski, who declined. "There is no UFO problem," he said.

Roach described a luminous object he had seen and photographed himself. He said he had not been able to identify it.

Menzel jumped up. "It was an airplane, of course!" he said.

"Don," replied Roach sadly and calmly, "I am always impressed with how easily you can solve a problem without having all the data!"

He also related his radio conversation with a forest ranger in California who was clearly terrified by what he saw.

"An incompetent observer," commented Morrison, "the testimony is worthless."

Hynek says that as long as Menzel was in the room, nothing good could come from the meeting. He also learned that Condon was definitely out of the picture. When Frank Roach went to tell him about his own sighting he said:

"I'm sorry Frank, but I'm really too tired."

Allen, on the other hand, is in splendid shape. He gave a lecture on UFOs at the Planetarium in Prague. A man came to him furtively after the talk and presented him with a stack of newspaper clippings from the local press. But he added that it was the military, not the Czech scientists, who did all the investigations.

Allen now realizes that my departure is imminent and that nothing will change my mind.

"Will you promise me something?" he asked. "You've often said that you would drop everything if you could work on this problem full-time, right?"

"That's still true."

"Will you make this commitment, that if I can get the right level of support to create a real research project on this topic, you will come back from Europe and help me run it?"

Yes, I told him, I would make that commitment. I don't think anything would keep me away from such a project, if it were seriously financed and built on a solid basis of hard science. I would know how to run it. My objectives are set. I already know what I would do the first day, the first week, the first month, the first year. And I'm not talking about mere computer statistics, either.

Chicago. Friday 8 September 1967.

Tomorrow Olivier and I leave for Paris. For the first time in our lives Janine and I will be separated by the ocean.

Kazantsev has sent the first copy of *Young Technology* carrying our joint article. I took it to Mittman and together we managed to translate a few paragraphs. This is quite a coup, the first article in the Soviet press which is open to the reality of UFOs:[22]

> The authors of the present article believe that the hypothesis that UFOs are machines built and used by an alien culture must be considered scientifically.

This article in a magazine printed at over one and a half million copies was immediately picked up by the trade union newspaper, *Trud*, which has an even higher circulation.

When I saw Hynek I gave him a copy, telling him that according to Kazantsev this issue of *Young Technology* had quickly achieved the status of a collector's item in the Soviet Union.

I have bought a golden frame. I have placed into it a beautiful reproduction of *The Lady and the Unicorn* tapestry which I intend to give to Allen as an ultimate gift, in Fred's presence, before I leave. Between the picture and the cardboard I have inserted the Pentacle letter. That document, after all, comes from his personal files, and I have no right to keep it from him, nor do I want to carry it to France with me. I will make Hynek read the document, then we will replace it under the painting. Let him decide what he will do with it. In this form, beautifully framed yet completely invisible, he can just hang it on a wall and forget it.

Part Four

MAGONIA

16

Paris. Rue de la Mésange. Tuesday 10 October 1967.

No one was waiting for us at Orly. We enjoyed keeping to ourselves this little secret of our return to France. We smiled at each other, our hands touching, feeling young and full of expectations. We arrived early at the apartment on rue de la Mésange. We waited at the door for my mother to come back from the Mouffetard street market. We helped her carry her grocery bags upstairs.

Now Annick has come over to pick up Janine, and the two sisters have driven off to Normandy where our son is. I have spent the day walking through Paris alone, looking intensely at every small detail. I felt a sense of pleasurable devotion when I saw the Île Saint Louis again. Let anyone who hasn't loved this city make fun of my weakness.

At the foot of the Petit-Pont a flat barge carrying an old steam crane on its ancient deck was unloading sand. Inside the steel cabin a man was rummaging in the fire with an iron hook. From the quay a worker was sending down a load of beer bottles along a cable.

Paris. Rue de la Mésange. Thursday 12 October 1967.

I have lost track of time. For the last three days I have been walking all over Paris. I am eager for Janine to come back and help me look for a place to live. I have been buying books, books and more books.

Paris. Rue de la Mésange. Friday 13 October 1967.

Janine comes back tomorrow. I have spent the afternoon with my uncle Maurice in his quiet apartment on rue du Cherche-Midi, behind the church of Saint-Sulpice, admiring his work and his dreams.

Before giving Shell formal notice of my acceptance of their offer, I have decided to talk to a few other French companies involved in the development of sophisticated computer programs, to make sure I'm not missing anything. Today I met with the head of software for Control

Data France. I tried to convey to him my interest in the new generation of languages that are appearing in the States. Would his bosses in Minneapolis allow his group to develop new compilers? I asked. He did not know how to answer my question.

An American friend has sent me a copy of an amazing article about UFOs published in the very authoritative *Science* magazine last month. Authored by William Markowitz, a physics professor at Marquette University, it contains such idiotic remarks as "the published reports generally describe objects about five to one hundred meters in diameter, which land and lift-off without the use of launching pads or gantries. No similarity to the giant undertaking of a launching from Cape Kennedy has ever been reported." He deduces from this that UFOs cannot be under intelligent non-human control if the laws of physics are valid![1]

Paris. Rue de la Mésange. Friday 20 October 1967.

I should describe an experience I had with the Plan Calcul today that has left me puzzled and amused. This Plan is a conglomeration of government-funded projects. Some of them are little more than disguised handouts to the big companies, others are research programs administered by various agencies, notably the Institute for Research in Informatics and Automatics, known as IRIA.[2] I tried to reach their people all day Tuesday. Since my mother doesn't have a telephone (there is a three-year waiting list to get one!) this meant going down four flights of stairs and walking three blocks to the Post Office every hour or so to try and get through. At the Ministry I was told that IRIA had just been assigned a building near Versailles. One look at the map and we felt enthusiastic: it would be great for Janine and Olivier to live there, West of Paris, so close to the countryside and the forest. The problem was the phone line: always busy, no way to reach anyone.

"The Plan Calcul is in the midst of such a boom," I thought, "that the switchboard must be literally swamped with requests."

So I bravely decided to do things the American way—going there in person to meet their top people even if I had to camp on their doorstep to be admitted among the throng of their admirers.

At the railroad station I bought a sandwich and tried the phone line again from a public cabin: still busy. So I let the little train take me away through the Western suburbs. To my left was a young fellow who set aside *L'Humanité*, the Communist daily, to roll himself a cigarette. In front of me an older gentleman was reading the conservative *Le Figaro*. I relaxed, feeling at home again.

From the train station in Versailles I called the Plan Calcul once more. Still busy! I demanded a check of the line. It was functioning normally, I was told. I went out to the fresh air and the sunlight. It was a fine October afternoon. The woman who was selling newspapers didn't know where Plan Calcul headquarters were, but she did know how to get to the town in question. Take the B bus to the end of the line, then you walk.

And suddenly it is as if nothing had changed since my childhood. It is as if I were eleven or twelve again. I am in the B bus which climbs up the rue de la Paroisse. The whole world is simple, a charming mosaic of little cottages with their gardens and their rough stone walls and their roses, their stone crosses and their plaster angels. Silence. Like a time machine plodding back to the distant past, the B bus turns into the Boulevard du Roi and rests at Le Chesnay. I am the only passenger left. It's lunchtime in this quiet, conservative suburb. I jump out.

"Rocquencourt? It's over there," the driver says with a vague sweep of the hand, showing me a narrow alley between two hedges, ironworks, barking dogs, dusty sunshine.

He adds: "It's quite a ways away."

Who cares? I grab my briefcase, which contains a few reprints of my recently published American articles and some Infol listings and I walk, keeping a weary eye on the black clouds that now line the horizon. I am not far from my goal, I am told.

"Do you see those old barracks over there?" someone tells me.

He is directing me to the former headquarters of a company of firemen who have recently moved to more luxurious accommodations. Yes, that is IRIA, those wooden shacks. It looks like the set for "Stalag 13." The famous new research institute occupies three small, silent offices there. A secretary lifts her eyes from a romantic novel and greets me with a smile.

"You tried to call us on the telephone? Ah, poor Monsieur, we have just learned that we have been cut off for two days. Fortunately somebody has asked the Post Office to check our line, otherwise we would still be without communications."

It was me, calling from Versailles! So I have already made a contribution to France's Plan Calcul, by reconnecting their one and only telephone line. It is a measure of the intensity of their business that the managers did not even notice that they had been without a telephone for two days. No one will be in the office before 2:30 p.m., the secretary tells me, so I walk back to the village to get another sandwich, feeling silly in my dark business suit and tie.

It is a bit after three o'clock when the managers return from lunch. My presence is now announced with proper official formality. Through the open door I can hear a bureaucrat speaking to the woman who says: "It's someone who has just come back from the USA. . . ." I catch the abrupt reply:

"Let him talk to Louis."

It turns out that Louis is an underling who was there all the time, of course, but nobody thought of him. He is quite pleasant and hospitable but I have already made up my mind about the whole meeting. This way of treating people, of handling outsiders, these offices that stink of bureaucratic neglect, of obsolete ideas, represent everything I fled in 1962. I ask a few questions, to be polite, to pretend I am interested in what they do.

"What computer hardware will Plan Calcul be using?"

"That hasn't been decided yet."

They have created this institute to fill the gap left by the universities which prove incapable of turning out good software research, but they are putting the same hopeless bureaucrats in charge of it. Plan Calcul is only the old tired coterie preserving the same petty privileges under a new label.

I left a few reprints of my American articles with Louis. He got up politely when I took my leave and he saw me to the front steps. I felt so light as I walked away, so happy to realize that I would never work there, that I started whistling along the hedges and the rosebushes of suburbia.

Paris. Rue de la Mésange. Monday 23 October 1967.

I feel a fierce need to attack new things, to look for the undiscovered. The grave and deep nature of this world must be knowable. It is a waste of one's life to look for anything else.

I have brought up the physiological effects of UFOs in a discussion with my brother the doctor. He doubts that the instances of paralysis reported in cases like the Valensole landing two years ago represent genuine paralysis through action on the nervous system. To him they sound more like psychological effects. Masse's sleepiness after his sighting could be a psychological reaction too, he tells me.

Paris. Rue de la Mésange. Wednesday 25 October 1967.

By taking another week to extend my network of contacts I have also taken some risks. I feel exhausted and we are rapidly running out of cash. What I am gaining, on the other hand, is a good overview of the French software world. There is practically no research activity in my current specialty of data-base management. Mr. Lepoisson, a man who could have been my boss if I had joined Thomson and who will also head up some part of Plan Calcul, tells me that EDF, the giant French electrical utility company, is about to create a massive index of its technical publications. They have decided to do it the hard way, under a conventional framework. I wish them good luck.

I only spent half an hour with Mr. Lepoisson at Thomson headquarters. I found the old building just as I imagined it, with layers of fading paint on the walls and centuries of bureaucratic passivity peeling off the ceiling. Technical directors talk to you indifferently in boring, drab offices where nothing happens, where nothing could possibly happen. They wear their brown pants and the butterfly tie cousin Gustave had offered them on their wedding day twenty-five years earlier. When I mentioned the new techniques for large data-base management Mr. Lepoisson shrugged and replied that it was a trivial problem which had already been solved by his management consultants. He did offer me a job in the end, but when I reacted in shock at the low salary figure he replied with a fine superior smirk that young men often returned from America with exaggerated notions of their own worth to society. . . . I certainly would never find elsewhere a salary much higher than what he had quoted me, he assured me.

I answered calmly that Shell offered me fifty percent more. For the first time, his reaction was genuine. It was bitterness mixed with contained rage.

I have called Plan Calcul again, just to close the loop.

"The problem in your case, *mon cher Vallee*, is that you have not been introduced by one of our Directors."

"But I saw Monsieur Lepoisson yesterday afternoon. . . ."

That changed everything. People suddenly spoke to me as if I were actually related to them rather than an alien from across the sea, an unknown quantity. But they only became truly friendly when I hinted, for fun, that I might agree to work without pay for a while.

On Friday, to complete my explorations, I will meet with the director of the scientific computing center for Armaments at the Montrouge Fort. Today Janine and I visited a two-bedroom cottage which is for rent in Saint-Germain-en-Laye.

Paris. Rue de la Mésange. Thursday 26 October 1967.

Aimé Michel and I went rummaging through the flea market near the Panthéon this morning. We looked at dusty books and old fabrics exuding the wondrous stench of bygone centuries. We spoke of the ills of the nation, the weaknesses of old man De Gaulle and the rigidity of French science. We came back to rue de la Mésange for lunch. I read to him passages from Paracelsus' *Book of the Sylphs*, which left him fairly stunned because of their similarity to the UFO entities he has been describing in his books.

Three letters have arrived from abroad: Olavo Fontes wants me to go and work in Brazil, a little too late. Charles Bowen is all excited at my sudden departure from Chicago and suspects it is hiding great truths. Gordon Creighton is urging him to jump into a plane and come to Paris to interview me and learn "my secrets." Indeed Creighton, who is a very subtle man, has understood that it was not simply a vague nostalgia that made me come back to Europe, but some specific discovery that had disgusted me with America. They want me to continue contributing to the *Flying Saucer Review*.[3] The third letter comes from Jim McDonald, who is wasting my time.

We will take the cottage in Saint-Germain. Janine has told me we will have another child in a few months. She likes this place with the narrow green shutters and the walled garden, where Olivier and the baby will be able to play safely. The forest is two blocks away, at the end of the street.

Paris. Rue de la Mésange. Friday 27 October 1967.

After all those visits to pompous bureaucrats, the Armaments guy was dynamic and realistic, a refreshing surprise. I told him that my visit was only made out of courtesy, since I had already decided to accept another position. He said I should keep in mind the possibility of working with them if I ever wanted to pursue self-organizing systems and artificial intelligence in data-bases.

I went to rue de Berri, near the Arc de Triomphe, to tell the managers at Shell that I would take the job they offered me in their computer data-base project.

Hynek writes to me that he has travelled to Columbus to see Pentacle and his team at Battelle. But it seems that instead of confronting him, he meekly inquired about the contents of the old letter, and naturally he was rebuffed. He says it himself: "I don't know any more now than I did before." And he adds:

> There are, very simply, two explanations, and I could defend both of them. Of course I didn't take the famous letter, but I quoted enough from memory to get them very worried. They insisted on the fact that their cards had been destroyed and there was no listing, and all that had taken place with the approval of ATIC.

Technically Allen is right, of course. The letter could have simply meant that Battelle didn't want to come before the Robertson meeting until all the facts were in. The Air Force was in a hurry to get rid of the problem, didn't give a damn about scientific truth, and went ahead with the panel. But it is more likely that there was indeed another project, a truly secret one, and that the panel served as a smokescreen, to keep civilian scientists away from a serious study of the observations.

Paris. Rue de la Mésange. Sunday 29 October 1967.

I have found Aimé Michel rather dogmatic, attached to old ideas. He has not developed their implications further:

"We are witnessing the first example of contact between two societies in which the inferior group, namely mankind, will not have died as a result," he says. "This contact with the extraterrestrials will become open when all of us on earth believe in Them. This is why they show themselves a little bit at a time: it is up to us to break down scientific and official opposition to their reality."

He is in danger of becoming an evangelist for the Saucer cause rather than an analyst trying to find the truth.

"How do we know that the true purpose of the 1897 wave wasn't to be rediscovered in 1965," he asked, "just like the laws given to Moses on Mount Sinai may have been designed precisely to make the Jews unable to become assimilated into other cultures, a fact that has given us Karl Marx, Sigmund Freud, Albert Einstein?"

I argued that we could only gain greater insight by listening more closely to what the witnesses are telling us. I reminded him of the Paracelsus piece I had read to him last Thursday. What about all the forgotten accounts of Little People, of Elementals, of Leprechauns? If these beings are part of the same phenomenon we see now, what does that mean for their nature? Are we necessarily dealing with extraterrestrials? And isn't it premature to assume, as he does, that we are the inferior culture?

Paris. Rue de la Mésange. Tuesday 31 October 1967.

We took Olivier to Saint-Germain to show him the little house and to sign the rental agreement. I wish I could earn enough money to buy such a place, to give Janine and my son the security of a shelter of our own, the assurance of a little warmth. I would also like to be able to buy more books, to have more time to write. But this is a good beginning, a new adventure. Living it through Olivier's fresh eyes makes it even more exciting.

Some evenings, when I have just closed some book like Anatole France's *Life of Henry Brulard*, the spirit who watches over dreamers like me sits down next to my bed and says:

"Will you always demand new topics? I had already turned your head with forbidden science, with paralyzing visitors and cosmic enigmas. You have gone beyond them. Now it is elves you need, dark forests, a whole landscape of sulphurous witchery. You seek occult destinies, you want to know the meaning of ancient archetypes.... Your old boss Paul Muller was right, my friend, you do think too much."

I feel like an Adept on some mystical retreat, ready to cut off relations with the outside world, like the Rosicrucians of old who were required to retire from public view periodically, for years at a time. To be sociable I really should answer Bowen, Fontes and Hynek. But I am not inclined to do it at the moment. This retreating movement is a deep imperative in the alchemy of my existence. I feel that here in Saint-Germain I am renewing a very old link. To live consciously, as someone who is spiritually awake, someone who chooses to direct his life instead of just experiencing it as a series of external events: such is my ambition.

Paris. Rue de la Mésange. Monday 6 November 1967.

This is my third day of work, creating a version of the Infol language that will run on the mainframes of Shell. The company has two large computers in the basement of its majestic building on rue de Berri, both of them expensive Univac computers. The large room is quite a sight, with its long rows of tape drives blinking, the operators efficiently gliding between the consoles and the assistants wheeling out stacks of reports as soon as they come out of the high-speed printers. All the data-bases are manipulated by hundreds of programs written in Cobol, the traditional business language. There is no facility for direct interrogation of any of these items, and every change requires weeks of planning, months of implementation. A high-level language such as Infol could make it possible to get information in and out quickly, to establish correlations, to keep client lists, contracts, marketing data, economic tables up-to-date with much less effort. But Infol has been written for another type of computer. So my first task is to rewrite the compiler.

I have written to Ben Mittman on the subject of software and to Tikhonov on the subject of UFOs. Hynek is finishing his article for *Playboy*, where he will make the dubious claim that the Russians are ahead of the United States in UFO research. He thinks this is a particularly clever ploy: is there a "UFO gap" like there is a "bomber gap" or a "missile gap?" he asks. I think the people he is trying to impress in Washington will simply shrug. The despots in Moscow will frown, and they may clamp down on what little research was beginning to flourish around Kazantsev and Zigel. I have warned Tikhonov that they should not expect too much from the Condon committee. I did not elaborate. He is smart enough to read between the lines.

Coming out of my alchemical silence, I also wrote to Hynek on the Pentacle letter, pointing out that "it throws into doubt all the statements made afterwards (by the Air Force), all the reports from official sources. . . ." And again I bring up the topic of the missing photographs. Why is the U.S. Government hiding these records that Allen himself has never seen, yet are carried in the Blue Book case index? Where the hell are they kept?

Paris. Rue de la Mésange. Sunday 12 November 1967.

We are getting ready to move to Saint-Germain. I have this strange feeling, that I have died and that another "me" is getting ready to slip in from the icy depths of the galaxy, to take the place of the being who had so religiously kept this Journal until now. In Van Vogt's haunting novel *The Adventures of Null-A*, successive layers of being are revealed as the main character keeps dying, only to be replaced by more sophisticated versions of himself kept in faraway sarcophagi under the guardianship of wise men. That is the way I feel, except that I have not met any wise men in a long, long time.

Berthold Brecht has said: "He who knows the truth and doesn't proclaim it is a murderer." But what can be said, then, of the man who knows the truth and proclaims it? Isn't he a greater criminal? Or a greater fool?

Saint-Germain. Monday 27 November 1967.

The workers leave the office at 5:30 p.m. I'm hungry as I go down to the wet streets. At Saint-Lazare station I swallow a big cup of coffee at the counter, with a crusty *croissant.* There are tickets to be torn off from the weekly travel card. Every day, in order for me to go from our house to the station, I must give two tickets to the bus driver. For the train, another little rectangle. To eat some sauteed veal with noodles at the Shell cafeteria, eight little coupons. What kind of life is this, a life measured in little pieces of cardboard, discarded one at a time?

If there is a "real" reality, it lies elsewhere. Not in Moscow, Paris or Chicago, not on this poor planet. It lies in a place where there are no coupons, no tickets, a place where people can think anything they like.

Saint-Germain. Tuesday 28 November 1967.

The company draws everything it can out of us, day after useless day. Yet we accomplish nothing. I feel very tired. Today we had a formal meeting about corporate files. It was a ludicrous waste of energy. It reminded me of the days when I was computing satellite orbits in Meudon, and when I spent long periods just listening to the dying drum before walking over to restart the dynamo, cursing the whole system. Like the Meudon astronomers, the managers here simply don't believe that computers can manipulate their data in new ways.

For example, a well-educated man I will call Martin is in charge of the file of service stations. He is quite pleased with his way of doing things: if he knows the access number of a particular service station, the machine will regurgitate everything there is to know about it, including the owner's name and the color of the geraniums in front of the pumps. But if all he knows is that thc owner's name was Dupont in 1965, he has to call two secretaries into the office, give them sharp pencils, and let them read the entire printout of the file until they find this Dupont. So why can't the computer itself look for D-U-P-O-N-T, I ask?

"Ah, but *Non, Monsieur Vallee*, that would be a name search, don't you see, and that would be, how shall we say, *linguistic....* Computers are for numerical operations, and everybody knows that they cannot think."

There follow a few citations from Kierkegaard and some allusions to the concept of the Self in Sartre, and Monsieur Martin rests his case. Yet I never said the Univac in our basement could think, I only said that I could make it look for a name just as easily as it could search for a number, but nobody will listen, it's too scary. These people only know about advanced computer techniques through the garbled accounts of American research they read in fake intellectual journals like *Le Monde* or *L'Express*. They mix up recent reports on artificial intelligence, on automatic language translation, on software trends.

I hate those long draining meetings where executives parade and posture, resolving nothing! All the real decisions are made informally in the hallways, between two doors. The director who hired me for this department, a gifted man who had a clear vision of the inadequacies of current computer techniques, is about to be gently pushed out. The concept of a modern data-base system for this company is too revolutionary, it threatens too many ideas.

In previous years I always awaited winter with such joy! Now I do not feel any enthusiasm. On the contrary, I am apprehensive about the drizzle and the fog of every new day, the push-and-shove of every trip in the crowded bus, the crowded train, the crowded metro. There is so little for me to do when I finally get to work in that magnificent office building near the Arc de Triomphe!

Saint-Germain. Tuesday 5 December 1967.

I have not been completely forgotten by my friends after all. Some strange noises still filter into my retreat from the outside world. A man named Andrew Tomas who founded the earliest civilian UFO research group in Australia is now in Paris following a two-month stay in Moscow, where he was doing library research. In our phone conversation he presses me to meet with him, saying that he has a message for me from Kazantsev. Tomas is hinting that the Moscow group may soon be disbanding.

Allen has sent me his *Playboy* article.[4] The text gives an impression of confusion both in his goals and in his methods. He keeps apologizing needlessly. And this unholy association of Hynek's lofty astronomical arguments with the magazine's glossy pictures of luscious wenches in various states of sexual availability is not going to help his cause with his conservative colleagues back at Harvard.

Saint-Germain. Wednesday 13 December 1967.

These are bizarre times. The joy of Paris is gone. The men and women we see in the streets are bitter, unhappy. We work hard on our little house: painting, replacing broken window panes. Janine is sewing. I have glued a large map of the Paris area over one wall of the little room that serves as my office. I am cutting and nailing shelves, re-aligning my precious books, playing with Olivier, building castles and extravagant cars out of Lego blocks.

The social situation is getting tense in Paris. Electrical and gas workers are on strike. My respect for Shell as a company is growing, however. They run a very mature, intelligent organization, except in the computer field, where the technology eludes them completely. I continue to implement software among the indifference of my managers, who have no intention of ever using these new tools.

I am also concerned about the apparent change in Aimé, who five years ago was full of ideas for research, hypotheses to check, studies to conduct. Today he refuses to admit that we are dealing with a problem whose nature we simply do not understand. He seems to be hoping for the ultimate landing, a grandiose and apocalyptic manifestation that will prove to the world that he was right. Does he expect the Cosmos to disgorge a whole tribe of Martians just to vindicate us?

Saint-Germain. Saturday 16 December 1967.

Aimé Michel wants to publish an article in *Planète* to expose the Condon committee. Such an article would be counter-productive, I argued. Instead we should give the public a positive outlook on the question, and some tangible facts. He complained about the deviousness of the United States.

"Why is the U.S. always blamed for everything?" I asked him. "I don't even believe the French Air Force is telling the truth about UFO research here."

"Well, we have Colonel Clérouin," he insisted.

"A puppet. Has he ever told you anything reliable?"

He seemed annoyed. "Who has the control of the files in France, then, in your opinion?" he asked.

"Probably an organization that reports directly to De Gaulle."

He whistles: "Those are not people you can approach just like that. I ought to talk to Bergier. Perhaps he will have an idea."

He went home, taking with him a bunch of notes about the Prague astronomy meeting, the Condon committee and the shadowy role played by Bob Low.

Saint-Germain. Sunday 17 December 1967.

I went to Neuilly today to visit Andrew Tomas, an Australian researcher of Russian origin who is in Paris for several months, finishing a manuscript.[5] It turns out that he was in Moscow when my article with Kazantsev was published and when the Stoliarov committee was created. Tikhonov and Kazantsev have asked him to brief me confidentially on the true state of affairs, but they don't want me to repeat anything to Hynek, whose credibility in their eyes has been shot down by his *Playboy* article about a UFO gap. This puts me in an awkward position.

As I already knew, Kazantsev and Zigel have created an "Org-Komitet," an organizational committee, with Tikhonov as secretary and Stoliarov as President. This committee was under the House of Astronautics and Aeronautics, a structure made up of retired pilots, young people and aerospace enthusiasts, not professional engineers or "real" cosmonauts.

The Stoliarov committee had an office in one of the Kremlin buildings. It only needed two signatures to become official. One of these signatures had to be given by the Air Minister, so Kazantsev and Zigel duly presented themselves before him and explained their purpose.

"Very interesting," said the official. "How can I help your research?"

"One thing that would be useful," proposed Zigel, pressing his luck, "would be to ask your men if they have seen anything in the air that hasn't been reported."

"Excellent idea, Comrades! It's such a simple thing."

The Minister pushed a button and a staff officer came in, snapping to attention.

"Order to all Air bases in the Union of Socialist Soviet Republics," the Minister started to dictate. "Report at once any observations of unidentified flying objects made by your personnel." Zigel and Kazantsev left the Kremlin elated, secure in the promise that they would be kept informed.

Forty-eight hours later their committee had ceased to exist. The Minister had been swamped with fifteen thousand observations, making the officials realize just how many sightings have happened. Military secrecy fell over the whole business.

Andrew Tomas told me all that in English, as we sipped the excellent coffee prepared by his landlady and as we looked over a small frozen courtyard where a pigeon alighted from time to time, looking for bread crumbs. He added some details about the impact of the publication of my article with Kazantsev:

"Do you realize what happened when it was picked up by *Trud*, the labor union newspaper? *Trud* has a circulation of twenty-two million daily copies. Word quickly got around that it included an article about UFOs, and that issue disappeared from the streets in a very short time."

He also told me a long, intricate story of a London woman who is said to have intercepted a Morse message between an extraterrestrial and a British physicist. She went secretly to the appointment, saw the whole thing and revealed her presence. The rest of it is an undocumented tale with the usual garbage about Aliens who come from very far away and learn our language while they sleep during their long trip. It happened in 1962 and the woman's name was Mrs. Marchwood. I listened politely to all this, took my leave and went to catch the bus that would bring me back to my computer.

Saint-Germain. Christmas night 1967.

Ten years ago I was beginning this Diary. My father was living his last days. Only a vague presence of him is still left: a certain strong tenderness I find in nature around me reminds me of him. Other beings have come and blessed my life: sweet Janine and Olivier. As I write I can hear my son in the next room, telling long stories to his teddy bear.

We are little people, with a small rented house, a little garden. We do not even own a car, or a telephone. We keep our secrets well. Nobody notices our happiness.

This afternoon I took a long walk alone towards the Marly forest. In Fourqueux I climbed all the way up to the windmill. From there I could see the whole of Paris and its suburbs. The big heavy bulk of the black clouds was rushing across the gray sky. The tepid wind blew wet on thick crusty layers of dead leaves, the beaten earth, the trembling clusters of mistletoe in tall black trees. I could imagine myself in a few years, exploring this little corner of the planet with my children, taking them for Sunday walks, drawing charts for them.

Yet all of a sudden something snapped and rebelled within me. I faced the harsh wind that pulled tears from my eyes. Have I not plotted out the map of Mars for Project Mariner? Here I am, planning to become comfortably buried in the dreary life of a French bourgeois, drawing little charts of some sleepy suburb: the creek that flows towards Chambourcy, the last little sliver of woods near Saint-Nom-la-Bretèche.... Have I forgotten Syrtis Major and Sinus Meridiani? I must choose: in which time, on which scale do I want to live? Back in the United States, Saturn rockets are climbing straight up in the sky. And here I am, wondering if I will ever own a little cottage of my own some day, how I can save enough money to buy a car, in two or three years....

So many people have asked before me, "Who am I?" I can only answer "I am you" when I see the elderly man who stopped me in the street to ask me which day of the week it was, the tired worker, the learned bishop, they all have a brother within me. From this crowd I seem to carry inside myself rises a simple, corny, inebriating feeling of love for the whole planet.

On weekday evenings, when I come back from Paris, the blue bus from the station stops on the square, and I get out, hungry and frozen with fatigue. A few office workers get out at the same time as I do. The bus leaves us there. On the little square I am worthless, harmless, alone.

Saint-Germain. Sunday 31 December 1967.

The present should not be tolerated. In the tiny home office I have built in our spare bedroom, I have gathered a whole world of maps and books, and fanciful toy machines with blinking colored lights. They encourage me to go on with my work in the gray morning.

I am reading Camille Flammarion's *Popular Astronomy* for 1881:

> At last we have reached the scientific era that every friend of progress was wishing for . . . the shadows of the night are vanishing gradually. Clarity comes into the souls of men. This is a manifest, eloquent, indisputable sign of the current state of minds and of their aspiration towards true science, positive science; towards true philosophy, scientific philosophy. It is sweet indeed to see such noble tendencies win over our great human family as it makes its slow progression.

Perhaps Flammarion's naively hopeful words should be contrasted with a more recent text by Jacques Bergier which appeared in *Planète*:[6]

> Our modern societies, with their foundation on democratic principles and their free circulation of information, may in fact have delegated essential responsibilities to a handful of men whose knowledge and power are kept hidden from the man in the street.
>
> Consider revolutionary warfare, where the leader is a shopowner or a farm worker; consider science and technology, where language has become an esoteric jargon, where those who have the true keys of civilization remain strictly anonymous; consider the field of propaganda, where it is impossible to locate those who are plotting new collective psychoses. And let us not even talk about the police . . . the very nature of things is drawing us towards a *cryptocratic* State.

Poor Camille Flammarion! His view of progress could not have anticipated the White Storks of classified science! Tonight the souls of man are as dark as the long shameful guns which have briefly interrupted their thunder in Vietnam: New Year's Eve. In our little house near the forest we will sleep quietly, the three of us, soon to be four in number. No great feast for us. We need no wine, no friends tonight. Only the intimacy of our love.

Saint-Germain. Thursday 18 January 1968.

"What is the role of the Delegation for Informatique?" I asked a member of this exalted body which oversees Plan Calcul. By meeting with people like him I am keeping my network of contacts alive in case I decide to change jobs.

"We report directly to the Prime Minister," he began with a competent tone that made tricolor flags wave in my mind in anticipation of great deeds. "This gives us authority..." he smiled ironically before continuing: "...to write to the other departments to exercise certain pressures. You see, the stationery of the Prime Minister always makes a big impression on other people. So we use it to influence their decisions when they are about to buy a computer. We try to make sure it is a French machine."

I cut short our meeting and I walked away with disappointment bordering on disgust. What they call French machines are nothing but refurbished designs from the United States on which a nationalized company, the old Machines Bull, does little more than paste a blue, white and red label. There are many smart computer people in France. With some decent management behind them, European hardware experts could put together a world-class computer. Instead, taxpayers' money is being used to finance the marketing mistakes of State-controlled companies whose history has been marked by a series of failures that would have destroyed any normal business long ago.

It so happens that I currently share my office with one of those excellent French specialists. I call him Mouse-Face behind his back because he is slight and sly, with a cynical sense of humor. A few months ago he was developing software at Bull. They had planned an ambitious line of machines designed to compete against IBM at lower prices. After years of waste and politicking they have accomplished little more than producing computers that have lower performance, higher costs and poorer software. Time-sharing, which is the wave of the future in computer applications, is not supported by Bull while IBM is already introducing it on its newest machine, the 360 model 67. Who will buy the Bull product? Five or six administrations, unless they have enough clout to resist the pressures that are placed on them by the Delegation for Informatique writing threatening letters on Prime Minister stationery.

A letter from Hynek says he is thinking of coming to Paris in March, when he will be travelling to yet another astronomy meeting in Prague. He may bring Fred Beckman.

Saint-Germain. Monday 22 January 1968.

Aimé Michel came to visit me in Paris this afternoon. We went to a nearby bistro. I have given him Kazantsev's address, and they have started an exciting correspondence. Now Aimé has written to one of his contacts that he is now in touch with both the U.S. and the U.S.S.R. on the UFO question, and that this "puts him in an awkward situation as a French citizen." Hence he would like to disclose this to a higher governmental level.

It is a childish tactic—the French call it *faire l'âne pour avoir du son*, or "pretend you are an ass in order to get some rye." But given the current state of affairs, it just might work and get him in contact with some of the French Government agencies that have not yet tipped their hand. Indeed his friend has answered: "I can put you in touch with the SDECE," adding cryptically, "That might enable you to expand your files of sightings."[7] Does that mean that the French secret service, the SDECE, is maintaining its own UFO files? An officer is supposed to call him soon.

Saint-Germain. Friday 26 January 1968.

"My name is Commander Granger," said the man as he shook Aimé Michel's hand yesterday. He was young, blond-haired, with an open face, "the kind of fellow to whom you feel like telling your whole life story." Well-trained: he took no notes.

"If I understand you right, you belong to a small group that is in an interesting position because it has contacts both with the Americans and with the Russians."

"Strictly on a personal basis, I should add," said Aimé Michel, "since we know very well that in France nobody takes this topic seriously."

"You are wrong to think such a thing!" said Granger. "There are some high-level people who follow the issue. Myself, I have always suspected there was a link between UFOs and military reconnaissance satellites."

Aimé Michel gave him his most astonished look: "Satellites? In 1947?"

"Well no . . . obviously, not back in 1947."

"Frankly, we have always hesitated to talk to French authorities, because we felt that other nations would find out about such contacts right away."

Granger was piqued: "Well, Monsieur," he said with dignity, "it is not for other nations that I work."

"And if you did, would you tell me?" asked Aimé with a smile.

He then told Granger about the precarious situation of Project Blue Book (a situation every ufologist already knows) and the predictable Condon committee fiasco. He was struck by Granger's lack of knowledge of the simple administrative channels followed by UFO reports not only in the U.S. but in France. For example he thought that the Gendarmerie had orders to report every sighting. His real interest was in the area of satellites, especially infrared satellites. He didn't care about UFOs at all, concluded Aimé, who had nothing to tell him about satellites, so he brought the conversation back to his own interests:

"Back in 1953 the U.S. Government gathered together five eggheads who leafed quickly through a few sighting reports during two afternoons and concluded there was nothing to the whole business. This conclusion was immediately accepted by the Air Force without any discussion. Yet the military intelligence services of the same U.S. Government had been studying the same reports for five years without finding any explanation. Now, you're in that business, *mon Commandant:* doesn't this seem a little bit odd to you?"

Granger just laughed.

"Anyway, we are not asking anything of you," concluded Aimé. "We only want to be sure that we are not acting stupidly in our correspondence with the Russians. Your presence here is a guarantee that we have fulfilled our responsibility as citizens, by informing you."

"What would you like for the future?"

"Again, we are not asking for anything. And we have nothing to offer, unfortunately, since France is not interested in the field. . . ."

"Well, if both the Americans and the Russians are doing something, we've got to do something too, of course."

"Some of the best specialists in the field are people like Plantier, Guérin, Garreau, Vallee and myself, who are French, but we work without any official recognition or support. We see three possible scenarios.

First, you could find me a job with some Ministry where I would have an administrative cover to spend some serious time on the problem. Second, you could help us formulate some guidelines that the Government could implement to improve information collection, such as an order for the Gendarmerie to report any local sighting known to them. Third possibility, a small group of scientists could be discreetly created to continue research."

Granger promised to report on this conversation and to let Aimé Michel know through Rocard if his superiors felt the contact should be pursued.

Saint-Germain. Saturday 17 February 1968.

As I return from a three-day management training seminar organized by Shell in Rambouillet, Janine shows me a surprising number of items that have arrived in the mail: a complimentary copy of Ted Bloecher's interesting book about the 1947 wave, dozens of letters (a physicist wants some details from recent observations for an article he is sending to *Science*, another man proposes the theory that UFOs are linked to geological fault lines) and two UFO magazines from America. Why are people so slow in forgetting me?

To dispel my feeling of distress I often walk through Paris in the evening, pausing to observe and to absorb. Yesterday in the gardens behind Notre-Dame I followed the course of the whitish clouds racing along the rugged stone roofs in the cold night. Last week, at the Beaubourg plateau, I rested my back against an old wall where posters torn from the decaying plaster were beating in the wind. At the edge of tears I saw the moon racing; its reflections on the uneven cobblestone answered my despair: what am I doing here? The promoters who are about to tear down Les Halles promise to build fine gardens for us some day; we will have beautiful walkways with unbreakable plastic steps, multi-level cultural centers made of concrete slabs. We race away, the moon and I.

Saint-Germain. Thursday 22 February 1968.

In the middle of the afternoon I came home, skipping the second part of an "information seminar" organized by a consulting group on the flimsy topic of decision tables. Shell had delegated me to this assembly of friendly young technocrats. They gave us plastic briefcases.

I now realize with a shock that, if the French lag behind the Americans, it is not because we are poorer or less well equipped. Many leading French companies would be regarded as large firms on the American scale. My employer, for example, owns late-model, multi-million-dollar computers, housed in a magnificent facility two blocks away from the Arc de Triomphe, in some of the most expensive real estate in the world. We have a large systems staff that would do honor to any U.S. enterprise. The gap is neither financial nor technical. The gap is a mental and cultural one. Two months ago I would have fiercely denied this. But all the presentations I heard today were based on timid interpolations of half-understood American models, and speculation about American developments—like Monsieur Martin's refusal to let our computers perform "linguistic" searches.

In the newspapers are pictures of the destruction of the Hué Citadel in Vietnam by Lyndon Johnson's guns. That image is in every mind: the sepulchres of venerated medieval rulers blown up by U.S. bombs; ancient cultures razed by the blind and pitiless advance of that American civilization which is condemned to destroy the past, that American civilization to which we belong, even if it is fashionable among French intellectuals to deny any connection with it.

Every morning when we wake up and turn on the radio we find that the world in which we try to live has been covered with a little more blood.

In Chicago, no matter how dull things were, I had the feeling that I was making a contribution that mattered, that others would work better and faster thanks to my own efforts, that their lives would change. Here I have one of the most desirable technical jobs available in France, but I am not helping others make progress. This morning I heard a manager make a little speech in which he thundered against those Frenchmen who refused to work below their level. "They demand an attractive job, as if jobs should be attractive!" Everybody would be happier, he argued, if they simply continued to labor at those things they were trained for.

Paris. Thursday 7 March 1968.

This morning I saw Aimé Michel again. One of his friends has heard that the French Air Force was about to set up a real UFO investigation group.

"You ought to be able to check this with Clérouin," I told him.

"It may just be a trial balloon. Did I tell you what I heard? The French military did a counter-investigation of the Socorro landing: They concluded the object was a postal plane, a mail-carrier!"

"We are wasting our time with those turkeys. They swallow anything the Pentagon is telling them. The Russians are not doing any better, you know. I have just heard from Kazantsev. Maybe he wrote to you too? A committee within the Soviet Academy of Sciences has met to discuss the issue. They have concluded that the anti-science propaganda that surrounded the problem was aimed at sowing trouble among the proletariat."

Saint-Germain. Saturday 9 March 1968.

Great news: Aimé Michel now confirms the fact that a research committee is about to be created by the French Air Force. The order has come down from none other than General Ailleret, who is De Gaulle's Chief of Staff. In addition, Granger has sent Aimé a note to set up another meeting in Paris, "as long as his presence was not interpreted as official approval."

Saint-Germain. Sunday 10 March 1968.

General Ailleret is dead. His aircraft crashed five minutes after it took off from the island of Réunion. Nothing is known about the circumstances. A nurse who was on board is said to be the only survivor.[8]

Saint-Germain. Monday 11 March 1968.

A telegram has arrived from Chicago: Fred Beckman will be here on Wednesday morning along with Hynek. Andrew Tomas writes to me that Tikhonov and Kazantsev miss my letters, "but now that the news is bad perhaps it is better to remain silent." He adds:

> I have before me the 29 February issue of *Pravda.* Three members of the Academy of Sciences condemn UFOs, following a recent report against "saucer propaganda" by Artsimovitch. They say there isn't a single fact that proves their reality. They quote Menzel, and they claim that U.S. scientists are unanimous in saying the subject is pure hogwash.

Tomas points out that in any other country such an article would not be cause for concern, but in the Soviet Union it means that our friends had better be careful.

Saint-Germain. Wednesday 13 March 1968.

Hynek and Beckman's flight from Chicago was four hours late. I only saw Allen between two planes, on his way to the meeting in Prague, but he promised to stay longer on the way back. Fred remains in Paris. Maman has agreed to give him the guest bedroom in her apartment, since he doesn't have much money to spend on a hotel. Janine and I have spoken with him all evening. We came away with the feeling that no progress had been made in Chicago since we left.

Saint-Germain. Friday 15 March 1968.

Yesterday morning Fred and I met with Andrew Tomas. Afterwards we walked over to an exhibition of naive American art. Paris is gray and gloomy, the news of the world is worrisome: financial markets have been in turmoil for the last three weeks, the monetary system seems to be crumbling, the London gold market is closed.

Saint-Germain. Sunday 17 March 1968.

Hynek arrived at noon, all excited to be in Paris again. He wanted to live "like a native," so I took him to a small hotel near Place Monge. While he was getting settled there I had lunch with Maman and Fred, and I went out again an hour later to pick up Hynek. Upon returning we witnessed this wonderful scene: Fred still at the table, seated in front of an assortment of cheeses, encircled by five old bottles, his lips moist with Armagnac and his eyes glistening with pleasure.

Hynek told us about the turmoil he had just witnessed in Prague, the extraordinary changes taking place in Czechoslovakia, the hopes for reform. With visible regret Fred finally rose from the table. I took them both to show them the ancient Roman Arènes, and later we went to see the Gauquelins a few streets away. Allen, who had courageously seconded my recommendation to bring their book to the American public, stayed with them to discuss the "Mars Effect."[9]

Saint-Germain. Monday 18 March 1968.

Hynek acknowledges he feels an emotional need to get even with Menzel, Condon, Klass and the rest of the skeptics. His colleagues' attitude towards him is changing to the point of contempt, and this pains him. He is not taken seriously among astronomers any more. Perhaps that is more a reaction to his style, his brash articles in magazines like *Playboy*, than to his ideas. But does he really have a choice, given the fact that scientific journals are refusing to print anything, anything at all on the subject? Even our own *Challenge to Science*, which is a dry and factual scientific monograph, got no review in the academic press.

At the recent ceremony for the tenth anniversary of *Sputnik* they could not avoid inviting him to speak, but the organizers took him aside and told him to be sure not to mention UFOs. He felt so humiliated that he almost stalked out of the room. He only stayed there out of consideration for his wife and his former staff members, who had come to see him again. But he had a heavy heart.

He also told me about his meeting with Pentacle. He did not see him alone, as I had recommended, but in the presence of four of his Battelle colleagues. When he started reading from his notes, Pentacle snatched the paper out of his hand and said it was an old story, it was all over and forgotten. Pentacle got rid of him as fast as possible, but he did not return his notes. Always fearful of confrontations, Hynek left Battelle with his tail between his legs. Such a violent reaction may in fact indicate that something important is going on. Why should Pentacle worry so much about a simple letter written fifteen years ago?

I took Allen to Les Halles for supper. He was terrified when he found himself among the crowded little streets, in the midst of so many odd and suspicious characters, and he remained nervous in spite of my attempts to explain what made the area so unique and so wonderful. I argued that poverty should not necessarily be equated with crime or squalor, that the poorer areas of Paris could have more dignity and shine with a more special light than the grandest of palaces.

After I converted him to my views, with much quotation from Victor Hugo, I had to confess to him that the newest crop of French technocrats was about to raze this wonderful den of decadence to build a cultural center and a shopping area out of plastic and aluminum. At least, after Les

Halles are gone, I will be able to keep the memory of Allen Hynek walking with me through these dark and convoluted passages from another age. I told him of the esoteric significance of the Church of Saint-Merri, the alchemical history of the Square of the Innocents and the gargoyles of the Tour Saint-Jacques which look down on the Great Work of Nicolas Flamel. Our conversations continued as we passed among the shops, the bars and the haunts where the prettiest prostitutes in Paris used to pirouette on the sidewalks. All that has changed. To the laughter of Paris, prudish Madame De Gaulle ("Auntie Yvonne") has decided a few months ago that Catholic morality demanded the "cleaning up" of this traditional garden of indulgence. Acting on behalf of every pious and pure person in France, Madame la Générale has made strict new rules to keep *Les Girls* in their hotels, out of the streets.

We had dinner in a little café where we sat next to truckdrivers and elegant aristocrats coming out of the theater.

Saint-Germain. Wednesday 20 March 1968.

Hynek went back to see the Gauquelins to discuss astrology and destiny while I had a long talk with Fred. I told him frankly I saw little hope for the future of UFO research. Perhaps we should write a theoretical piece, Fred and I, with his friends from Argonne Laboratory, to outline what we think should be done. But in trying to publish it we would surely run into the Condon committee brick wall.

Saint-Germain. Thursday 21 March 1968.

A wasted day. We met in the morning. Hynek has spent most of last night visiting every tourist trap in Paris with Mrs. Ackerman and he is very tired. Fred is not faring much better. They both seem exhausted and ill.

There are mounting social tensions around us. Yesterday rioting students broke the windows of American Express on rue Scribe. Peace talks between the U.S. and North Vietnam are supposed to start soon in Paris. Vaguely discussing these developments, we ordered some chocolate and Hynek started writing a letter. He was so distracted that we could not get a single meaningful word out of him. He only told us that tomorrow he had to make sure to get to Orly early enough to buy a few toys and some duty-free booze. I became discouraged and left. There was so much we should have discussed!

Saint-Germain. Saturday 23 March 1968.

They are gone. I felt bitter when I came to meet them at the hotel yesterday. As we waited for the cab to the airport Hynek asked:

"Are you still getting sighting reports from the public?"

"What would I do with them?" I said.

He seemed taken aback. "How do you mean?"

"Well, we don't have any way of studying them or of following up, do we?"

"It's important to keep these reports," he mumbled vaguely, "because it's science."

He must have realized he had just said something silly because he added:

"In the coming year I want to build up a file with a number of cases that are air-tight, something I can throw against the Condon Report."

"Isn't it a little late for that?" I asked. "We have been talking about 'air-tight' cases for years. The Condon Report is going to be a public relations fortress built up by the military, the press people, the scientific establishment, accompanied by a fanfare of editorials in major newspapers and the scientific press. How do you expect to counter all that?"

If he tries to respond to the report with his own arguments, Hynek will be greeted not just with skepticism but with irritation, no matter what good cases he cites as evidence.

Perhaps he wants to correct the bad impression left by his careless attitude yesterday. He now tries to get a serious conversation going, realizing that we only have a few more minutes together, and it may be months or even years before we see each other again. But how could we hold a serious conversation now, when the taxi has been called and could show up at any minute?

"Jacques, you gave me an excellent idea the other day. If I have a chance to answer Condon before Congress, I will say that his report is biased and incomplete because he didn't take the Pentacle episode into account. If they ask me to explain that, I will demand to go into executive session and I'll tell them the truth about the Battelle work."

I shrugged. "Maybe it would work."

We are not making any progress. Hynek is clearly afraid to go back and confront Pentacle. I will be very surprised if he demands to see the

Drury film.

"What should I say if the Air Force tells me they didn't make a copy of it before returning it to the Australians?" he asks.

"Laugh in their face. Demand to see their correspondence with the Royal Australian Air Force."

"And if they say everything has been destroyed?"

"Go back to your UN contacts. Ask them to send an official request to the RAAF to get the original. It's worth trying."

"You have an answer for everything."

"Your cab is here, Allen."

Fred shook my hand warmly, promising that he would "think about the things we discussed."

I was sad to see them drive away. During this visit I have had too few glimpses of the other Hynek, the profound and clever mind who appreciates the limitations and the biases of science. Before leaving he did remind me of the commitment I made in Evanston: that I would drop everything, return to the United States and be his lieutenant if he ever managed to obtain support for a "real" research effort on this subject. I reassured him on that point. Yes, I would gladly drop everything, set aside my computer science career, go anywhere to work on such a project on a full-time basis. But it would have to be for real. Not another Blue Book, not another Condon circus, not some academic pipe dream.

Allen is always surrounded with rich acquaintances attracted by his celebrity, his media charisma. They keep hinting they might give some funds for a private effort some day, but real support never materializes. These are wealthy people who want to be entertained but who understand little about the constraints and the hardships of real science work: idle widows who like to parade with a highly visible professor in their entourage, or industrialists with their own pet theory of extraterrestrial life who want someone to confirm their private fantasies. But Hynek is the eternal optimist, and his personal magnetism may eventually motivate some of these contacts to fund a real effort. So far, however, nothing has happened. No wise man has offered support, no secret agency has emerged from the shadows, no tycoon of industry has flown to Chicago with his key advisers to help us assemble a team that could crack this mystery.

Saint-Germain. Sunday 24 March 1968.

We woke up in a gorgeous new world today. I am sorry Hynek missed it. All he saw of Paris was grayness, haze, gloom. The French Spring suddenly arrived last night, putting a sweet and heady fragrance in the air, golden sunshine everywhere. At the end of the day a brief, gentle rain washed the earth, creating fresh smells, a breeze like the breath of newborn angels.

Olivier left for Caen today with my mother. We came back from the station, Janine and I, mystically close, carried through the streets by the same thoughts, the same love. The sun sweeps the forest where leaves are sprouting, and when we lie in bed it comes indiscreetly through the white drapes with the pink medallions and kisses us. A hush hangs over us, the expectation of new life about to appear, the treasured time of birth. Mine is such a secondary part in this sacred process. . . . I envy Janine, who is initiated to such mysteries.

Saint-Germain. Thursday 28 March 1968.

We are faced with a technical dilemma in our use of computers. The pressures of business are forcing our company to install remote terminals in the South of the country to allow our regional offices to reach the central mainframe to update financial and marketing files in real time. Accordingly the firm bought some American equipment, but the French Government exerted its usual arm-twisting in the name of Plan Calcul, and demanded that we install an equal number of "French" terminals, a thinly disguised subsidy to various local firms whose presidents probably went to the same Grandes Ecoles as the Minister. We had to do it, because they blackmailed us by hinting they would deny us the telephone circuits we required. The result is that we now carry twice as much equipment as we should, at twice what it would cost a U.S. firm to do the same job. And the French terminals keep giving us trouble: many of them don't work at all and two of them have actually caught fire. I must be missing something. In what way does this help the French industry?

Saint-Germain. Friday 29 March 1968.

Aimé Michel has sent me a confusing letter, full of news and of ideas. He has initiated a number of experiments with a woman clairvoyant who claims to be able to "read" the personality of individuals from a simple photograph in a sealed envelope. She has already produced a psychic profile of Allen, but it is so general it could apply to any man with a beard.

His plan is to continue these sessions, including some standard targets to calibrate the medium herself, and eventually to expand these psychic "readings" to the UFO pilots themselves, whoever they are. It's an interesting project.

Saint-Germain. Wednesday 3 April 1968.

Yesterday I was supposed to attend a seminar about a new programming language. I left the office early, intending to browse through some of the occult bookstores in the little streets around Odéon before going to the seminar. Near the Seine I found a shop I had never seen before. It was full of fascinating books. The man who owned it said he could locate for me any item in demonology, science fiction and other sulphurous subjects. He was willing to discuss sylphs and salamanders, elves and the ghosts of Flammarion's time.

Afraid that I would buy too many books if I lingered there, I left the shop after a while. I walked along rue Saint-André-des-Arts and paid a surprise visit to my uncle Maurice on rue du Cherche-Midi. Suddenly, talking to him began to appear much more important than listening to a group of consultants pushing some new software scheme.

Every visit to my old uncle is a discovery, a marvel. A self-made man without an advanced education, he took classes in draftsmanship when he luckily came out of World War I with little more than a foot injury, and later he went to dental school, which has enabled him to make a good living. But he is most adept at precision mechanical and electronics assembly. Always building some new device in his three-room apartment, his most recent idea is to analyze his dreams by recording anything he might say while he is asleep.

In his living room, with the window wide open over a courtyard whose walls are draped with ivy, he played some of the tapes for me with their curious disconnected sentences, their words out of context. We also discussed astronomy. He is responsible for my interests in that science, and the telescope I have often used to scan the night sky was a gift from him. He always has something new and unexpected to say, even at seventy-six years of age.

A light rain started falling. The wind was biting as I walked back along the Seine.

Saint-Germain. Thursday 11 April 1968.

I spent the morning solving a tricky computer problem of remote interrogation of data-bases. I found a simple and elegant method to accomplish it. I returned from lunch fairly pleased with myself, with two hours to kill before my new software came back from compilation. My colleagues were only thinking of one thing, and that was the planning for their annual Summer vacation. So the time had come, I decreed, to execute a long-delayed project: visiting the Bibliothèque Nationale to browse through the masses of documents having to do with apparitions in general and Elementals in particular, a literature that presents some striking parallels with modern UFO lore. I have never had the leisure to do serious research on this formidable and complicated topic.

I had never visited the Palais-Royal and the area of the Bourse where Stendahl had lived, where Colette died. At the Nationale I filled out the customary application forms. From there I walked all the way to the Châtelet. Once I had reached the Seine I decided to see Notre-Dame in the tender green of the trees, and from there I started loitering along the Quays, where I found Tyrrell's book on Apparitions. Reaching the Institute on the Left Bank, I cut across the old streets to Odéon, still on foot, finding more bookstores everywhere, and I finally sat at the terrace of a bistro to rest my tired legs and eat something before returning to my computer.

Along the way I kept wondering, what is the point of staying in Saint-Germain? We do not feel that we belong in this snobbish town where neighbor doesn't talk to neighbor. Added to this are the poor services, the need to walk all the way to the post office in order to use the phone, the bottles of butane gas to be dragged home for the antiquated heating system.... We had not noticed any of this when we moved in, so pleased were we with the forest and the charm of our little doll house.

Research is what I want to do. What I seek above all is new knowledge, new understanding and eventually new wisdom. All day I have been ruminating about these topics, and about the long-term meaning of these computers of ours. They are changing industry and science, but will they ever change Saint-Germain, where every house is surrounded by a high wall?

Saint-Germain. Friday 12 April 1968.

A journalist has asked General Nguyen Van Hinh what the French Air Force was doing to investigate UFOs.

"Nothing about this is secret or even confidential," said the General, who added some interesting administrative information. There is a "Prospective Bureau" within the Air Force where advanced topics are considered, including "numerous documents about UFOs" coming from bases of the four military regions, and "related phenomena reported by civilian or military crews, the Gendarmes, individual citizens and the press." Perhaps it be possible for us to work with them.

Unidentified flying objects appear spontaneously, often out of thin air. Yet they take the form of physical, perfectly tangible machines. Close to them the witnesses often see short humanoids with a large cranium, as well as human beings identical to us. But what about cases when these "entities" appear by themselves, without a craft? What about the cases of 1897, the landings of the Airship in the Midwest? What about the beings seen by honest folk in the days of Paracelsus? The great doctor had confessed his puzzlement before such stories.

Medieval folklore also mentions other reports: beings of light seen in the clearings, at dusk, close to "shining tents." They are rare indeed, because the witnesses were afraid of being accused of witchcraft by the exorcist, just as our modern witnesses are afraid of being accused of "obscurantism" or irrationality. Let's not even mention the cases of Fatima or Ezekiel: what we have here, whether we like it or not, are paranormal apparitions. I am reluctant to come to this conclusion: I have been trying to avoid it all these years. Yet I cannot refuse to consider it indefinitely. I have classified, analyzed, eliminated everything else.

My own sighting of 1955 in Pontoise is still present in my mind. I would be ashamed of the human race if we did not respond in some honest, authentic way. I also feel it is crucial to place the evidence on record, whatever it is, so that people will be informed in case of new developments. Aimé Michel is right when he says that it's not enough to write books. But we disagree on what to do next. He wants to address the public, to hasten the moment of ultimate contact.... I do not follow this line of reasoning. I believe that the skepticism of the bosses of science is irrelevant. They are the same people who, ten years ago, were equally skeptical of electronic computers, of artificial satellites. History has shown they are powerless to stop new ideas. The real problem is that we do not have enough reliable data on the nature of the phenomenon. We have no proof that we are dealing with extraterrestrials in the sense the ufologists imagine.

Yet the so-called "saucers" have too much impact on our culture not to be controlled by some intelligence. It is hard to imagine that it is only the result of some psychic projection coming out of mankind itself, as the psychologists suggest.

Saint-Germain. Tuesday 16 April 1968.

Aimé Michel is back in Paris. I continue to have a deep feeling of friendship and admiration for him. His motivations include a need for vindication in the face of those who deny him the few resources he requires. He is not getting much help from the few powerful friends he does have, even though Rocard is one of the finest minds in French science.[10]

An interesting letter has arrived from Hynek. He is trying to find out what happened to the Drury film after all. He also tells me that John Fuller is about to publish an article in *Look* to expose the Condon committee as a major scandal. Anything can happen now.

Saint-Germain. Tuesday 23 April 1968.

The U.S. Post Office has just forwarded to me a letter from good old Professor Menzel, who remains obsessed with flying saucers. Speaking of Father Gill's remarkable sighting in New Guinea, which I quoted in *Anatomy of a Phenomenon*,[11] he writes:

> Gill was afflicted with myopia, probably a severe case, and failed to have his glasses on. Squinting severely distorted what he saw and, at best, the view was that of an out-of-focus Foucault test of his own eye. To verify my theory, I increased my own myopia by using a circle lens of 0.75 diopters and looked at a distant, bright porch light. As I squinted, my own eyelashes, irregularities in the layer of liquid on the surface, and simple diffraction made a beautiful UFO, with the legs on the bottom and people on the top.

The argument is childish: What about the few dozen other witnesses who were with Father Gill on that day? Were they all myopic too?

Saint-Germain. Thursday 9 May 1968.

I have resumed my correspondence with Don Hanlon, with whom I feel I can communicate on the level of mutual trust. He writes to me:

> The last time I was in Chicago, Allen, Bill and Fred mentioned that you had been rather silent recently, and in fact they thought your activities were suspicious. I expect Allen's visit has clarified all that.

This throws a strange new light on Hynek's visit to Paris: Could it be that he and Fred came to France to find out if our little secrets were still safe? Now I feel sorely disappointed in their lack of trust.

John Fuller has just exposed the Condon fiasco in *Look.* He reveals the evidence of a deliberate bias discovered by Saunders, Levine and Mary-Lou. It is worse than a bias, it is a smoking gun. An official University of Colorado memo by Bob Low, dated August 9th, 1966, discussed the research proposal they were getting ready to submit to the Air Force in the following terms:

> *The trick would be* to describe the project so that, to the public, it would appear a totally objective study but, to the scientific community, it would present the image of a group of nonbelievers trying their best to be objective but having an almost zero expectation of finding a saucer.

Now I have learned how the exposé unfolded. It appears that a meeting took place in Boulder, gathering all the critics of the Condon committee including Hynek and McDonald. But it is only after Hynek's departure that Saunders told McDonald about the memo, which had been discovered by staffer Roy Craig. Jim caught fire, charged ahead and used the memo immediately, in his typical "elephant in a china shop" fashion. This precipitated a crisis: Saunders and Levine were blown and fired from the project "for incompetence."

I find it interesting that the key players did not take Hynek into their confidence. He only learned of the memo through the article in *Look.* A general feeling of mistrust hangs over the whole scene. This is typical of American ufology; why can't these people ever work together? Fred, who perhaps could enlighten me about the situation in Chicago, has not written to us since he came to Paris.

Last Monday I joined an interesting meeting of Shell specialists in The Hague as a representative of *Shell Française*. The gathering was organized to define strategic priorities for the Group in the software arena.

One of the top Dutch managers set the stage by explaining that there was growing international demand for better software within our sister companies. The meeting lasted two days and concluded with a fine dinner at an executive's home in a suburb of The Hague. Our host went around the table to ask each one of us what we felt was the best investment the company could make in the computer field. The unanimous answer was "a generalized data-base management system." In other words, exactly the kind of software I have been developing in Paris. Another scientist in Holland is working along the same lines. Our host looked very stern and there was silence.

Then he stated, "The Group is aware of this requirement. But we are not in the software business. We are in the oil business. All we can do is put pressure on Univac and IBM to develop such software." He is wrong, of course: the computer companies are only interested in selling "iron," not in developing new fancy tools for us.

We bowed our heads, as befitted young and well-educated Europeans when the boss had spoken. But there was a young Texan with us, and he felt no such constraint.

"Well now," he said loudly, "down there in Houston we spend about sixty million bucks a year on programming. I reckon we're in the software business."

I could have kissed the guy.

When I got ready to fly back from Holland I suddenly realized at the airport gate that I had foolishly locked my passport inside my suitcase.

"Well, don't you have some kind of identification?" asked a courteous Dutch officer, sensing my growing panic.

I fumbled in my wallet and pulled out my Shell personnel card. When he saw it he almost snapped to attention and motioned me to board. After all, anybody can get a French passport, but not anybody can be on the international staff of Shell.

Upon my return I found a deep change in the air. There is new turmoil in Paris: the students are restless, skirmishes have erupted with the police. There are random strikes in various industrial sectors. France is sinking into a deep malaise.

17

Saint-Germain. Monday 13 May 1968.

A General Strike has begun in Paris. I was in the cavernous computer room at Shell when the central power went off. Our two mainframes, with their flickering lights on the large consoles, and the rows of tape drives with their green and red panels glowing in the darkness, presented a magnificent sight. Calculations in progress were not interrupted: emergency power supplies kicked in immediately.

The amazing new fact is that the students and the younger workers in revolt are ignoring the directives issued by their own Unions, which they accuse of being just as bureaucratic, antiquated and rotten as the structures they were supposed to fight. The absurdity of the Regime has finally given birth to this surrealistic student movement that claims it will expose and tear down the fancy stage of French politics. So far it has done no such thing: it is purely an anarchistic uprising, without any constructive vision. Can it grow and become significant enough to play a role in the future? Where will it feed itself? Are we going to witness another historic upheaval ten years after the Thirteenth of May 1958, when the Public Salvation Committees of the Far Right precipitated a crisis secretly manipulated by the General's men to bring him back to power? Perhaps this is the end of his reign. Yet the images deployed in this new, spontaneous storm are naive, idealistic and pompous, like the circle of red flags they formed around the Arc de Triomphe.

6:10 p.m. The big demonstration is over, according to the radio news I just heard. But it was only a beginning, everyone senses it. We keep listening with a thrilling sense of anticipation mixed with alarm. Suddenly the heart of Paris is beating in new ways. Something extraordinary is indeed in the air, even if it still expresses itself in tired old clichés. Everything now depends on the reactions of the factory workers, who are heavily unionized. Their leaders are trying to take control away from the wildcat strikers who are challenging them.

6:30 p.m. The bulk of the demonstrators has just reached Denfer-Rochereau, near Paris Observatory (around the corner from Guérin's office), where they are supposed to disperse. Many people are going home, but the hardcore revolutionaries and some "anarchist" leaders call for renewed demonstrations. Student leaders have met with the chief of staff of the Prime Minister, who conceded all the minor demands that did not challenge the policies of Georges Pompidou. He denied all the others. But the students don't really seem to know what they want. They are calling for change, any change. The truth is that most Frenchmen are utterly fed up. I thought it was just me, a symptom of my own uneasy attempts to find my place again in the old country.

Saint-Germain. Saturday 18 May 1968.

Last night the Saint-Lazare train station presented the spectacle of a huge steel carcass where the heads of the suburbanites quivered like caviar in a box as they waited for a few rare trains. Another surprise strike had disorganized the schedules, emptied the ticket offices, erased the wide departure boards.

Today the entire nation is in shock. The rioting students have changed all the rules. Other groups are following their example, learning from them. Employees are on strike at Renault, the trains are rare and people wonder: are these the growth pains of a new French society, or simply a rough moment as we are shaken up by the same waves that are upsetting the planet, demanding new freedom everywhere? The Czechs have their own revolt, American students are up in arms, China is in the grips of the Red Guards.

Should we try to tell the students who have taken over the Sorbonne that much of the turmoil of our age is precipitated by the new information machines? That this is part of an evolution of the planet's nervous system, beyond all political considerations? They would only laugh at us. They equate computers with what they call the "system of economic oppression" they want to destroy. They don't realize that the world has already bypassed such ideas just as it has bypassed the old bosses they want to send into retirement.

Saint-Germain. Sunday 19 May 1968.

Vague visions of our future: we will certainly not remain in France. We will move again, perhaps to Northern California, to the San Francisco area. Yet we are in no hurry to leave. Janine is due to give birth in a few days. And I still have research to do, my own private research in the dusty stacks of the huge National Library, if not computer research for French industry.

Saint-Germain. Monday 20 May 1968.

The time for Janine's delivery is very close, so I am reluctant to leave her side for fear of getting stuck in the huge traffic jams that block every road in and around Paris. Gasoline is drastically rationed. This morning, on the square in front of the Château, I waited for the Army trucks that are supposed to replace the striking trains. The convoy had been stopped somewhere, it never arrived.

The woman who sold newspapers was worried:

"It is going to get better soon, won't it, Monsieur?"

No, it isn't going to get better, because this particular movement is uncontrollable, unpredictable, unpremeditated, spontaneous. The leadership itself doesn't know what it wants or where it is going.

All we know is that the example set by the students is spreading everywhere in France: many workers have now seized control of their own factories, *locking out both the management and the Union leadership!* The strikers themselves are producing and distributing gas and electricity. There was a run on the banks this morning, and gasoline distribution is about to pass under the complete control of refinery workers.

8:20 p.m. This is fascinating. I was wrong, I misread the movement around me. An extraordinary upheaval is indeed in progress now. On Wednesday night, in the small town of Cléon in Normandy where a small firm turns out auto parts for Renault, two hundred young workers locked themselves inside the factory after hours. They kept the director in his office. On Thursday the younger employees of the major factory at Flins had heard of the action and the wildcat strike spread there: at the lunch break they locked the gates. When the Unions attempted to take over leadership of the movement, they were violently rebuffed by the workers themselves.

In an altogether different field, namely medical research, the same inspiration has found sudden resonance: the younger staff is fed up with the *Mandarins*, the old tyrants who run the major hospital departments like medieval fiefdoms. Thus Professor Soulié at Broussais hospital has just been sent home by a vote of his nurses and of the doctors on his staff. This is not simply a political movement, this is not ordinary labor discontent. We are seeing a reshaping of French society. Can it be effective? Can it last? Only the cobblestones of Paris know the answer.

Saint-Germain. Tuesday 21 May 1968.

12:30 a.m. Janine my love, I think of you in your hospital room. Tonight I came home alone. I was able to reach my brother on the phone in spite of all the turmoil. Always ready to help, he came over and picked up Olivier. I said good-bye to his little sleepy face as we rested him on the back seat of the car. And you my soul, you are in pain now, you suffer alone and I feel helpless.

2:30 a.m. Catherine was born an hour ago. When I reached the Saint-Germain hospital Janine was still in the delivery room. How tired and happy you were, my darling, when I kissed you! Our little girl looked at us both with her blue eyes, pulled on her ear, and spat.

Saint-Germain. Thursday 23 May 1968.

10:15 p.m. The labor Unions have tried to recapture the leadership of the social movement again, like a drunken cowboy trying to control his stampeding horse. Yesterday the CGT Union, under pressure from its Communist bosses, strongly disavowed the student demonstrations.

This afternoon I attended an in-house meeting for the Shell management and staff. Our president made no attempt to hide the seriousness of the situation and the grave concerns of French industrialists, which border on panic. Indeed, all our production units have stopped work, he said. We only have a few days' worth of gasoline consumption in inventory and seventy days' worth of fuel in the refineries, three-quarters of which has been processed. But no one could get to it without crossing the line.

A very recent event illustrates the unprecedented nature of the strike. The Shell refinery workers committee received an urgent message from the striking workers at the main Saint-Gobain glass factory: they were running short on fuel and requested to be resupplied. The message, which bypassed the management of both companies, was very urgent indeed: when the temperature of a glass oven falls under a certain threshold the whole vitrification process stops. It is not a matter of re-heating the oven when the fuel supply is resumed. Instead, the entire factory has to be rebuilt around a brand new oven. This meant that unless Saint-Gobain could be quickly refueled, this major firm would be unable to produce any glass for a year or more.

What were the Shell workers supposed to do? They debated the wisdom of keeping Saint-Gobain in business. The majority expressed the view that since France would soon be under a Workers' Republic, it should have a first-rate glass industry. Accordingly, a special convoy of tanker trucks was organized. They opened the gates, and they kept Saint-Gobain in business.

The management of Shell is ready to negotiate an end to the strike. The stockholders want the company to get back to work at any price. The Unions agree: they would like nothing better than a settlement. The CGT quotes the words of an old Stalinist boss of the forties, Maurice Thorez, who said "one must know how to end a strike." But the strike is still going on, whether the Unions like it or not.

The historical reality is far different from anything these people have seen before. This evening a new, seemingly irresistible movement is springing out of the Latin Quarter. As I write this I am listening to the radio. Barricades are going up everywhere. The largest one is at the corner of Boulevard Saint-Michel and rue de l'Ecole de Médecine.

On the old Place Maubert, barrages have been built out of wooden crates. Indeed, Annick and I drove through the square at four this afternoon on our way to see Maman and Olivier. It was full of these wooden crates because the striking garbage men are no longer picking them up after the morning produce market. Now the students are setting them on fire to slow the advance of the riot police.

We found Olivier playing at the Arènes de Lutèce, the ancient Gallo-Roman circus behind rue Monge where I took Hynek a few weeks ago. It was filled with kids released by local schools where the teachers, too, have gone on strike. Old ladies, as always, were sitting on the benches to catch a few rays of the sun while they were knitting among the confused youngsters on forced holiday.

When we came back from rue de la Mésange with my son, who had stayed there since Catherine's birth, we found a curious, colorful, excited crowd in the streets: it was the same explosive atmosphere I remembered from 1958, with buses full of riot police parked on the bridges, blocking access to the Right Bank. The sun was lingering in the cold sky. The crowd was restless.

At the Sorbonne and at the Odéon there are non-stop public arguments. "The Power has been taken over by Imagination," claim the new slogans. "It is forbidden to forbid." It's all a crucial experiment. I had assumed too soon that the recipe for change had been lost here. The creative genius of the French has been temporarily freed up at last following the bankruptcy of the bureaucrats.

Yesterday morning I went into a bistro for coffee and croissants. There are certain things which never go on strike in France, food and drink prime among them! At the table next to me a group of middle-aged workers were arguing loudly about "the events" around a bottle of red wine. It wasn't a question of right or left, of belonging to the Unions or not, someone said. It was the young against the ancient order: the young workers were our hope, and the young managers too, against De Gaulle and the Bosses, but also against the Unions, the Dinosaurs, "all the old encrusted fools," proclaimed one man.

"Listen to the young of today," he added, a cigaret dangling over his lower lip. "Do you understand what they say?"

"My daughter is fourteen," another one broke in, "believe me if you want, I can't even help her do her homework."

"There shouldn't be anyone over forty in the Government."

Last Friday, coming back from Le Bourget airport, the cab driver had told me the same thing in more direct terms:

"The young, the students, they have shown that they had the balls less mushy than all the old fogies who run the Unions."

Not everyone in France agrees with this view, of course. I have heard a shopowner thundering against one of the student leaders, Daniel Cohn-Bendit:

"As if we needed more Germans in our country, and a Jew too!"

Cohn-Bendit is the leader of the March uprising in Nanterre. Currently in Germany, he is barred from crossing the border into France. The students have gone back into the streets with signs that read: "We are all German Jews!" It seems that every time the demonstrations settle down or run out of steam, the Government or the Police do something utterly stupid that ignites the riots again.

Saint-Germain. Friday 24 May 1968.

De Gaulle has condescended to speak, giving the people the benefit of his beautiful rationalizations. What else could he do? It is disheartening to witness this old soldier promising to the young France of 1968 eternal happiness in his tired arms while his riot police savagely beat up the kids of Paris at every opportunity using their wooden sticks and throwing tear-gas grenades.

The entire country is at a standstill. It is an awesome era, a suspended time as a modern nation stops dead in its track. We are lost in the time dimension, drifting along. Power is up for grabs.

8:45 p.m. New barricades are being erected. I am listening to a press correspondent who watches the scene from an apartment window in the Latin Quarter. The demonstrators have attacked the police. Trucks are overturned. The first grenade explodes. There are barricades around the Prefecture in Lyon. Riots have begun in Nantes.

10:50 p.m. Small groups are milling around aimlessly in Paris, without any specific objective. One such group made a move towards the Bourse, then wandered off . . . everyone wants to be down in the street tonight, to take part in the big change. But a change for what? To build a new world, claim the demonstrators. What kind of world? Under whose direction? No one is able to answer that.

11:10 p.m. The students have erected a real barricade on the Left Bank, made up of cobblestones, of iron grates, of overturned cars. There is a fight in progress on Place Saint-Michel. Thousands of demonstrators reach rue Soufflot with red flags and black flags side by side. Others have decided to return to the Bourse, the stock market area. In Lyon, more barricades are going up. Dozens of people have been wounded by the police, the radio reports scenes of looting.

11:15 p.m. High barricades are rising all over the Latin Quarter. At La Nation people, young and old, have gone out of their apartments to debate with the striking students around a heap of burning crates.

I wish I could go over into the city and join in this fantastic happening, but this is Janine's last night at the hospital. Tomorrow we bring

Catherine home, and my first responsibility is with them. So far Saint-Germain has not been affected by the turmoil. But this upheaval I follow minute by minute over the radio has taught me how little I understood, how poorly anyone had measured the depth of the French *malaise* of these last few months. I do not believe those who speak out now are especially qualified or competent, but history has thrust them forward. Not to be in the street tonight is to miss a great lesson, an opportunity to witness a raw process that is reshaping the entire nation in a few days.

11:45 p.m. The radio reports that the riot has turned ugly on Place Saint-Michel, that tear-gas grenades are exploding. A woman is screaming at an upper-story window. There are reports of people wounded on the Right Bank too. In Lyon a police official has been killed by a runaway truck loaded with stones. There are fights in the streets of Bordeaux.

Saint-Germain. Saturday 25 May 1968.

Now there is talk of reforms to come. The government states that yesterday's riots were due to "the mob." Yet the general strike continues. Gasoline reserves are getting very low. We have difficulty finding such basic items as mineral water, or milk for Catherine's bottles. People have panicked. They are stocking up on sugar, oil and coffee.

Janine and Catherine are home with me tonight, and I feel very happy. My thoughts go back to our life in Chicago, and I dream of what we might find some day in California. If I really believed that something new was about to happen in France I would stay here to help build it. Erecting barricades is one thing, consolidating a victory and managing a new regime is another. I am just a bit too cynical to think that the current movement is capable of deep changes.

A tremendous psychodrama is in progress. It is healthy and generous, but the outcome may be just a lot of wind and idealistic proclamations. The movement gives no evidence that it has a vision of the future it wants.

In the middle of all this I still get news about UFOs. Galia writes to us from Moscow that one of her friends plans to visit Paris soon, and will bring us some information. New developments have also happened in Brazil in the affair of the "Lead Masks" which took place in August 1966. There is still no explanation for the large luminous object seen above the hill of Morro do Vintèm by a society matron and her children around the time when two technicians died. But the two victims have been identified as electronic technicians Miguel Viana and Manuel Ferreira da Cruz. They had been seen by several witnesses the evening before their death.

Paris. Wednesday 5 June 1968.

Today I was able to walk around Paris at last. All traffic has come to a standstill. The strikes have frozen the city. In the Latin Quarter, strikingly beautiful Anarchist women dressed in black hand out leaflets: they are calling for the ultimate uprising. On the boulevards a happy crowd goes about its business on foot. At Les Halles the confusion is complete. The naughty girls that "Auntie Yvonne," De Gaulle's sour and bigot wife, had tried to ban from the streets have reconquered their sidewalks. They victoriously swing their purses and display their generous cleavages among the debris of the rotting mess.

I ordered a café crème at the Capitole, my old hangout of student days. It is from this bistro, one evening in 1962, that I called Aimé Michel to tell him about the statistical relationship between his files and the catalogues of Guy Quincy. It is this remarkably robust correlation that gave us the impetus to start a global analysis of the problem. I have come back to this area to locate a shop which is run by a family who has relatives in Russia. One of these relatives, a girl named Marina, has just arrived from Moscow. She is the friend of Galia mentioned in her last letter. I am hoping she will tell me what is going over there, and will explain to me the things a Russian citizen cannot write in a letter.

Marina is a smiling, attractive girl in her mid-twenties, loaded with presents for us from Galia. She tells me that the Moscow group has not disbanded, contrary to what has been said in the West. Indeed they now have two hundred members, with secondary groups in Kiev, Minsk, Tallin and Novosibirsk. They are still hoping to receive a formal authorization to become a State Institute for the study of unidentified flying objects.

Saint-Germain. Saturday 15 June 1968.

The great movement is over. The strikes have ended. Nothing has changed on the surface, even if the recent upheaval has revealed deep crevices underneath, gaping chasms, profound fault lines. I have resumed my attempts to get a job with an organization interested in advanced computer techniques. I find a friendly reception everywhere. People like the fact that I know the United States well. I could go back to America on their behalf and bring back new products they could distribute here under license. But they all say the same thing: no research. Research is too risky, too expensive. Only the Americans can do it, they think. Wait for IBM.

My work on rue de Berri remains easy and flexible. All the ambitious plans made last year by the managers who hired me have been shelved or postponed. Talk of a "data-base group" was far too scary. Instead we have set up a committee on "basic data," which meets regularly to inventory the files and argues endlessly about abstract structures, but does not implement anything.

Sometimes I discuss this state of affairs with our field maintenance engineer. I discovered this mystery man one day as he was kneeling behind a tape drive, a screwdriver in hand. He comes from Kansas City and speaks no French. Univac has assigned him to the Shell account, so he practically lives in our computer room. This fellow is the man who keeps our center operating. He personally drives to Orly once in a while in order to pick up the new disk drives and the new printers he installs for us. Nobody takes the time to talk to him. He thinks the way we use his powerful machines is very funny.

I have the freedom to wander off into Paris while my programs are running. I spend as much time as I can at the Bibliothèque Nationale, reading, taking notes. I walk among the bookstores looking for works on hermetic traditions and folklore.

Last night, as I came home on the train loaded with rare esoteric volumes, the contours of a new book took shape in my mind. It would draw a formal parallel between the UFO phenomenon of today and the medieval traditions about elementals, elves and fairies. Indeed my current research shows these beings to be strikingly similar in their behavior to the alleged ufonauts. The UFO phenomenon, I will argue, is *folklore in the making*. No one has yet established the relationship between the landings of flying saucers and this mass of popular traditions about aerial beings down through the ages, their relationship to man, their role in mysticism. I am thinking of calling this book *Passport to Magonia*, after the magical country situated above the clouds, alluded to in *Comte de Gabalis* and in the works of blessed Saint Agobard, written in the ninth century.

I have resumed my correspondence with Bill Powers, whose theory of the mind is progressing rapidly. And I am secretly dreaming of a day when I could do advanced research again.

Brunville-par-Bayeux. Tuesday 16 July 1968.

We are back in the fat land of Normandy. At night we hear whispering, crackling sounds, as if some giant aristocratic lady strolled invisibly through the countryside in her great velvet dress. The weather is cold and the grass is damp. The ponds fill up after the briefest shower, the whitish morning fog hides the hills. At seven in the morning Madame Fleury, our neighbor, milks her cows in the pasture under our window. A train rumbles in the valley. Last night the sky was so pure it was like a dream.

From her cradle my darling Catherine smiles at the flames in the ancient fireplace, at the red lanterns of Bastille Day hanging from the wooden beams. Olivier watches Madame Fleury milk the cows and he plays in the fields with his cousin Eric. Janine's sister has bought and renovated this big rambling farmhouse which dates back to 1642. The stone walls are two feet in thickness, built like the ramparts of a fortress. We are ten minutes away from the coast. In this conservative landscape cattle and dairy products are the only industries, wealth comes from the land and the poor are silent.

There are many temptations for us here: we could buy an old house like this one, overlooking the fat pastures; we could dream our lives, watching the great beautiful sky, raising a family of Norman kids as my paternal ancestors did, close to the earth. There is nothing wrong with such a plan. But we will not do it.

France has begun sinking into a new phase of the Gaullist regime, a paradise for the new rich, the *parvenus*, the promoters, the technocrats. The Think Tank companies, the purveyors of "gray matter," are making renewed offers to me: their idea is not to do research, but to grab new American products and to appear as great heroes by being the first ones to apply them here. It is a very stupid bet. It means that those French researchers who are capable of independent thought must leave if they want to flourish and create.

Whatever was left of the dreams that exploded so brilliantly in early May has now collapsed. The people who were endlessly talking on every street corner have run out of things to say. The Unions and the *Patrons* sat down at a big long table, shook hands and agreed. Everybody got a raise.

Any hope of a European vision is gone for now. There is no open avenue of dialogue, no concrete way to start implementing a strong Europe that could counter-balance the two major powers.

New elections have brought an overwhelming Gaullist majority back to power. Pompidou is gone. The General has appointed Couve de Murville as prime minister.

We are spending a fortnight here in Normandy, enjoying the easy hospitality of Janine's family, dunking great *tartines* of bread and butter into huge bowls of coffee mixed with hot milk. When we return to Paris I will go on buying books, saving what I can save of the old treasures, egotistically assembling an esoteric library to support my future work. And we will move West again, with fewer illusions this time.

I have typed up about half of *Passport to Magonia.* Compiling and analyzing this material, I see that we cannot hope to understand UFOs with the methods of physics and statistics alone. We are facing a form of intelligence that cannot simply be tracked with radars and cameras. It was already here before science was invented; it will still be here when science becomes superseded by other ways of gaining knowledge. Our soul is the only tool that is of any use in this search. Our soul, alas! What is left of it?

Brunville-par-Bayeux. Wednesday 24 July 1968.

At sixteen I knew more about myself than I do now. Certainly I have learned much about other people, other countries, other ideas, but in the midst of all that jumbled knowledge I am in danger of losing the secret state of mind which enabled me to measure how truly small we were on a cosmic scale. Today, as I walked back from Bayeux through the fields, I had the feeling I was resuming the use of a part of my mind which had not worked for ten years. Indeed I have written, classified, coordinated, invented, programmed, managed and implemented. But I have failed to contemplate.

My greatest treasure: our children, Janine's love, and that which is still childish in myself. I have spent the last three days in the middle of a grassy field building a plywood airplane for the kids, ten feet long, yellow and bright red, complete with a big propeller that spins proudly by itself in the wind, to the astonishment of the local farmers. I am young and a bit crazy, and that observation reassures me when I see what wise and serious people around me are doing.

The waste I witness in France makes me sad. Those who don't have an independent source of wealth are suffocating economically. I've done some calculations: if we took our savings of the last few years and we invested them here, we could buy a car, get a telephone, make a down payment on a small apartment. But we would quickly sink under the taxes and the fees those few possessions would bring down on us, although my salary puts me among the relatively affluent layers of the middle class. The French State penalizes you as soon as you start trying to live. The truly wealthy, of course, have many convenient ways to get around such frustrating problems.

On the stage of French society certain characters survive from one scene to the next, increasingly contrasted as if they were illuminated by the setting sun.

A celebration is planned here tomorrow. We expect doctors and lawyers with their wives and mistresses. No great inspiration rises from their ranks, no great sin, no great dark passion either. American novelists, who take such voyeuristic pleasure in describing experiments in immorality, would eavesdrop in vain around these alcoves. In many ways this is still the France of François Mauriac, not the France of François Villon. At dusk I can hear the dinner bell at a nearby manor.

On Wednesday I will be back in Paris, in my tedious office. Through the window I will see the vast gray courtyard, little rusty roofs spotted with pigeon shit and a few offices on the other side, where people constantly mill around, looking busy.

I have decided to use what is left of the Summer to complete the first draft of *Passport to Magonia.* After the holidays I will quit my job. A French Think Tank has made a firm offer to me. It is run by a man with a reputation as a clever politician who managed to buy the Control Data machine for which the Americans had denied an export license a few years ago.[12] But the job I am offered involves no new research. They all sing the same cowardly song.

I am tired of going through Saint-Lazare station every morning and every evening, with the realization that thirty years from now, if I stay in France, I will still be taking the same train, seeing the same people.

Brunville-par-Bayeux. Friday 26 July 1968.

When the media strike was over, when television journalists were forced to return to work, a political commentator capitulated with these words:

"If I were twenty-five I would leave this country. But I am only a poor old man. Let us acknowledge our defeat. Let us rejoin the ranks of the mediocrity."

Coming back to France last year would have been one of the dumbest decisions of my life, if it weren't for the opportunity it afforded me to witness first-hand the historic upheaval of May, and to gather unique documentation for *Passport to Magonia.* Patiently I make the rounds of all the firms that use big computers. I find the same attitude everywhere. When you cannot do research you cannot do development either. Soon you lose the ability to do simple engineering.

Such failures are unimportant to French business leaders. Political contacts are everything here. Knowledge of technology and the ability to develop new products are irrelevant skills for the ruling class. They don't tell you: "We have here the man who developed the first Algol compiler, and the genius who solved the transfer equation for neutrons in nuclear reactors." Instead they try to impress you by bragging: "We have a fellow who knows everyone at the Department of Public Works, and one of our directors went to Polytechnique with the Minister of Health, and So-and-So has a father who has influence over the military budget." In the last few months I have found a solution to a significant problem in the management of data-bases: how to design large information systems that are self-organizing, reordering their data as a function of the questions the users are asking. Nobody seems very interested. Politics, not technology, is the name of their game.

At our office the director for computing recently invited me to an elegant and select lunch with a small group of in-house scientists. I discovered that our executive dining room was one of the finest tables in Paris. We were there to greet Paul Rech, a French expatriate who is an operations research specialist at Shell Development in California, near San Francisco. As we were popping champagne corks one of our brilliant mathematicians expressed his pleasure at Rech's visit:

"We have had some very interesting talks," he reported to our boss.

"Perhaps for you. But I doubt if Mr. Rech is going to learn very much among us."

The comment was flippant and unfair: our applied mathematics group has recently succeeded in solving a basic problem of optimization in integer variables that will save the company a good part of the cost of moving its railroad cars around, a problem that had stumped their American counterparts. Three months of savings represent the entire yearly budget of the applied mathematics group. Tomorrow the same computer program could save a similar percentage for the world-wide fleet of supertankers, millions of dollars a month.

Paul Rech, with whom I spoke afterwards and discovered a real sympathy, understood very well what was going on:

"It's typical, the reaction of your executive. He doesn't recognize how much his own organization is worth."

He seemed pleased to find someone with whom he could discuss America. I explained the events of May to him. We left the building and we walked across Paris to have coffee near Saint-Germain-des-Prés. The weather was warm. We sat outside at a terrace and watched the barricades which are still standing: it will be months before the work crews can clear the streets of rubble and return the Left Bank to normal traffic.

Paul told me about California, and I listened eagerly. We discussed computers, those strange machines from America that are responsible for the crumbling of the old structures in Europe. I remember Guérin writing to me "a good researcher behind a telescope will always be more economical than a machine," a terribly misleading statement. Today's computers are still only embryos, the first prototypes of a new social nervous system.

We have seen enough of France. There seems to be no cause we can serve here, no future in which we can participate, nothing we can build. In French society at large my skills are useless. To remain here is professional and intellectual suicide. To leave is to be reborn. With a lingering regret, though. France is a mistress who still inspires lust as one leaves her bed at dawn. I am especially sad about the failure of the dreams of May. That France of the Possible we briefly saw flash before us will never materialize.

18

Edinburgh. Wednesday 14 August 1968.

Tea for two in Edinburgh, where an international computer science meeting is being held. Janine and I are making new plans, even as we meet old friends again. The conference was a perfect excuse to see Ben Mittman from Northwestern, and Jean Baudot who came over from Montréal. I had dinner with Bill Olle, the man who directed the implementation team for Infol, the first generalized data-base management language. It turned out he had heard that I had written a compiler for his language in six months.

By the time we were ready for dessert he had offered me a job in his new group at RCA. I hesitated, reviewing all the negative points: RCA is a newcomer to the computer business. Can they really compete head-on against IBM? More specifically, can their new Spectra computers take on the IBM 360 Series? Their facilities are in New Jersey, a region of America that has little attraction for me. It is of California I dream.

On the positive side, here is the chance to work with a team that thinks about the same problems at the same level I do. I would rejoin my intellectual family, the brotherhood of software developers that forms a sort of Freemasonry. New Jersey could be a stepping stone to other places.

After the Congress itself we took the time to climb Arthur's Seat. Janine bravely went up the side of the old volcano in her Parisian dress and high heel shoes, a scarf over her head to protect her hair from the mists of Scotland. From the top we could see the city and its fairy-tale castle, the little lakes and the Firth of Forth.

We have made new friends in the bookstores of Edinburgh: here, in the country of the Little People and the Good Neighbors of legend, we spent hours rummaging through treasures like *Celtic Magazine, Scots Lore*, the history of the Picts, and a collection of the early volumes of *Knowledge* magazine. My documentation for *Passport to Magonia* has taken another step forward.

Saint-Germain. Friday 23 August 1968.

The short roots that still tied me to the French soil are withering. Yet I cannot deny I still react to a poetic feeling in the air of Paris which moves me to delight. I have many friends in the crowd that walks along the Boulevards, and I love the shades and fragrances of the countryside. What I cannot stand in France is the greed and the spite of the upper class, even if I might have approached that exclusive circle some day.

I have gone back to the think tank that had offered me a job. If I did stay in Paris and joined their company, would I be free to lead a software research team into uncharted territory? The manager told me he was personally skeptical about generalized data-base systems: they were still beyond the horizon of the possible, as far as he was concerned. He wasn't even sure they were needed.

"Then, as you can see, it's better for me to go back to the States. I can always come back here when the need for such systems has become obvious to everybody!"

"*Mon pauvre ami*," he answered condescendingly, "if you leave you will never come back here. You won't be able to find a place again. All the good *situations* will have been taken."

It was my turn to be flabbergasted.

"Taken by whom?"

"Well . . . by those who, in France, will begin such applications. . . ."

People like him live by a simple rule: there is only one world and the center of it is Paris. Nothing matters except as judged by the standards of Paris, and no life is desirable unless lived in Paris. There are only a few "good situations," as he says, and these choice jobs are held in reserve for those who know how to maneuver cleverly within the system. There is no thought, as in America, of being rewarded for creating opportunities for others. Their idea is to grab the greatest possible number of privileges and then to slam the door shut, blocking the way for others, keeping them as far as possible from success, even if that means restricting one's own landscape, narrowing one's vision, sacrificing one's own range of opportunities, postponing one's achievements.

Such tactics don't work in the computer field because the technology itself will sweep aside its own leaders.

Now I can either leave Paris without looking back, or I can stay in

France as a member of the technocracy and never think about California again. There is also a third alternative, a compromise which many French intellectuals have chosen. It can be painted a variety of colors. Some like to brood in Left Bank cafés, saying: "Yes, I am a *raté*, a failure, but I am staying here to save what can be saved of civilization." Others claim: "I stay for the sake of my old parents," or "for my children," "for my insect collection," "for my little house in Aquitaine to which so many family memories are attached."

Another delusionary compromise is the revolutionary dream: "I stay here to help build a better world once this rotten system falls apart."

There is also the erotic fantasy: "My work isn't important to me, I only live for sexual freedom. What would the girls of Paris do without me?"

These people are spoiled. They have understood nothing of the upheavals of May, they have missed the fact that the whole social matrix has rattled and shifted under their very feet. No one can control the world any more. Every nation wants freedom, even if that means facing the tanks, as we just saw in Prague.

"Play for the Czechs!" demanded the public at a recent London concert given by a Soviet orchestra. After playing the *Symphony for a New World* the first violinist, a Russian, left the stage in tears. In Prague itself the same evening, a Red Army soldier committed suicide inside the University complex his unit had just seized. Ironically, while the world is passionately reaching for liberalization, the U.S. is about to put into the White House a conservative politician, Richard Nixon.

At the office, near the Champs-Elysées, the passing of hours does not concern me any more. This time flowing away around me is not mine. Last week, I spent a few days of poignant tenderness at my mother's apartment while Janine was back for a visit to her parents in Normandy. Maman lives beyond the Panthéon, in a six-storey building without an elevator. We spoke late into the night. Above and below us children played and cried, people came and went, doors closed with a thump.

Night came majestically in the apartment, draping its shadows over a desk my father built, over the red carpet and a fine black cabinet enhanced with copper angels and a gold border, and over my mother's paintings on the walls. She doesn't seem to change. She is a schoolgirl, delighted with all the adventures of everyday life as well as the grandiose claims of modern science, for which she has not developed any skepticism. Nothing unusual or funny escapes her sharp, ironic blue eyes. She attends an evening English class where she talks to other students fifty years younger than her, dispelling their sadness and their concerns for the future with her contagious enthusiasm. Recently she took my brother's two sons and his daughter to a science show about meteorites. Never did a better grandmother fall from the sky.

Paris. Thursday 12 September 1968.

This morning I stopped for coffee at a sidewalk terrace on the Champs-Elysées and I went over our reasons for leaving America last year: a need to refresh our emotional beings, to find new poetry, to breathe the passions of Europe, to break up the routine of work in Chicago. I have satisfied all this, witness this manuscript I have just completed. *Passport to Magonia* is a book written in English and destined for the American public, yet I could never have written it in America.

I walked back by way of the *Herald Tribune*, where I successfully tracked down a reference to an aerial object seen in 1908. Then I came back to this office to stare at the ugly courtyard below my window. Day after tomorrow I will be in Bill Olle's office in New Jersey for a series of interviews with his team.

Later the same day.

Every Frenchman has a virtual placeholder, a little rectangular invisible box which represents his place in society. It comes into existence at the moment of birth and moves under its own power throughout life. Its path is defined by one's parents, their social position, their contacts. This past year I have discovered that it isn't important to actually reside inside one's own box. You can step off the plane at Orly and find some perfect strangers readily speaking to you as if you were their lackey, while others will treat you immediately as their boss. I could come back again twenty years from now and find that my box has accumulated seniority and privileges as well as restrictions; whether or not I am here, whether or not I work, love, experience and think, succeed or fail, it will follow a career path carefully charted for it by unknown forces. Some people find this reassuring. They will actually go on strike to demand such "security." But to me it is a terrifying vision.

It is only during the May insurrection that I have seen the French really living, with an unpredictable, passionate life, full of virtual powers. For a short time they felt inebriated by it, they delighted in their new-found freedom. Then they got tired and scared of their own audacity, and they sought shelter again under the skirts of their old General. A few contemptuous words from him and higher wages for everybody was all it took to force the system back into the normalcy of mediocrity.

I could easily stay here to manage a small technical project in poetic surroundings, but I know very well that I will not do it. Surprisingly, it is in today's *Figaro*, a conservative paper, that I find a striking image under the pen of Valéry Giscard d'Estaing:

> France is a hexagon filled with pyramids. Like the Mexican pyramids, ours have multiple levels. Our social structures, thanks to their remarkable vigor, have succeeded in placing the same social class on the same levels of the various pyramids: it is the same people who manage, who execute or who suffer (whether they happen to be) in Administration, in Education, or in Industry. This historical monument is condemned by the forward rush of contemporary events.

It is a very appropriate metaphor. Yet I see no evidence that the pyramids are about to crumble. On the contrary, they seem more stable than ever. From their summits the high priests are heard chanting every day at sunrise.

Cherry Hill, New Jersey. Sunday 15 September 1968.

The plane to Philadelphia circled Boston at sunset. No cloud in the sky. The Atlantic coast was below us, spreading its woods, forests and lakes, Manhattan looming ahead.

Now it's a quiet Sunday in the Garden State. From my hotel window I can see a perfect lawn and pretty flowers. Beyond the tall trees, long quiet limousines seem to float along the expressway. Bill Olle has promised to call me before noon. How wonderful it is to be in a country where the telephone actually works! Yesterday I called Allen Hynek. He told me he had in his hand my very latest statistics based on the Air Force files, which I had sent him a few days ago. He warmly recommended that I take the RCA job:

"You can't imagine all the things that have happened since you left! I just got back from Boulder.... It would take too long to tell you. Have you seen the latest *Playboy?* Did I tell you I am writing a report about Project Blue Book ...?"

Next I heard the voice of Don Hanlon, sensible and vibrant. He has been discharged honorably from the Navy, but he just broke his right hand in a boxing match in Chicago.

Saint-Germain. Monday 23 September 1968.

Last week I met the whole RCA data-base team in Cherry Hill. I discovered that most of them were Europeans. Bill Olle himself has a doctorate in astrophysics from Manchester in England. There is a fellow named Burkhardt who left IBM-France when they stopped working on advanced software. Others have come from Sweden, from Holland. It would be a lot of fun to work with them.

The decisive meeting for me involved a discussion with Alonzo Grace, an engineering manager for RCA, who explained to me their computer strategy:

"In past years this company has made the key discoveries in color television, and we operate some of the largest communications networks in the world. IBM has no such experience to its credit, it has a lot to learn in those areas. Future projections indicate that ten years from now the most important issues in the computer marketplace will happen to be exactly in our domain of expertise: networking and display of large quantities of data at a distance through communications devices. We are not there yet, of course, but RCA can use its expertise to take a shortcut to that future world, well ahead of IBM. We will intercept them at the pass, so to speak."

It is a brilliant strategy, but will they have the resources and the will to execute it decisively?

Paris. Thursday 26 September 1968.

Today a dull gray light comes down the well. It throws an uncertain reflection on the bluish office walls. I do not raise my eyes. To avoid seeing the plastic partitions covered with dust and grime.

On Tuesday afternoon the fat black telephone on my desk, which looks like a regulation revolver, rang suddenly. Bill Olle's joyous, energetic voice intruded into my dull reality. He gave me the final details of the RCA offer.

My manager must suspect that I am looking for another job because he took pains to explain to me that my technical proposals "were under serious consideration."

He tried to make me feel better, as managers always do, by giving me the usual bromide: "You are not a number here, we treat you as an individual. You are part of our big family." He added that I should be patient, by 1972 the technical developments I was advocating would probably become a reality. "So maybe I should mark my calendar, and make plans to come back in four years?" I thought to myself.

Now I can hear the director of the programming section in the next office, cutting his fingernails: tick, tick, tick. After that he will take up personnel management issues: salary reviews, performance evaluations. If time permits he may even tackle some technical problems in codification, in which he is something of an expert. He dreams endlessly of universal identification systems. A good, harmless fellow, the kind who periodically proposes to the Académie Française to reform the spelling of every word in the dictionary, in the name of rationalism.

This morning I broke the news to my mother that we would go back to the States. She did not appear very surprised. At noon I will meet Janine to buy a tapestry we have seen in a shop window. We plan to take just a few precious things with us. I have found a copy of Dom Calmet's *Dissertation on Apparitions.*

Saint-Germain. Saturday 28 September 1968.

Marcel Granger expresses the mood of many Frenchmen when he writes in *Adieu à Machonville:*[13]

> No more green spaces, no more old mossy houses, no more fishermen with their lines, no more games of *pétanque*, and the Beaujolais no longer tastes the same. In the streets people don't argue any more, there is no time. On Sundays we are bored to death. The Saône river stinks of shit and despair.

With the eyes of a mere traveller on the earth I look around me. Is it right to leave again? On one side is the great creative wind of freedom, the immense potential of America. When I look for something to put in the balance on the French side, what comes to mind is not science or art, which have become mere servants of the State, but the humble street scenes. That old woman I passed on rue de la Verrerie, for instance, whose hand was shaking so much she could hardly hold her grocery bag. I do not think of friends, they are so few: in one year here I have gotten to really know only a handful of new people, and only one has invited me to his apartment. I had never realized how stuffy French society could be. Our elegant neighbors in Saint-Germain only meet one another in church on Sunday morning. The ladies wear white gloves. The little girls carry clusters of marigolds.

The delightful temptation of the return to childhood is ever-present here. In the disquieting evening I look at the ancient crumbling walls, the stately cathedrals, and I feel reassured, even if I have long ago left behind the childhood faith that impressed me with its Latin hymns, its stained glass Saints. . . . Some of that respectful awareness of a higher level of being lingers within me, in a mystical tone that still echoes through *Passport to Magonia.* I needed the vibrant little room I turned into an office, opening on a garden covered with dew, with the presence of centuries around me and the light touch of elfin hands on my shoulders and the weight of scholarship around me, even if the Latin cantatas have been replaced by the Moody Blues' *Nights in White Satin,* which Janine and I are running endlessly on our little record player.

Saint-Germain. Sunday 29 September 1968.

I have written to Aimé Michel to tell him our decision:

> I am leaving for good this time. I sincerely hope that you're happy and that you have been able to save some of that ancient culture which, for some silly reason, remains important to both of us. You guard it fiercely on your mountain like a spirit watching over a treasure. Perhaps you have found that higher lucidity I sometimes grasp briefly, and which is compatible with happiness. You are genuine and true, like flying saucers. Maybe that's why people don't believe in you any more than they believe in them.
>
> Now I am rather glad that Rocard did nothing. It is better not to have unrealistic hopes.

He replies:

> Ah *merde*, I expected this from the beginning, that you would split again! This country, which is the smartest in the world at the basic level of the average plumber, the farmer, the mailman, has managed to acquire the most pretentious, cynical and rotten elite one can imagine.
>
> There is still a little hope, however, on the UFO side. I am coming to Paris on Saturday. The *Barbouzes*[14] want to see me. I am told they are fed up with the continuing accumulation of UFO cases, and they are starting to have serious doubts about the good faith of their American friends and colleagues. They are wondering if they haven't been fooled from the beginning.

Shell Building. Thursday 3 October 1968.

At the *Mandragore*, where I found a two-volume set of Lenglet-Dufresnoy, the owner is an expert on Gothic fantasies with demonological overtones. However, like many occultists, he is ill-informed on UFOs, an observation that led me some years ago to abandon the Rosicrucian organizations that had attracted me as a teen-ager. His first reaction, worthy of a Rationalist, is to doubt the testimony of the witnesses, whom he calls hoaxers and drunkards. It is easier to mock them than to confess they may have seen something none of us understands.

I was still pondering this observation on the skepticism of occultists when I reached Vivien's bookstore on rue Mazarine. The shop was closed. A pale young man all dressed in black velvet was staring at the books in the window. We struck up a conversation, and I said that I was interested in the history of apparitions. He told me, as in a dream:

"A long time ago, I might have been thirteen, I saw several small beings in the countryside, at the edge of a clearing, a whole band of little men, all dressed up."

Did he notice anything else? Any unusual light? Did he look for traces? No. He pushed the incident out of his mind. How tall were the gnomes? He put his hand flat in the air, less than one foot off the ground. No, there were no UFOs in the vicinity. He was not interested in flying saucers, he said, only in Magic. "Transforming the being, isolating the spirit in its tower," he insisted feverishly, with the eyes of a cornered animal.

Shell Building. Monday 7 October 1968.

Aimé Michel spent the day with us in Saint-Germain yesterday. Once again, some significant developments have "almost" happened.

I saw the true Aimé Michel again—alive, witty, eloquent, curious to examine our little treasures: Rener's paintings, the seventeenth- and eighteenth-century books I have been gathering, and the *Archives of the Invisible College* I had wanted to show him for a long time.

He told me all about his latest meeting with the *Barbouzes.* They weren't the same ones he had seen before, the colleagues of Commander Granger. Instead the people he has just met were from a different service, which I didn't even know existed. Our mutual friend Jacques Bergier[15] who has extensive knowledge of this shadowy world had given him pointers to establish the contact. They confirmed to Aimé that they had some embarrassingly detailed sightings in their own files. They have finally come to the conclusion that what they heard from their American colleagues was pure bullshit and that Project Blue Book was simple window-dressing. They wanted to talk confidentially to someone who had researched the phenomenon.

"Could I bring one of these gentlemen discreetly to your house for a good long talk?" asked Aimé.

"I suppose so," I said with some hesitation, "but only if it's clear they are not on another fishing expedition. And I won't betray any confidence."

"Of course, of course," Aimé said, "I wouldn't expect you to do that. I know you well enough."

He was having dinner that same evening with Bergier, who has just left *Planète* after a violent political disagreement with Louis Pauwels, who is drifting to the right. Bergier is a ferocious anti-fascist and has the scars to prove it: during World War II he was deported and tortured by the Gestapo, who rightly suspected that he was a spy. In fact Jacques Bergier played a major role in the Marco Polo network of high-tech espionage which located Peenemunde, the Nazi rocket base.

When Aimé reported our conversation, and my reactions, Bergier deflated his hopes. He had made a few more inquiries. There was nothing genuine behind such conversations with the *Barbouzes,* he said.

"Don't waste Vallee's time. These people want to see you because they are paid to be curious; of course they want to talk. But they are not serious."

Paris. Tuesday 8 October 1968.

Yesterday, in Dorbon's shop, the owner allowed me to consult his card index because I am a good customer. It was a clever move on his part; many of his books are not on the shelves but in the back of the shop. I went away carrying Papus' *Magie Pratique*, Flammarion's *Invisible World* and Murray's *Dieu des Sorcières*. He also told me of a forthcoming sale at auction of the library of Stanislas de Guaita.[16]

Same day, afternoon.

Bill Olle has called me. This saves me from having to walk over to the post office and stand in line to send him a telegram. The devil of a man has even found a house for us in New Jersey: three bedrooms and two bathrooms.

The bad weather has gone away. Clouds race through the sky, coming from the West. Suddenly I find them fascinating. I lean out the window to watch them, those clouds flying in formation over Paris.

Saint-Germain. Thursday 17 October 1968.

Aimé Michel tells me "I'll never be able to leave for America." He says this with despair in his voice. He is the most "French" mind I know, yet he would go away tomorrow if he could. At work my resignation is official. A string of colleagues come through my office, and they all say the same thing: nobody can get anything done here, anything new.

Saint-Germain. Saturday 19 October 1968.

Yesterday afternoon, *Chez Doucet* near the Luxembourg gardens, I had coffee with my old uncle.

"And your mother?" he says softly. "She must be sad that you're leaving again."

"She expected it, you know. And we will be back often. To see her, to see you."

"It's not the same thing."

When I grew up here I was firmly convinced that class structure was something the English had, with their Establishment boys in their proper jackets and their Oxford ties and the races at Ascot. But we French had had the Revolution. Hadn't we proclaimed the Universal Rights of Man? We were all equal citizens.

Returning from the United States after five years was a major shock. It opened my eyes to the fact that there are still Lords and indeed there are servants in France. The scientists, the people who run the technology, the cadres are today's *larbins*. Stylish, liveried lackeys to be sure, well-fed and well-clothed, but they do what they are told, while the Lords above them have access to another world of subtle, silent attentions, a world full of privileges where there is so much money the very word need never be mentioned.

The only passion that could keep me here is the luxury of rare books. In the Reading Room at the Nationale I can assemble Bodin and Wierus around me and call a meeting with Paracelsus and Gorres, then I sit back and listen to their arguments back and forth as if we were sitting at home around a fireplace. I can attend their passionate debates about the reality of the soul, the paranormal, the question of evil. It is a rich and delightful human company. I see my books as merely an extension of theirs. They are the standard, the only critics I acknowledge. Our continuing research beyond the centuries is the only activity that matters.

I saw some fascinating items in Leconte's bookstore. He is the expert for De Guaita's library of rare magical books which will be sold at auction on Friday. Among the collection is a seventeenth-century manuscript about Chiromancy. On each page is the contour of two hands, where a researcher has carefully copied the features of the palm he was examining, along with the destiny, name and birthdate of its owner. Lecomte says he has never seen anything like it in forty years. Under one of the pairs of hands I read the note *"Il fut esgorgé en 1616"* (He had his throat cut in 1616). Naturally such treasures are beyond my feeble financial resources. I did buy Wier and Bodin from Dorbon recently, stretching the limits of my purse.

Saint-Germain. Sunday 20 October 1968.

It is hard to believe that in two weeks we will be back in the States. The weather is perfect, sweet, magnificent. It would be nice to just stay here and walk through quaint villages. France is a great and wonderful country, and the most pleasant place in the world if you don't have to bother earning a living.

Paris. Friday 25 October 1968.

At the auction house on rue Drouot I watched as several million dollars changed hands. All the books of Stanislas de Guaita were sold in a few hours. They represented a collection of occult works which exceeded in rarity even the library of wealthy attorney Maitre Maurice Garçon, according to the rare-book expert who ran the sale.

As I left the auction house in mid-afternoon I found myself in the same Metro car as a frail woman whom I had just seen win the bidding for a hand-painted fifteenth-century *Book of Hours*. She had paid approximately a million dollars for it. She went out at Miromesnil, very calmly, brushing against the afternoon crowd with the small volume in its red cardboard case under her arm, as casually as if she were carrying the latest romance novel.

Paris. Rue de la Mésange. Saturday 2 November 1968.

For the last few days I have been waking up with a nauseated feeling. Neither reading good books nor taking leisurely walks can dissipate this anguish. My stomach rebels, my legs are wobbly, my brain is confused: Could it be that my body is telling me it wants to stay here? On the morning when the movers came I felt weak. Yet I dutifully went to the office and I was immediately caught up in a blizzard of paperwork. Janine is in Normandy for a few days. We have no home left. It will be good to listen to the four jet engines screaming behind us when the Boeing takes off.

19

Willingboro. Saturday 16 November 1968.

Big empty house. Janine has gone to the hairdresser. Little Catherine is slowly falling asleep. Olivier draws quietly in the kitchen, the only room where we have any chairs in this new American house we're renting. This evening we are expected at Bill Olle's home for dinner. On Monday I fly to New York as my new employer's representative at a Standards meeting on data structures. Suddenly I am back at the cutting edge. I cannot dream of a better environment for my work, although there is something missing of course, after the intensity of Paris, in this quiet little development. Willingboro is a modest suburb near Cherry Hill, on the New Jersey side of Philadelphia. The house we are renting faces a small wood, the last undeveloped area in this community. A year from now it may well be gone, but we will probably have moved away by then.

Our friendly neighbors have already introduced themselves. They rake the leaves religiously.

Luxuries: on Tuesday we will have a car once again. I have not driven a car in a year. The phone is already connected, with the apologies of the company for being late by one day. In the good American tradition I have bought all kinds of insurance we don't need. What little furniture we had is on the way from France. As soon as we have someone here to take care of the children Janine is eager to get back to work, to feel alive and productive again.

Willingboro. Sunday 17 November 1968.

Don tells me he is now working as a photographer for an architectural firm. According to him, UFO research is dead in the water. More work is actually being done on old sightings than on current ones, which are being ridiculed.

Sitting in the large room on the second floor that will become our library and office, I dream of things to come. If people mock the UFO subject, this must be a propitious time for good research. People always laugh at important things, to release tension, to regain control over what scares them.

Willingboro. Wednesday 20 November 1968.

The Condon Report is two months away from publication. Tom Ratchford, of the Air Force Office of Scientific Research, is said to be reviewing the final draft. Hynek doesn't have access to a copy.

Yesterday I called Harvey Plotnick at Regnery to get his reaction to *Passport to Magonia*: he told me enthusiastically that he wanted to have it out in hard cover in the Fall of next year. I continue to be intrigued by the fact that I had to go back to Europe to write that poetic celebration of the

Little People, that acknowledgment of our *Good Neighbors*, a glimpse of an order of consciousness into which every age has projected its own fantasies, and whose existence modern man persists in denying. Could it be that events and thoughts, and even books, are a function of location? Are new ideas and new images to be fished out, like sardines, only at certain predetermined spots?

I have difficulty judging the quality of what I write, even when Janine's opinion is favorable. If she frowns, I always know exactly what to do: the manuscript goes into the trash without another look.

Willingboro. Saturday 7 December 1968.

Hynek has made another trip to Dayton, where some changes have taken place. He had taken with him his letter of resignation, ready to slap it on the desk. But he found in the new commander a very different man from what he had imagined, a man who listened to him, he said, when he pointed out the Air Force could still save face in this whole business and conduct a good scientific study. But the chief scientist doesn't want to do anything until the Condon Report comes out.

David Saunders' book (entitled *UFOs? Yes!*) is well-written, according to Hynek, who has seen an advance copy. Now that he sees what little progress Saunders' splinter group has accomplished, however, he is glad that he followed my advice and stayed away from them.

"It's just so much philosophical talk, generalities," he told me on the phone about Saunders' theory of UFOs. "Nothing like the kind of in-depth work we really need."

Willingboro. Wednesday 29 January 1969.

Winter has brought cold nights and ice storms. The suburbs of Philadelphia offer no beauty, only an absurd steel bridge, and a huge refinery next to the river. The dirt and stink of New Jersey are not much better. And up this road, an hour and a half away, New York City and all its monstrosity. The icy roads are dangerous. Forty cars have just collided together South of the city. Schools are closed. Janine and I are both staying home.

I am using this welcome free time to review my files, compiling the first catalogue of landing reports, which I envision as covering an entire century (1868 to 1968). I have mailed a copy of the manuscript to Don to get his advice. I feel alive. Man only moves forward through the changes he creates. What can one learn when one's back is to the wall? That's the way I always felt in France: my back held to the wall by the forces of convention. The letters we get tell us others feel the same way. Even my old conservative uncle writes "we must be forgiving to our country." Perhaps we have always forgiven too much.

The new generation growing up in Europe seems unable to correct the mistakes of the older one. If they choose revolution, it will be a long hopeless gray thing where souls will be wasted, as in the Soviet Union. If they turn to nationalism it will be rancid and ugly. If they set their economy in motion again they are in danger of turning Europe into a giant version of New Jersey, with quiet pools of poison on either side of dull expressways that link refineries, nuclear plants, chemical factories. You can already drive from the suburbs of Brussels to those of Paris or Milan and see what I mean.

I feel as one does at the end of a very long road, my love: tired in all my bones, the brain shaken by the vibration of the engines, the howling of the wind.

In my work I have rejoined my peers at last. Our software factory near Riverton, a small town fifteen minutes away from here, gathers the elite systems group in a facility no customer ever visits. They are kept focused on the task of turning out massive software tools. In the basement of the large building we operate six *Spectra 70* computers, including the very first prototype, wired by hand. The new operating systems, the compilers, the application packages are being tested day and night on these machines.

Next week I fly to Chicago, where I expect to see Ben Mittman again. Janine has a programming job with RCA too, in the group which manages the manufacturing schedule for the computer plants. I worry when she drives off on these roads. We will not spend another winter in New Jersey.

Willingboro. Sunday 9 February 1969.

Hellish winter hits, wind-blown snow drifts in tough blows of white draped over the dark naked wood. The house shakes and vibrates. The air is taut, twisting around like the skin of a drenched balloon lost in a storm.

I spent three days in Chicago and came back drained: three days of lectures and arguments. My friends offered me a position on the Faculty as they had two years ago. I declined again, in spite of my respect for most of the people there, but mindful of what happened with the Dean when we were faced with a real challenge. Hynek does not understand why academic life doesn't appeal to me. He cannot imagine anything better.

"You really should be a professor somewhere, Jacques."

A poisonous kind of beauty rises from Chicago, from the mud of dilapidated Clark Street, from all the ruined buildings downtown, from the empty lots shiny with shards of broken glass, from the drunks in the doorways. At the bus stop an old insane black woman planted in the snow laughs alone like a wicked witch. From the top of the Palmolive building the beam of a searchlight looks like a crazy Menzelian experiment, drawing ghostly disks in the fog. Beyond it the carcass of the Hancock building carries a few lighted offices deep into the cloud itself.

"My answer to society, here it is!" an old cab driver tells me, sweeping theatrically the inside of his car. Following his gesture I see nothing but a grayish photograph of his face on the dashboard, a serial number in big black figures, the meter with its red digits, a pool of mud under my feet where the shoes of his previous clients have dragged in the Chicago winter. Not much of an answer.

And my own answer to society, what will it be? It should begin with the expression of human emotions. Last year the turmoil of Paris was softened by sensuality even at the height of the crisis. But America doesn't encourage emotions. Even in love this greedy Freudian continent does seem uniformly cold and mechanical, I thought as I looked at Chicago, searching old memories. I do treasure a few exceptions, but they came by pure chance, when the cold of the night brought lives together by the light of a fireplace for a brief moment of tenderness.

Is paranormal research just another attempt at a "response to society," like the cab driver's little world, or is it a genuine way to transcend ordinary reality?

Last Saturday we had a reunion of the old *Invisible College.* Hynek was there, Fred and Don too. Bill Powers came in, jovial, with a new beard and plans for a huge research proposal that would cost two million dollars.

"It would go over like a lead balloon," grumbled Allen.

A man named Donald Flickinger, who works as an agent for the Treasury Department, has sent a curious confidential report to Hynek: in 1961 it seems that four hunters saw an object near Minot, Minnesota. There were several figures near the landed craft. One of the hunters fired, and at that point the report becomes very confused and dream-like. One of the pilots of the object is said to have fallen.

"Why did you do that?" another supposed ufonaut screamed *in English.*

The witnesses came home disoriented, having lost four hours somewhere. They swore to one another they would say nothing to their families. The next day, however, official-looking men came and picked up the fellow who had fired the gun. How did they find out who he was? The mystery continues to deepen, with no solution in sight.

Willingboro. Sunday 16 February 1969.

The end of the week turned out to be full of beauty, with the sun in the clear sky and the cold, crisp air in our lungs. At the computer plant the whole software team met to prepare the testing and the release of our data-base system, called User Language-1. I love the energy of this group.

A young French girl from Normandy has joined our household. Maud is nineteen, a redhead, curiously jaded about everything she sees in America. But she cares for the children, they love her and this restores some measure of freedom to us.

On Friday Janine and I spent the evening with a journalist from Philadelphia who knows the witchcraft traditions of Bucks County. In Upper Black Eddy, he told us, a woman named Mary Manners Hammerstein received witch Sybil Leek in 1964. She is said to lead a coven associated with spiritualist circles in Philadelphia and with a man who is the inventor of a saucer detector: thus the Delaware Valley is rich in deviltry to this day, even if Dow Chemical and RCA Electronics own increasingly large sections of the land. Nearby, in Everittstown, an elf-like ufonaut was seen in November 1957: he came out of a saucer and tried to steal a dog! Was there a symbolic message in this display? The Russians had just orbited a satellite with a dog named Laika on board.

Yesterday we went on an excursion to Princeton with Maud and the children. Catherine is almost standing up by herself now. And Olivier has started to ask sharp questions for which we have no answers.

"What will there be, *after the days?*" I wish I knew.

Sunday night: classical music. I read *La Mystique* by Gorres, a five-volume treatise on everything in the higher universe, from angels to demons. There is adventure in the passing of time, an inspiration in the air, the spark of love in your eyes.

Washington. Tuesday 18 February 1969.

It is because of a retired French officer named General Lochard that I am here. He represents AFNOR, the august French standard "Association for Normalization," at an international meeting about the vocabulary of computing for which he is Chairman. Sending me there as part of the American delegation was a diabolical plot on the part of Pete Ingerman, the man who is in charge of standards for RCA.

A few weeks ago Pete came into my cubicle, playing with his black beard and eyeing me with a conspiratorial air.

"How would you like a job as a double agent?" he asked.

"What's the assignment?"

"This company has a problem with standards. Not just this company, but the whole computer industry. See, we have to have an international standard for data processing vocabulary that can be used in every application."

"That shouldn't be too hard," I said. "There are only three or four companies that run the whole business anyway. If they really want a U.S. standard they can do it in a few months."

"You're right, but the point is, they don't want a U.S. standard. Instead they want an international standard that can be applied in this country but can also be used in selling machines all over the world. That means getting the ISO, the International Standards Organization, to standardize the terms used in computer technology throughout the planet."

"Is there a problem with the ISO?"

"It has two official languages, French and English...."

I began to see what was coming.

"The French want to protect their own language, which is fine with everybody, but they keep delaying the proceedings until they have a perfect French equivalent for every term. The last international vocabulary the ISO standardized was that of Telephony. Do you know how long it took?"

I had no idea.

"Fourteen years. Now you see the problem? In fourteen years all the terms we use in data processing will have become obsolete. So my idea is this: you work with us drafting the American definitions, and you think through the French equivalents. When we get together for the international sessions you sit with us behind the star-spangled banner. And when the Chairman, who is an old French general, tells us he cannot accept our definitions because they have no French equivalent, you get up and you read them to him in his own language!"

That sounded like a lot of fun, but more importantly it was an exceptional chance to meet some of the pioneers of the software field, so I agreed enthusiastically.

Willingboro. Friday 21 February 1969.

Maud, the young French girl who has recently joined us to take care of the children, is a stunning redhead who is quite ready to have fun. Yet when she goes out with the local boys she comes back scared and utterly disgusted with their games. She tells us stories of drag racing in their parents' cars along the deadly expressway, of mindless groping on the back seat. She finds all that stupid and debasing, yet that constitutes young love in the United States, the rite of passage.

America the Great Teaser. There is plenty of apparent, obvious sensuality at every street corner, of course: the advertising media thrive on suggestive images, naked women and subtle seduction. But these empty promises are just the tantalizing side of a great equation of hypocrisy: there is no depth, no sophistication of feeling, no complexity of the heart around us.

Later the same day.

At the National Bureau of Standards the building seems too big, the rooms a succession of empty caverns in the gigantic Federal style which is uncomfortably reminiscent of Stalin's architecture and of Hitler's Berlin. From the gigantic windows one can see the plains of Maryland, white and gray, monotonous.

I am not comfortable on the East Coast. From Boston to Washington I can hear nothing calling me, no detail with which I can identify.

Willingboro. Sunday 2 March 1969.

I called Hynek today. The new commanding officer at the Foreign Technology Division, Colonel Winebrenner, has taken him to lunch in great style, but Allen still gets into regular arguments with Quintanilla. There is no reaction yet to the Condon Report, which is pretty much what everyone expected: a despicable snow job, padded with an irrelevant section about radar. Jim McDonald is expected to write a formal critique on behalf of NICAP. Always the same old maneuvers: These people have understood nothing, nothing at all.

Willingboro. Wednesday 5 March 1969.

I watch the beauty of the sky when the sun sets in the fields, beyond the thundering expressways I leave behind as I come home from work. Bill Olle has put me in charge of a five-man testing group. I only leave the factory to walk around after lunch with a Swedish friend who is discovering the Delaware Valley with as much astonishment as I do.

When I come home I take little Catherine in my arms. She laughs. She grabs my glasses and plays with them. Together we watch the sunset behind our wild little wood. Soon the greedy developer will come back with his bulldozers and raze everything in sight. Why should he refrain from building more houses, putting in more streets, leveraging his investment in the land, selling, selling and selling? This is a planned community, the kind of town of which contemporary Americans dream. Naturally, it cannot tolerate one little undeveloped wood. Didn't anybody tell the planners that people will eventually go crazy if they are not allowed at least one tiny corner of wilderness?

My own life is serene. We will go away as soon as the first trees are felled. I have written to Rener, the Belgian artist friend of Serge Hutin, to buy some of his latest paintings. I am carried forward by heady anticipation, the same feeling I had in Evanston when I worked closely with Allen. I believe Man has an infinite ability to create.

Our company keeps recruiting good programmers from Europe. I talk to them as they get off the plane. They all say the same thing: it is not possible to be professionally alive and creative over there, in a tiny stifling world crumbling under its moldy crust.

Willingboro. Sunday 9 March 1969.

The things I love: a breeze in the foliage, Janine's smile, my children, the sunset, the quaint intricacies of language (old French, Elizabethan English), the efforts of men, especially those who failed, who feared, those who searched relentlessly through the mud.

Who are they, these people who foolishly thought they had found the key to dreams and destiny? The visionaries: Paracelsus, Nostradamus, Joan. The geniuses: Galileo, Copernicus, Bruno. The dreamers: Cyrano, Swift, Rener, Hynek, Villon, Aimé Michel. The lovers: Casanova, Baudelaire, Rimbaud. Not a single empire-builder among them; not a single servant to the politics of his time. And none of them was ever helped by those in power with either honors or money. So let us not talk about the State protecting the Arts and the Sciences. Beauty, the creative force, runs deep and invisible under the crust of history, under the pedestals of power and wealth.

All those places I love, too: tiny, quiet, modest rue de la Mésange, where my mother lives; Quai de l'Horloge along the Seine; and the streets where no one ever goes any more, in the villages of despair a long way from Paris: rue des Etanets in Pontoise; the carriage doors with their large nailheads where no horses come through. Rivers of memory flow towards oblivion.

We took a Sunday drive into the country today. The road followed the Delaware River, leading to the northern hills. Beyond Frenchtown we found a rugged land, impressive, alien and dark. A steep incline took us to the crest of Everittstown. A few farms showed up in the desolate landscape covered with snow. The Alleghenies rose on the Pennsylvania side. Ten more miles and the atmosphere, the people changed again. Here the little man dressed in green tried to steal John Trasco's dog. Yet we are only one hour away from New York City. Why should our Visitors love such forlorn, isolated spots? When dusk arrives and the crust of freezing snow crackles under the car tires I begin to experience the same fears described by John Keel in his books, I understand the primeval terror he writes about.

RCA Factory. Riverton. Friday 21 March 1969.

Curious Spring. The developers have started to ravage our little wood sooner than I thought. Nonetheless the excitement I feel here doesn't go away. Janine is visiting friends in Florida. How should I spend my time? Should I leave the children with Maud and go explore the bookstores of Philadelphia? Try to make new friends? It hardly seems worthwhile. The remaining months here will be short.

Willingboro. Thursday 3 April 1969.

Five pages of corrections and additional references for *Passport to Magonia:* I mailed the whole thing to Regnery today. The full text of the landings catalogue, all 923 cases, is ready. The phone lines are active among Don, Allen, Fred and me. Alleged pictures of little men with long noses have been published by the media. They look like obvious fakes. Keel's stories, which have the urgency of terror, are taken too seriously by many believers. Don believes that Jerome Clark, a young ufologist from Chicago, has become so convinced that an extraterrestrial invasion was imminent that he has been driven close to a breakdown.

Once again Fred is angry against Allen, who has just returned from an investigative trip to California with a suitcase full of magnetic tapes so poorly recorded they are inaudible. At one Air Force base where he planned to interview some of the pilots he was sent away with the terse comment:

"Project Blue Book? That doesn't exist any more!"

Willingboro. Monday 7 April 1969.

At my suggestion Fred invited Don to the University of Chicago to show him what scientific research was really about: Don is quite gifted but he has grown up in the streets with the teen-age gangs of Chicago and he has only seen the cold, official, boring face of science: the books, the public television shows that never tell the actual truth. Fred says he doubts if Don was very impressed or moved by what he saw in his lab. They did talk about magic, secret societies and the films of Crowley disciple Kenneth Anger.[17] As they came out both of their cars had parking tickets on the windshield. They were signed by an officer named A. Crowley!

Willingboro. Sunday 4 May 1969.

Michel and Françoise Gauquelin visited us yesterday. They brought greetings from Allen, with whom they had spent two days in Evanston. We took them to see Princeton. They were surprised to see the campus, with students sprawled on the grass typing their term papers on portable machines, and girls in flowery shorts sitting under the trees, annotating books on quantum mechanics. We were a long way from the gray Sorbonne indeed. The big people in Princeton are the highly paid consultants who regularly take the train to Manhattan or Washington to sell their ideas to Blue Chip companies and Government agencies. Yet even the RCA Labs, which are famous for their numerous past discoveries in radar, radio and color television, suffer from a high level of compartmentalization.

Michel and Françoise were in Los Angeles when the news came that De Gaulle was leaving power. They were as surprised as we were by the political commentaries, which show an amazing level of fascination for *Le Général* even among people who hated him profoundly. As if De Gaulle had been anything other than the concentrated reflection of French collective shortcomings. The Providential power that seemed to propel him was little more than a clever magic trick, an illusion utilizing the people's own carelessness, their relinquishing of responsibility for themselves. Some commentators are foolish enough to doubt that the scepter will be passed on to Pompidou.

Olivier, who detests visitors, astonished the Gauquelins by precipitating a family crisis in the course of which he coldly smashed a glass against the wall. This daring terrorist act demanded a certain strength of character, since punishment was inevitable. He ran away into his room. When I followed him there, however, everything was in place and no kid was visible. The window was open: I looked out and decided he could not have run away fast enough to disappear from the landscape. I concentrated on the room itself and eventually found the culprit hiding under his bed.

Willingboro. Friday 9 May 1969.

Computer technology is being swept forward, all around us, by an irresistible current. The latest concept is that of simultaneous, interactive use of the machines by many users in a mode called "time-sharing." It will make possible a whole range of new applications because it will no longer be necessary to have computers used by a single person at a time. One will not even need to go over to the computing center itself to access the machine. RCA is advancing in that direction, although with great difficulty. So is IBM, whose latest machine is capable of being utilized in that mode, but good software is still lacking. The computer field is exploding: opportunities abound for those who can seize them. Programmers are in short supply everywhere.

Aware that I could accomplish far more than what I am doing now, I have sent a letter-proposal to Rowena Swanson at the Air Force. I would develop an automated documentation service that could serve as a test-bed for artificial intelligence research. Perhaps I could run such "experiments" at Princeton. I am scheduled to give a lecture there on Wednesday at the invitation of Saul Amarel, a well-known software pioneer.

Our Riverton group works hard but the RCA computer effort in general is not going well. The grand strategy which called for beating IBM in communications and sophisticated human interaction is not being executed. Our marketing department is convinced that we must remain able at any time to respond to new product introductions by "Big Blue." But the implication is that, like the French, we always follow rather than lead. How can we ever intercept IBM's technical path in the distant future if we are doomed to react to every move they made last week? Those are the things the systems group discusses emotionally at the cafeteria.

A case in point is a very advanced implementation language called IL-1 that some of our brightest developers, including my friend Max Smith, have defined. The systems group wants to release it. Initially it was envisioned that all our software would be written in that language, which would provide a tremendous productivity tool to specialized teams like ours. Instead, marketing has now decreed that IL-1 must not differ significantly from a new IBM language called PL-1, which is supposed to be standardized "any day" for general use in the IBM user base, replacing Fortran, Cobol and even Assembler in one big blow. But PL-1 does not contain the advanced programming features we need, so we have been forced to revert to writing all our software in expensive, cumbersome assembly code.[18]

It is the private opinion of our little group that IBM has no intention at all of ever standardizing PL-1 in spite of all this hype; that our own marketing geniuses are wasting their time following that carrot dangled in front of them. Good intelligence from systems programmers within IBM itself tells us that their own shop is doing all their development in assembler, just like us, contrary to claims that PL-1 is the universal language of the future. That means we cannot use our best secret weapons. And without secret weapons RCA will probably be forced out of the field.

Science is no longer the single major factor of human progress envisioned by Camille Flammarion. Yet digital computers are truly magical tools, deeply changing the world, destroying entire sections of the old mental framework in the process. The power of a few groups of systems programmers, hidden away in the back woods, exceeds everything people had imagined. The growing sophistication of computers combined with media conditioning on a large scale can only mean one thing: the secretive technocrats with high security clearances who shape the major decisions in the advanced nations will soon have the ultimate power in the world. Jacques Bergier calls them "cryptocrats."

At the Warwick Hotel in Philadelphia this afternoon I saw the members of the annual conference on automated documentation slowly falling asleep. I did have a chance to talk to Ben Mittman, who gave a talk about Infol. Through the windows of the twentieth floor I could see Locust Street. It was a glimpse of a foreign planet. Many things are wrong in this well-ordered world which keeps the blacks among blacks, the whites among whites, the poor among poor, and cops at all the crossroads. If no deeper current comes to provide new vision we will end up with a society made up of multiple Willingboros surrounded by an infinity of Harlems, and a few Princetons majestically riding on top of the hill. People will be locked and entertained within their own little spheres. They will be allowed to evolve slowly with the ebb and flow of directed fashion. The ruling class will be a refined group of old anonymous minds. Yet who could say that this depressing future would be any worse than our past, the odious history of mankind?

Willingboro. Saturday 17 May 1969.

What I will remember of this town: the sound of lawnmowers, the smell of weekend barbecues in the backyards, the retired military men and the retired insurance agents who are our neighbors, decent folks with their red checkered shirts and baseball caps, and the boredom of the curved streets. Psychologists have discovered that straight streets made people unhappy, so everything here curves like the halls of the *Queen Mary*, meeting other curving streets, all the way to the main highway. This is not where I want to spend the rest of my days.

On Wednesday I went to Princeton. I was kindly received and the seminar I gave on information retrieval met with polite applause. Then a man with intense eyes and silver hair took me aside. We sat on the benches in the lab next to the lecture hall.

He said, "There is a fundamental fallacy in artificial intelligence, and you're falling into it like everybody else."

"In what respect?" I asked with the feeling that this discussion was not going to conform to the usual exchange of generalities heard at most professional meetings.

"Artificial intelligence is trying to emulate nature, it wants to approximate what man does."

"What other inspiration is there?"

"Imitation of nature is bad engineering," he answered patiently. "For centuries inventors tried to fly by emulating birds, and they killed themselves uselessly. If you want to make something that flies, flapping your wings is not the way to do it. You bolt a 400-horsepower engine to a barn door, that's how you fly. You can look at birds forever and never discover this secret. You see, Mother Nature has never developed the Boeing 707. Why not? Because Nature didn't need anything that would fly that fast and that high. How would such an animal feed itself?"

"What does that have to do with artificial intelligence?"

"Simply that it tries to approximate man. If you take man's brain as a model and test of intelligence, you're making the same mistake as the old inventors flapping their wings. You don't realize that Mother Nature has never needed an intelligent animal and accordingly, she has never bothered to develop one!"

I could only greet this stunning thought with silence. He went on:

"When an intelligent entity is finally built, it will have evolved on principles very different from those of man's mind, and its level of intelligence will certainly not be measured by the fact that it can beat a chess champion or appear to carry out a conversation in English."

With his piercing eyes on me, I had a brief vision of what an intelligent machine might be. If Nature has never needed an intelligent animal and hasn't evolved one, I kept wondering, then what are we? In our feeble attempts to handle the information we call our life, can we trust the creations of our dreams? Are we perhaps nothing more than the process through which another form of intelligence is evolving?

We spoke of software operators, their role in language understanding and in pattern analysis; he gave me an example of how a recognition operator would do its work by successive matching approximations: "it's a bird, it's a plane, it's Superman!" This made me smile because Olivier is always insisting I should come back from work early in the afternoon to watch Superman with him on television. He tells me enthusiastically:

"Superman, if you kill him, you can't kill him!" Another nice example of logical operators in conflict.

Chances are slim of my doing serious work at the Princeton Research Labs of RCA. Their computer is already saturated and their overall management is shaky. So I sent my résumé to Stanford University, where a position is open for a senior information scientist. It is time for us to move. We have agreed that we should find a place where we could buy our own home, a house with an opening on the sky, a house that would live, breathe, commune with the night.

Washington. Monday 19 May 1969.

Another international standards meeting. Dinner and an evening of conversation with Saul Gorn, a professor from the Moore School of the University of Pennsylvania where ENIAC, the first fully operational computer, was conceived and built. Saul Gorn is one of the founders of the field of software theory. He was one of the first people, among Von Neumann's colleagues, who understood that the instructions given to the new machines were not just a series of electronic quantities represented by zeroes and ones but the beginning of an actual language, an "artificial" language, as opposed to English or French which have evolved "naturally" out of human usage.

We continued our discussion as we left the restaurant and walked through Dupont Circle. Blacks were playing drums, hippies sat on the lawn smoking pot, motorcycle gangs rode around making as much unmuffled noise as possible, Puerto Ricans strutted back and forth, each group closely conforming to the accepted rituals of their own tribe while pretending to rebel against society at large.

"No matter what you do, we can't prevent our kids from having the same values as the other kids when they go to school," said Saul Gorn.

"That doesn't bother me," I replied truthfully. "I know that my children will have new data, new knowledge and will be exposed to different values. I just want to make sure they have their own set of standards by which to judge all that."

Washington. Thursday 22 May 1969.

After four days of discussions and decisions presided over by the wild and multi-faceted mind of Pete Ingerman, I have come back to the hotel to rest and read. The evening will be spent with Rowena Swanson, one of the best-informed people in the small world of artificial intelligence. This is a world which functions as a secret society, selecting its members through unspoken rules.

Two interesting episodes took place during our deliberations in Washington. The first one came after we had just defined the term "language" as

> A set of signs, represented as symbols, used to convey information.

A problem arose when we tried to separate artificial languages, which computers are using, from natural languages. We agreed that natural languages arose from human usage, while artificial ones were based on machines, but we could not define it in such an "intuitive" way. If the mere fact of prescribing a language made it artificial, then Esperanto would be an artificial language, which confused the issue even more.

I posed the following problem to the group: two computers, each of a different make and each equipped with a communication device, are placed in a room where there is no human. They are programmed to establish contact with their environment. After some preliminaries, we would expect the computers to exchange some signals, acknowledge the contact and establish a joint protocol to send and receive messages. Eventually they would define a communication structure they both understood.

"Would that language be natural or artificial?" I asked.

The linguist expert in the group thought for a long time, got up, paced up and down, tore up his hair and finally said, "natural."

Later, General Lochard announced triumphantly that the French delegation had finally discovered an appropriate French term to translate "software." We all perked up, because the problem had been a vexing one for years. A prize has even been proposed in Paris for the first person who would solve it. With our group was a very distinguished middle-aged State Department interpreter of French who was on loan to us for the duration of the proceedings. He was dressed in a dark suit and tie and sat humbly in the back, behind the chairs of the delegates, waiting for his services to be required.

"Well, what is that fantastic new term you have found to translate *Software?*" asked Pete Ingerman.

"It is the word *Programmerie,*" General Lochard said proudly.

I was too stunned to react. But the State Department interpreter jumped up from his chair like a devil out his box and sang in a clear and enthusiastic voice:

> Quand un programmeur rit
> Dans la *Programmerie,*
> Tous les programmeurs rient
> Dans la *Programmerie!*

This was a witty play on an old French children's song mocking the Gendarmes:

> When one Gendarme laughs
> In the *Gendarmerie*
> Then all the Gendarmes laugh
> In the *Gendarmerie!*

He had simply replaced the word "Gendarme" with the word "programmer" and the word "Gendarmerie," a barracks for Gendarmes, with the proposed new term *Programmerie,* which thus became a barracks for programmers.

After this unseemly explosion he must have realized what he had done, because he turned as red as a ripe tomato and sat down in confusion among our laughter. Not only was it witty, but for an American it was an amazing demonstration of intimate mastery of French colloquialism, a lesson General Lochard was not likely to forget. In any case *programmerie* was never mentioned again, and the term *software*, itself an astute play on words in English, remains untranslatable into French.[19]

Willingboro. Sunday 25 May 1969.

Yesterday at 3 a.m. I saw a wonderful thing in the glow of my headlights as I arrived home from Washington: all the weeds are growing wildly on the lawn we have not mowed for ages. Our neighbors have been too polite to complain, but it is clear I am expected to do something about this jungle very soon.

Rowena Swanson, with whom I had dinner in Washington, saw little reason to get excited about the current state of artificial intelligence research.

"People are just endlessly re-doing what has been done before," she said as I filled her glass from the Chianti bottle.

"What about Stanford? Everybody is talking about their robot," I said. "Saul Gorn, who is very cynical, thinks its only purpose is to train a few students."

"I'm even more cynical than Saul," she said with a smirk. "The robot's only purpose is to spend a few million dollars of Pentagon money. There is a terrible lack of imagination everywhere."

She warned me about the West Coast:

"It's a tough world out there, nothing is ever stable in California. It's ruthless. Whenever they temporarily run out of money they just lay off all their engineers. They will have a big crisis if Washington ever gets tired of paying for all their high-tech toys."

We walked over to Georgetown, where a young idle rich crowd seemed to be waiting for something no one could define.

Now I am back home with my little family and we listen to the *Moody Blues:*

> The trees are drawing me near
> I've got to find out why....

We have celebrated Catherine's first birthday. Olivier was a real pest: he wanted no part of the cake, and locked himself in his room. The challenge for us is to keep re-affirming the structure around him without breaking his independent spirit: he will surely need it later. But this time it was not so easy to restore the structure. I had to get a screwdriver to take apart the doorknob before firmly putting him to bed.

Willingboro. Tuesday 27 May 1969.

A simple business trip has turned into an exhilarating visit to New York. There are flags over Rockefeller Center, strains of music beyond the open-air restaurants. Yet the financial markets are falling. On the marble steps outside the major banks, groups of men my age wearing three-piece suits argue in concern.

I have been invited to New York by the same kindly recruiter for Shell who had spoken to me two years ago. To my surprise he wanted to know if I would consider returning to Europe to work for them again. He had the same subtle Dutch elegance I found among the top executives in The Hague. He took me to lunch at his Club, complete with expensive carpets and leather armchairs, old books, white-haired butler. A middle-aged waitress with rosy cheeks served us food like a mother. The thrust of the discussion was a generous new offer. I said I was honored and pleased they had made this effort to contact me.

"I really have no interest in returning to Europe," I told him. "If I were to work for you again it would be in Emeryville, in the research company," I added, thinking of my friend Paul Rech who belongs to their operations research team in California. He promised to look into it. I was astonished to find how much work he had quietly done on my background. He has even read *Challenge to Science.*

In Manhattan I was struck by the sudden sensual itch in evidence everywhere. Pornography has become a major retail business. Miniskirts are in fashion. A pretty girl sits on a bench in the sun, displaying the flesh of her upper thighs. She unfolds a red cardboard that becomes a tanning mirror. A blind man passes in front of her with his dog, missing the whole spicy show.

Willingboro. Monday 2 June 1969.

Don Hanlon has just spent three days at our house, sleeping in my large office on the second floor. Our talks began among the old books, continued in the RCA computer room and concluded on Saturday as we walked under the night sky. We spoke of Hynek and Project Blue Book, of John Keel. But our major topic was esoteric theory and especially the Rosicrucians. I recently published a humorous piece in the form of a ufological horoscope. It included a short paragraph destined to Allen:

> You will learn with amazement that the U.S. Air Force has closed down Project Blue Book. You will be hired as a scientific consultant on a new project called Sitting Duck. Grow a beard.

Don, who was there when Hynek opened the letter, tells me it made him burst into laughter. Indeed, he now expects Blue Book to be closed for good. He believes the higher levels of the Air Force are privately divided over the UFO issue. His relationship with Quintanilla has reportedly deteriorated to the point where they can't stand to be in each other's presence. Allen, who is usually restrained and gentlemanly, refers to the Major in the most unflattering terms.

We debated the various schools of magic, Don being partial to Aleister Crowley's *Ordo Templi Orientis* to which avant-garde director Kenneth Anger also belongs. Any group that demands total belief, as the OTO does, and provides absolute symbols, seems meaningless to me. Magic should be the meta-science, not just a set of cooking recipes. I am not looking for a belief, I am not eager to witness cheap marvels and parlor tricks. What impressed me about the Rosicrucians was their belief that no human organization could be anything but a fragile support in this kind of work, a temporary resource, and that the true realm to which they felt allegiance was "of another level," beyond human life and physical reality itself.

According to Don, the most accomplished scholar of modern magick in America is a man named Israel Regardie who lives in Los Angeles. He once served as Crowley's secretary. Recently vandals have ransacked the old doctor's library, stealing Crowley's manuscripts. Some people suspect that California motorcycle gang members, like those depicted by Anger in his movie *Scorpio Rising*, may be involved in the despicable burglary.

Don also showed us Crowley's admirable Tarot cards, powerful images of genuine beauty.

It is out of the question for me to belong to a faith, a church or a sect, no matter how much intelligence its leaders demonstrate. I am looking for my own slow deliberate progression towards a personal spiritual truth. This means gradually transforming my own life, placing it into the right conditions. I agree with Allen when he says that a man who is capable of an active spiritual life can survive anywhere.

Now a terrible wind is blowing, a Summer storm. The trees howl and shake. The rain is about to begin.

I wonder what I will think when I reach the end of my life. I don't think I will miss the earth, mankind, this body.

Willingboro. Wednesday 4 June 1969.

I have just spoken to Fred Beckman in Chicago. He is very upset at Allen who, he says, is kidding himself:

"He seriously believes the Air Force is going to give him five people and a real budget to study the electromagnetic effects of UFOs. He is wasting his time. He just won't face the fact that the same Air Force has just spent half a million bucks to convince the world they didn't exist!"

Willingboro. Tuesday 24 June 1969.

Next Monday I fly to San Francisco for a job interview. I have had several phone conversations with Paul Armer who heads up the Stanford computation center, reputed to be the best in the world, the frontier in the software field.

Fred's voice on the phone registers a careful reserve which is very clear when I ask him to read the galleys of *Passport to Magonia.* I know that he does not agree with me that flying saucers are "folklore in the making," but he will read the text out of friendship. Hynek has left for Corralitos observatory, a small astronomical outpost with two domes operated by Northwestern in the New Mexico desert.

The other day, reading an old Scottish book on folklore, I came across a beautiful proverb: "There comes with Time what comes not with Weather." Curiously it cannot be literally translated into French, where time and weather are the same word, temps. I should mention this to Saul Gorn, who keeps a collection of linguistic anomalies. Indeed, as I get ready for my first trip to the West Coast, I wonder what will come with Time.

20

Stanford, California. Tuesday 1 July 1969. 00:20

The little red alarm clock is ticking away, still showing New Jersey time. The white Galaxy I rented is parked out in front of the Tiki Inn motel. I got lost twice before finding the little road through the hills that brought me back after my first visit of the Campus.

I had never flown over Utah and Nevada. I spent my time alternating between the views of the Great Salt Lake and the reading of Baudelaire, with whom I have become re-acquainted at thirty thousand feet. I have seen little of Palo Alto: uninspiring lines of motels and bars, a fancy hotel with plaster statues, the usual kitsch. But the air is soft and tender. I have time to look at the land, to consider where we might live if I got a firm job offer, if we moved here.

Stanford. Thursday 3 July 1969.

House-hunting in Half Moon Bay: the real estate agent whose office I had spotted in this little coastal town has died. I went to the closest coffee shop to think. I discovered with astonishment that the food was very good: real bread that tastes like bread, and butter! Oil and vinegar for the salad. Fresh vegetables.

One cannot understand the coastline just by looking at the map. One has to see the farms spread between the hills and the Pacific, and drive on the rough dirt road to the top of the cliffs that overlook the surf.

Very few people live along the coast. This wonderful landscape is empty; the big crowds are on the Bay side of the hills, where all the motels and the bars lie clustered together. Here there is nothing but weeds, blue wild flowers. Seagulls survey the rolling waves.

San Gregorio, the next little town down the coast, has a few wooden houses lost in the fog and the spray.

La Honda is a beautiful and heretical village in the tall drama of a pine forest high in the hills. What are those palm trees and those eucalyptus trees doing among the redwoods, with an occasional willow to increase the gorgeous confusion of the foliage? The weather was warm, the smells intoxicating. I stopped everywhere to take pictures for Janine. I went to another real estate office in Woodside. They showed me photographs of available houses. I promised to write.

I didn't have the feeling I was seeing the edge of the Western world when I suddenly reached the top of the cliff and faced the Pacific for the first time. Instead I felt a projection into the future, not twenty or thirty years, but a thousand, two thousand years. Everything suddenly seemed possible, everything was allowed. The feeling of total freedom that the rioting students of May '68 had sought in vain was there, in plain view. Everything else, even the details of everyday life, seemed irrelevant. A warm wind blew over, a touch of infinity. The same impression was present last night when I contemplated San Francisco from the top of Telegraph Hill. Only a sin, a mistake, temporary insanity on the part of the gods can explain such magical beauty, vibrant and clear, diaphanous.

Willingboro. Monday 7 July 1969.

The haunting beauty of the California coast stays in my mind now that I am back in New Jersey. It is unique because the landscape there is shaped by constant seismic force. Erosion has no grip on those hillsides.

The people I met at Stanford, however, were not in keeping with the awesome grandeur of their surroundings. Paul Armer turns out to be a Teddy Bear rather than a real manager. A pleasant man in social surroundings, he became somber and withdrawn when we spoke in his office. What problems are hiding there? I was interviewed by Ed Parker, a professor in the Department of Communications, which runs an information retrieval project into which the Federal Government has poured untold millions. There have been no results so far, only wordy research reports.

Another interview the next day, with Bill Miller this time, the vice-provost for computing. He pulled the strings, that much was clear. The description he gave me of his empire was chilling.

When I sat down at one of their terminals and ran a few tests I could see that the database software was a dismal failure, but one thing became obvious: the operating system itself was a real joy, the culmination of every specialist's technical dream. I would give anything to meet the people who built it. They are years ahead of everybody, in a field where six months constitutes an eternity.

Willingboro. Saturday 12 July 1969.

We are coming back from Princeton. One last attempt to explore alternatives, to compare. It was raining in Princeton. We ate some pancakes with Maud and the children. We visited a house for sale. The windows were nothing but narrow horizontal slits.

Now the rain has intensified, my shirt sticks to my back in the muggy weather, Catherine cries. The world has less meaning than ever.

The other day I was reading Baudelaire's *Le Voyage*, from *Flowers of Evil*, as the plane flew over Nevada:

> Yet the real travellers are only those who leave
> For the sake of leaving; light hearted, like balloons,
> They always follow their fate.
> Without knowing why, they always say: go on!
>
> Those whose desires take the shape of clouds
> And who dream, as recruits dreaming of the cannon,
> Of vast, changing, unknown raptures,
> Whose name the human spirit has never known!

I have shown Janine the pictures I brought back from Palo Alto.

Willingboro. Thursday 17 July 1969.

Our bedroom is dark, hot and humid in spite of the air conditioner, which makes an annoying sound. Janine is working on a dress. I still hesitate to move to California, fearful of getting caught in the parochial fights that are going on at Stanford. They remind me of the academic feuds I had known at Northwestern, only deeper and more vicious, because more serious money is involved. It is an atmosphere in which creativity may be a liability more than an asset.

A young mathematician who works for me on information structure problems told me he was not surprised at my fears of returning to a university. Academic life is frozen and dull in spite of all the pretentious airs that surround it, he said. If anyone raises a serious problem he is often sent down a blind alley. It is a micro-culture that cannot be threatened or forced into change. He points out that even the beautiful slogan of the Sorbonne students in May last year, which claimed so proudly that *It is forbidden to forbid!* was only a logical self-negation, an empty statement of utopia, with no real operational value.

We must move to California. We will have a door leading to the woods, a window over the sea, a trapdoor to the sky, "and thy dress to the evening wind."

Yesterday three well-trained pilots left for the moon. We watched the launch of *Apollo* on an employee's tiny black and white television set installed in the RCA employee cafeteria, a place which smells of dishwater. The rocket went straight up, looking like a child's pencil. But most of the American public, bored with the space race, was watching other channels.

I miss being away from my mother at times like these. I imagine her in Paris, listening to every news bulletin, buying every newspaper. It is one of the great dreams of her life which is taking shape now. I admire her enthusiasm. But I already see the world that will come after we plant our flag in the dust of the moon: I fear it will be a strange dreamless world. I worry that the entire sky may soon become filled in every direction with spy satellites, flying bombs, orbital barracks and the cosmic latrines of the new secret armies.

Will we ever come back to the true questions: who are we? What is that strange process we call thinking? What are the dimensions of our world? Perhaps we should make a machine to think about these questions for us, since we do not dare ask them any more. But there can be no such thing, of course: everybody knows that building a thinking machine is just an unrealistic dream, like going to the moon....

Willingboro. Sunday 27 July 1969.

Around us, Americans of all ages are increasingly using drugs for recreation because, as they stupidly say, "it makes them happy."

When I was growing up my parents' generation viewed a trip to the moon as a purely theoretical idea. It could only be the culminating achievement of a supremely enlightened scientific culture, something that would not be possible until men became immensely wise. The very fact of going to the lunar surface would be a signal of cultural eminence, an irreversible break with the ancient world of wars and human misery.

Human life would be forever changed into a golden age.

The other night we watched *Apollo XI* land smack in the middle of *Mare Tranquilitatis.* We saw Armstrong, then Aldrin get out and plant the flag in the moon soil. Yet nothing has changed. Our world is as divided, fearful and greedy as ever. People still need to escape into drugs. Why did I have the feeling that an enormous opportunity had just been wasted?

We must leave for California. I need a land of ecstasy, a secret Abbaye. I cannot live much longer on the East Coast. Now I hear Janine's little Opel driving into our garage. I am happy and I do not know how to measure my own happiness.

Willingboro. Wednesday 6 August 1969.

Fred called me from Chicago tonight.

"I was looking out the window over the empty lot across the street. I saw a man who was running back and forth with a huge butterfly net, under the light of the mercury lamps, and I thought: It's been a long time since I've called Jacques!"

Willingboro. Thursday 14 August 1969.

It is hard to believe that we are spending our last few days in New Jersey. We don't open the windows any more for fear of all the mosquitoes.

At the office my mathematician friend and I had a conversation about consciousness. I reported to him a new question posed last night by the fertile mind of five-year-old Olivier, who wanted to know "How did it all begin, the days?"

Olivier has a theory that the world started "when all the grownups were at the hospital," since that is where babies come from. A deduction of impeccable logic. My friend tells me that as a kid he was fascinated with the observation that human thought could think about itself. That aspect also captivated me: the identity of every "I" in its relationship to the whole being. Could it be that there is only one "I"?

Willingboro. Saturday 30 August 1969.

In the small room downstairs boxes of books are piling up again. Yesterday I took down all the shelves. Janine went out to buy our airline tickets. Her car is sold and I have officially resigned. We have a lot of practice in mobility. She feels like me, a traveller on the earth. To move is a magical experience, a distillation. One of my colleagues commented, when he learned the news: "I knew you wouldn't last very long here. I've been watching you work. It was like standing before the tiger's cage at the fair."

Carl Jung writes in *Alchemical Studies:*

> I had learned that the greatest and most important problems of life are fundamentally insoluble. They must be so, for they express the necessary polarity inherent in every self-regulating system. They can never be solved, but only outgrown.

What is the method making such developments possible?

> We must be able to let things happen in the psyche. For us, this is an art of which most people know nothing. Consciousness is forever interfering, helping, correcting and negating, never leaving the psychic processes to grow in peace. It would be simple enough, if only simplicity were not the most difficult of all things.[20]

I believe I have understood that lesson in "difficult simplicity." All the movement in the world, including scientific progress, is rooted in the unconscious. The dogmatic view of science as an Absolute is a pretentious notion that has survived from the days of aristocracy, an illusion carefully fed and maintained by a privileged elite because it legitimizes scientific power. That the Rationalists should be the first victims of this idealistic illusion is an admirable and touching fact.

Science often denies any new observation or idea that challenges the Absolute. The true progress has to take place underneath the official structure.

When Hynek and I first met in 1963 he wanted to scrupulously preserve academic formalism while bringing UFOs into the forefront of scientific debate. He was anxious to publish a major article on the subject in a recognized journal, a laudable goal indeed. For my part, I wanted to inform a broader public and to quickly reach the younger generation, the scientists of the future who, I hoped, would turn out to be more open-minded than the men I saw in power.

Today, six years later, Hynek still hasn't been invited to submit an article on UFOs for *Scientific American* or *Nature.* A forthcoming meeting of the American Association for the Advancement of Science is supposed to discuss the subject but it is already shaping up as a futile debate about philosophical generalities. I have decided not to attend.

Discoveries are often born out of legend, chance and chaos. The deeper psychic chains remain invisible, even if scholarly books later rationalize the process and demonstrate in hindsight its inescapable logic. The fabulous ocean of human unconscious and conscious dreaming bears the little raft of science over its foamy waves like a mere toy.

If folklore expert Hartmann is right when he says that "human imagination obeys fixed laws," then it ought to be possible to speed up or slow down the development of science in a given culture, not by the play of budgets but by spiritual means, or by exposing the target culture to novel images. Is that one of the grand mythical roles served by UFO phenomena?

Willingboro. Sunday 31 August 1969.

An ultimate image of New Jersey remains fixed in my mind, with the parking lot of the software factory as a backdrop. Olivier has come with me to help carry out some boxes to pack our books. Today I don't have time to let him play with the terminals connected to our machines. I see my son's little face behind me, framed by three big boxes he carries as he stumbles about. The thick, humid East Coast summer hangs around us. We can't see any blue sky: only powdery, heavily polluted air mixed with the ground dust and the sweaty asphalt. Summer has turned into an endless ordeal.

Last Thursday I flew back to California for more job interviews. This time I slept at the Hotel St. Francis in the center of San Francisco, courtesy of Shell Development. I saw Paul Rech again. He introduced me to one of his colleagues, systems manager Mike Kudlick. We were friends immediately. Yet there was a curious atmosphere of insecurity in their shop, because people are often "promoted" out of the area by the company, generally to Houston where Shell has its American headquarters. That isn't my idea of a promotion.

I paid another visit to Stanford and something interesting happened there. I had been asked to give a seminar on data-base management. They even asked me if I would like to join the Faculty. The talk was well received. I had hardly stepped down from the podium when a short, red-haired fellow with a short beard approached me, introduced himself as Rod Fredrickson, director of the Campus facility for computing, and asked me to come into his office. I suddenly realized that this was the man whose team had built the extraordinary operating system I had admired on my last visit. He closed the door behind us.

"Look," he said as he sat down next to his massive IBM terminal, "I'm a guy with a very big ego. I believe that if you give me a problem and you give me a computer I can solve that problem with that computer, period."

I looked at him, wondering what would come next. There wasn't much I could answer to that introduction.

"There is one exception to what I just said," he said with a chuckle, "and it's information retrieval. I don't understand it and I can't fix it when it goes wrong. All the projects that use data-bases around here are screwed up, including the Communications Department project which is eating up more and more time on my machine. You seem to know what you're doing in data-bases. If you want to come here instead of working for those guys I'll hire you as manager of information systems for the Campus. But if you haven't fixed the problem for me in six months you're fired."

"When do I start?" I asked him.

We struck a deal. Next week we move to California.

Willingboro. Monday 1 September 1969.

It makes me nervous to keep this diary, to maintain a regular catalogue of fantasies which are only a rough draft of future life. It would be much easier to write after the fact: then fevers and follies could be carefully built up or smoothed out, the sudden impulses could be explained, transitions could be skillfully managed.

What interests me here is the process by which thought bears upon the real world and eventually results in action. But thought can only be observed when it is caught in the inspired jumps and somersaults of everyday impressions. Too bad for the historian who naively believes he can describe reality by drawing an average line through such follies. It is the peaks and valleys of individual life that count, even if they don't lead anywhere. The doings of famous people are important, like the dates of major battles, but they don't give the essence of an era.

All I can do here is describe and report, capturing specific points of view, perspectives on temporal landscapes that are always contradictory, illogical, linked together by the single irrefutable fact of my own consciousness, my painful awareness.

Unfortunately I won't have much time to go on keeping this Journal once I start working at Stanford.

Stanford. Sunday 7 September 1969.

Wednesday was our last day in New Jersey. By nightfall we had emptied the house. Warm rain fell in the narrow night. We left the children with Maud in a motel while we drove the car to the movers. In the middle of their warehouse, incongruously, we could see our sofa, lost like a small rowboat among the battleships in the navy yard. Finally our taxi arrived to take us all to the airport. The rain had become intense. Several intersections were flooded. A stormy weather system covered the whole Atlantic coast that night.

We first flew into Chicago where Janine visited her brother while I met with Allen and Don. I spent Friday talking to Fred. Their world has not changed.

For the last three days I have been showing Janine what little I've learned of San Francisco and discovering new parts with her. We had a picnic with the children in a clearing at the foot of giant trees. We have started to visit houses all over the Peninsula.

This evening I saw Half Moon Bay again. A half-naked girl on horseback was riding on the beach, her breasts in the wind, a can of beer in her hand. Behind her the ocean in turmoil was throwing surfboards up high above the waves. We heard the laughter, the music of a very young world where everything is possible, everything could easily be destroyed and rebuilt, and that colorful name: Pacific.

Stanford. Sunday 14 September 1969.

I just had a night of strange dreams and sadness. I feel uprooted, caught up in a swirling current. Today we drive up to Mount Hamilton to visit Lick Observatory. In three weeks I fly back to France for the next International Standards meeting and another encounter with General Lochard. I am told that his French friends have counter-attacked: they have hired an American from IBM France to sit behind the *Tricolore* and work out the tougher definitions with me!

On my return we will settle in our new home at last. I will be able to resume my private research into the paranormal. But it will probably take much longer for my emotional being to find its bearings again.

It is only when I hold Janine in my arms that the world stops, opening up an immense landscape. She is still the girl in the simple dress with whom I used to listen to Mahalia Jackson songs in my tiny bedroom near Porte Champerret; she is the pretty psychologist from Lille who used to pick me up at the little observatory in her blue Renault all shiny with rain; the smart career woman in Chicago, carrying thick computer listings under her arm, and smiling at me through the snow and the wind; my companion for the whole journey.

Olivier is often upset these days. He has left his friends behind again, has no school to go to and runs around in narrow circles. At other moments, however, he is loving and very tender with us, no doubt sensing what we are going through. Yesterday he presented us with his treasure, a little stuffed donkey, for us to sleep with.

Yesterday we went to the beach and Catherine braved the tide, both little feet firmly planted in the cold sand, shiny waves crashing around her in silver mist. When the wind turned cold we climbed up the cliff. Perched on my shoulders she turned around and yelled poems of her own invention to the Pacific far below.

The first reviews of *Passport to Magonia* have appeared. Many UFO believers are upset because I question the "nuts and bolts" model: for them flying saucers can *only* be spacecraft sent here by some other civilization in space. "Vallee has gone off the deep end," I hear these people say, denying any parallel between reports of ufonauts and other strange visions in ages past. Many of them had refused to consider these same "ufonauts" until *Anatomy* was published three years ago. But things are not as simple as they would make them.

The success of the book testifies to the fact that some fraction of the public is seeking a more sophisticated statement of the problem: such scenes have always existed, they have always inspired our dreams, perhaps even our science.

The *Library Journal*, however, is strongly recommending *Magonia* as "the one book on the subject even the smallest collections should want to stock."

This brings me to the most recent meeting of our Invisible College at Hynek's house. Fred began by saying again that we must look under the bed. There must be a secret study in progress somewhere, he insisted.

"Do you say this because you have some new data, or are you just posing the question again?" I asked him.

"I don't have any new data," he confessed.

Bill Powers then said what he always says whenever someone raises the issue of a secret study of UFOs. "That can't be true. The story would leak out. We would meet their agents in the field. You shouldn't assume these guys are intelligent just because they work in Intelligence."

Fred answered what he always answers when Bill says this. "They may not need the same data we are seeking. Why should they bother with civilian sightings? They would have access to gun camera footage, radar records. They have the ability to put a smokescreen around the whole business. It wouldn't necessarily leak out."

We have heard both sides of this argument many times. Hynek broke in.

"If I know that another astronomer is using a telescope of eighty-inch aperture to look at the same stars I am investigating with a small Questar, what do I care? My own research satisfies me. Nothing guarantees that his results will be more scientifically significant than mine. Perhaps I am smarter than he is. Perhaps I am a better observer. Perhaps I am looking at some characteristic he hasn't identified yet."

"That's true," I said, "but it isn't a very convincing argument. All it says is that you don't want to look under the bed."

I did not bring up the Pentacle letter. Neither Don nor Bill were aware of its existence, which has remained a tightly guarded secret between Fred, Allen and me.

I have to agree with Fred when he says that the Air Force kept Hynek around only as long as he was silent. I came to Evanston six years ago and put pressure on him, urging him to change his stance. A string of important cases forced the issue. When he started talking, arguing for a new study, the Air Force simply pushed him aside. First they defused the issue by getting their most vocal opponents to testify before bogus Congressional Hearings; then they selected Ed Condon, a physicist who was about to retire, and he signed his name to a report which was a travesty of science, yet reassured the establishment. They used that report to bring about the liquidation of Hynek's position, but they were careful not to fire him.

Allen is now fifty-nine years old. He still goes to Dayton regularly and remains on the payroll as a part-time consultant. He never sees Quintanilla, who still works there with a lieutenant and a couple of secretaries. He is received personally by the commander, who hints he might put him on his own list of consultants some day. Colonel Winebrenner, a former military attaché with the U.S. Embassy in Moscow and in Prague, takes him to lunch at the officers' club.

"What do you talk about?" asks Fred.

"Oh, we talk about lots of things, ordinary things like the weather," answers Allen innocently. "We talk about the places we have both travelled to. Foreign foods, European cuisine. He snaps his fingers and old bottles of Château-Latour materialize on the table in front of us. He speaks to me in Czech. He even gave me a copy of a UFO novel, *The Fortec Conspiracy.*"

Hynek has been charmed and neutralized by the Air Force. That doesn't mean anything, and it especially doesn't mean there is an ongoing secret study. It is highly undesirable for the Air Force of any country to have the citizenry believe in the reality of a phenomenon against which our jet fighters are powerless.

But what kind of science is this, if it is only allowed to discuss those phenomena it can explain? There should be room for other phenomena, for another science. We have the opportunity to help with its birth.

Sadly, our discussion went nowhere. Hynek yawned, got up from his armchair and suggested we listen to a magnetic tape he had received from a UFO witness in the Northwest. I wasn't ready for another light in the sky. I said we should first agree on some sort of practical, tangible joint action.

"We don't get together that often any more; we should use the time as productively as possible," I said. "In any case Don and I are committed to continue working on the landings catalogue," I pointed out somewhat bitterly. "I assume we have at least your moral support. Allen and Fred could consolidate the contacts we have begun with scientists who have volunteered their help, to form a real information and investigation network."

"That's really what we need," Hynek says, "making the Invisible College visible. But it should never become a formal UFO group. That would defeat its purpose completely. Hell, it would be a disaster if it turned into another NICAP, or another APRO, with dues and memberships and a whole management structure, and people fighting for turf all the time."

We all agreed we wanted no part of such an organization.

The next day Fred insisted on driving me to O'Hare. He wanted a word with me in private. He showed me some very interesting photographs taken from an airplane.

"Do you know who took these? Allen did! But he hasn't recorded the place, the date or the time. . . ."

It turns out that Allen was aboard an airliner when he suddenly noticed a white object at his altitude, seemingly flying at the same speed as the plane. He made sure it wasn't a reflection and he convinced himself it must be some faraway cloud with an unusual shape. He pulled out his camera "to see how fast he could snap pictures." In all he took two pairs of stereoscopic photographs and gave it no more thought.

Fred only learned of this a few weeks later. By then Hynek had lost the negatives and one shot from every pair was missing. All that was left consisted of two enlargements, taken separately, showing a well-defined white object, the top rounded like a lens, the bottom cloudy and asymmetrical. Naturally the loss of the negatives makes it impossible to determine whether it was really a cloud or not. Fred is indignant: "Sometimes I have the feeling Allen just doesn't want to know," he says.[21]

Stanford. Sunday 21 September 1969.

Yesterday, as we were all horsing around in the motel suite, Olivier fell hard against the angle of the coffee table. Blood gushed out of his forehead. We drove to Stanford Emergency. An intern arranged his head under the light, closed the wound with a few stitches and told us there would be no worrisome consequences. Olivier had not even lost consciousness and did not seem in pain. Yet the incident made us all aware of how vulnerable and insecure we were. All our earthly possessions will have to stay in storage until we find a house. We live precariously in two rooms at the Tiki Inn on the edge of campus, eating every night with the student crowd at the nearby sandwich and pizza place, where we listen to rock music and put quarters in the Pong machine.

Today we drove over the San Mateo bridge to visit Berkeley. We had lunch in the foothills of Mount Diablo.

I am reading a book about World War II as seen by the simple folks of Normandy and I wonder: how many Americans my age could understand what that era meant? How many have ever discussed the war with an actual witness of it? How many, for that matter, have ever seen a horseshoe maker at work, a water mill in operation, a woman using a washboard in the river?

Stanford. Wednesday 24 September 1969.

We have found a fine place and we are trying to buy it, with the feeling that we are making an investment much beyond our means. It is a large wood-frame house in the hills of Belmont, so filled with light it seems magically suspended in mid-air over the landscape.

It is hard to pick a location in this land of microclimates and microcultures. We almost fell in love with a very similar house above Half Moon Bay, but it would have been engulfed in fog a good part of the year. We decided we wanted the sun.

Today is my thirtieth birthday. In one week I go back to France for another standards meeting, but this time I know I will be able to return to a real home.

Paris. Friday 3 October 1969.

The kind futility of France enfolds me once again. I am reconciled with it. I experience with delight this country that doesn't change, its amiable folks hopelessly dominated by an egotistical elite with an enormous ability to exploit those it should serve.

This afternoon I walked all over the Latin Quarter with my mother. The weather was warm and diaphanous, with that delicate expectation of Autumn that the light places in the eyes of the people, the kids playing on the sidewalks, the stained glass windows.

On rue Saint Jacques I paid about five dollars for *The Other Side of Nature* by Mrs. Crowe, but I was disappointed by the little bookshops along the Seine. Little of value is to be found among the *bouquinistes* any more, only silly watercolors and overpriced copies of the Kama-Sutra for sexually repressed tourists. At Vivien's bookstore the owner recognized me and shook my hand. I bought an old parapsychology book from him. It shows curious photographs of grave experimenters watching a table floating up into the air through the psychic action of medium Eusapia Paladino. One of them, a young man with a beard, is none other than astronomer Camille Flammarion.

At *La Mandragore*, a miracle! The exquisitely rare, complete three-volume set of *Les Farfadets*, the fanciful first-person story of a nineteenth-century man named Berbiguier who was convinced God had sent him to the earth with the specific mission to destroy the plague of these ugly little devils. The most curious thing about the book is that the act of publishing it seems to have cured the folly of its author, who spent his last years buying back and destroying as many copies of his work as he could, thus increasing the value of the few remaining ones.

Maman and I had coffee near the Odéon. We spoke about the era when spiritism was in vogue. She remembered that time well. She told me about the fascination of the Belle Epoque for ghosts and mediums. Her own father, a businessman, was a member of the French Astronomical Society and often attended experimental séances inspired by Camille Flammarion. Unfortunately, in the fashion of the time, a superb banquet usually preceded the scientific work, and the liquid spirits played a more extensive role than the paranormal ones.

After coffee we decided to pay a surprise visit to my uncle. He was putting the final touches on an equatorial mounting of his own design for a new telescope, although he can hardly hope to see the sky from the dark recesses of his apartment. Such details have never bothered the scientific enthusiast. We came back on foot at nightfall. Workers were digging up the square in front of Saint-Sulpice to build a large underground parking lot.

Few other nations have that ability for happiness the French possess, so deep and clear. It can be seen in the lovers' quiet abandon, in the smile of passersby. Kids take each other by the arm familiarly, friends embrace in happy ways. Why are Americans so afraid of touching each other?

Paris. Saturday 4 October 1969.

This city is a hot elixir of gasoline and melting asphalt, its smell a mixture of the burning metal of car engines, perspiration of crowds, effluvium of coffee and chocolate, perfume of elegant women and the ever-present whiff of dog crap. In this vast cauldron the fragrance is as heady and stupefying as the atmosphere of an opium den.

Something has changed in French social life. Is it because De Gaulle is finally gone? People do seem more free.

What was the probability that I would meet Granville on rue Monge, when I have only seen him once since the Jules Verne Prize? He was walking towards me, lost in thought, wearing a purple shirt and a gray sweater, carrying a grocery net. We laughed at this chance encounter, and we laughed even more when I attempted to explain what had happened to me: that I was spending my sixth year in the United States, but that I had lived in France for most of 1968; that I was here for a computer science meeting, but as a member of the American delegation; that I was still interested in astronomy but I hadn't worked at the telescope since 1963.... I took him to a Chinese restaurant nearby to catch up on our lives.

Paris. Sunday 5 October 1969.

The whole city was happy and carefree last night. Near the Tour Saint-Jacques whose stone chimaeras have finally been freed from the workers' scaffolding, it seemed that Paris had almost recovered from the loss of Les Halles. On the Beaubourg square a flea market for old metal objects was a paradise of rusty chains, iron lamps, unmatched wheels.

Paris. Saturday 11 October 1969.

Autumn has come, the subtle doorway into the new season has been crossed. I know it in the sudden depth of faraway engine sounds, in the lingering whistle of tires on wet pavement down there on rue Monge. Paris is a fine soft machine, recording and absorbing indifferently the changes of the millenium.

On Wednesday night I had dinner with Granville. In our conversation could be heard an echo of our old debates.

"We are the men of judgment," he said. And his own judgment on this superficially happy country was pitiless: he described to me the work of a certain ministerial commission on education, the endless studies for the reform of the school system, the despicable way retarded children are treated. His voice was that of someone who was appalled, discouraged, angry. Under the pretense of vague experiments nothing is done. French education, he said, was sinking into chaos.

Nor is there any real move towards greater moral freedom: only the annual national orgy at Saint-Tropez beach, and in Paris a few groups of hippies who are carefully preserved, like the Mona Lisa, to show the world how daring and creative we all are. In reality the true feelings of yesteryear are gone; the old intellectual circles are broken; the enthusiastic social movements are dying, he claims, as unlikely to be heard again as the plaintive guitar in our little café.

"They have created an Old Town, just as your Americans did in Chicago: the snobs from Neuilly come over every evening to see what the snobs from Passy are wearing. Golden adolescents in Jaguars pretend to reject the conservative world around them and sip their whisky while perfumed minions parade back and forth on the sidewalk. Tourist buses from Dusseldorf gape at all this and are amazed. Is it the same in America?"

No, it is not the same, I said. I described the insane tom-tom of an African priest in Washington, the lawn on Dupont Circle where a drugged-out girl breastfed her baby while the Black Panthers made speeches about setting the entire continent on fire. Something drastically new and unpredictable is happening in America, I told him.

Paris. Sunday 12 October 1969.

Aimé Michel suddenly appeared at my bedside at ten this morning while I was quietly reading Berbiguier. He is turning fifty, but has lost nothing of his enthusiasm and wit. We called up Guérin and we all went to lunch together. Guérin believed that a vast cover-up explained Blue Book, the Air Force, the Condon Report and everything else. His theory was told in a breathless tone but it left me unconvinced. I think there is indeed a cover-up, but it isn't that simple.

In Guérin's old office at the Astrophysical Institute where everything began, the very same basement office he had ten years ago, we argued about flying saucers. We walked down the same dusty hallways crowded with wooden crates which testified to bureaucratic carelessness. Around us the heavy benches, the antique electrical instruments and the bulletin boards where Union demands were pinned, threatening immediate strike, added a smell of constrained rationalism to the place, with that special lingering odor of crushed ideals and mental sweat without which no French research institution could ever attain excellence.

We laughed, we argued about every camp in ufology. We debated the case of "Doctor X," the health official from Provence who watched in amazement last October as he saw two large disks merge into one. Aimé gave us the unpublished details, which involve levitations and strange healings. And he mentioned the red triangle which appeared on the man's abdomen and on that of his child, for which his colleagues in dermatology have no explanation.

Guérin has not changed much. He is kind and warm, always excitable, open to all arguments presented in good faith, capable of uncommon depth in his analyses, even if he sometimes falls into dogmatic attitudes. Perhaps he lacks knowledge of the larger world, but he is pure of heart, a rare feature in this city of false intellectuals and consummate politicians.

Tonight the shopowners are demonstrating. Once more the Latin Quarter is full of cops in riot gear. The scene reminds me of the days of Charonne. Have they learned nothing?

Paris. Wednesday 15 October 1969.

At thirty I can leave behind the memories, the hope and the rage that took me away from the darkness of Champerret and the drizzle of Lille, from the night dew of Meudon. We left France, burning to see other crowds, other lives. Why deny that I am very happy?

Yet our research has not answered the questions which started it. I still don't know what I saw in the sky over Pontoise in 1955. I still cannot tell what game various governments may be playing with UFOs.

People wear so many masks! That much I have discovered. Our words are masks, just as the daily newspaper is a mask journalists of different persuasions put over current events and the facts of life. Gestures, habits, customs are masks. Official science is the mask we put over knowledge so that no one will glimpse the terrible grimace or the tantalizing, seductive beauty of the unknown. Doctrines and philosophies are the masks that fit over changing, unpredictable, iridescent human thought. Academies are the rigid masks we impose on the face of creativity and of discovery.

There are very few times when we see the world without its masks, the real world of man and things rather than the carnival of our own delusions. My childhood visions of the war bore the unmistakable stamp of authenticity; so did my parents; Aimé Michel drinking Chinese tea and speaking about Liszt; the evenings at Bryn Mawr with Allen Hynek,

Fred and Don; the naked girl of Half Moon Bay; dawn coming over us; our silences.... Those are the few exceptions.

Such images condense into a single sphere that shrinks to a tiny point in arbitrary space, that space we call mankind. Then a haze comes over the magic mirror, my consciousness returns to drown the vision into the trivial present again.

Stanford. Saturday 13 December 1969.

The plane has taken me back to our house on the hillside, well-sheltered from the cold foggy Western wind. From our redwood deck we can see the Southern part of San Francisco Bay, from Mount Diablo to Mount Hamilton. So this is where we are going to live.

At Stanford the windows of my office in the computing center are always open. There is a flow of fresh air, heavy with the fragrance of a giant eucalyptus tree nearby.

Our weekend excursions around the Bay leave us with postcard visions of horses drinking from ponds in the hills, fiery sunsets beyond the broken fences of old ranches, campfire smoke in the steep forests. Why does it all seem so unreal?

Everyone knows such scenes belong in movies. Reality is supposed to be made of rain and blood, mud and grease.

I am aware of a faraway future, with no other country than the whole planet, no other nourishment than Janine's smile, no other vision than this: a tremor in the great skies above, a cloud we catch as we fly, high over the stormy currents. What is this diary, this record of a few passing years? Only the iridescent foam left behind on the beach when the big waves of existence arise, swirl in tumult and go away as we will go away, my love.

EPILOGUE

The passage of time is highly corrosive. Not only does it erase from our memory many facts, dates and figures, but it erodes even our perception of those human beings who have made an impact on our lives, and it distorts our view of ourselves. A diary, kept religiously enough, is a formidable weapon against this erosion. But it also makes our shortcomings more obvious, our failures more plain. A source of humbling experience, it puts even our proudest achievements in the perspective of far greater accomplishments by others.

This volume would be flawed and incomplete if the story was simply allowed to stop abruptly at the dawn of the seventies. Nearly a quarter century has elapsed since the last entry of this diary was written. It is natural for my readers to ask what became of the protagonists, what findings were made in the intervening period, and how did the events described here determine the present state of the UFO problem and influence its future. These issues can be addressed under five major headings.

+ + +

The ***first issue*** *involves the over-arching question of the reality and possible nature of unidentified flying objects. The sad truth is that they remain as much a mystery today as they were in the sixties.*

The major cases I had recorded in the Journal as they were unfolding—Socorro, the Michigan "Marsh Gas," Monticello, the Hill investigation and many others—were followed by equally sensational events in the seventies and eighties. The new cases, such as the encounter involving two fishermen at Pascagoula, or the Travis Walton abduction, captured the headlines and repeatedly sent Allen Hynek before the bright lights of the media, only to be forgotten a few weeks later. These cases augmented the database but they elicited no new pattern. On the contrary, it seemed that the UFOs took a sadistic pleasure in sending us confusing signals.

A new computer analysis of historical trends, compiled in the mid-seventies, led me to plot a striking graph of "waves" of activity that was anything but periodic. Fred Beckman and Dr. Price-Williams (of UCLA) pointed out that it resembled a schedule of reinforcement typical of a learning or training process: the phenomenon was more akin to a control system than to an exploratory task force of alien travellers.

There are many control systems around us. Some are part of nature: ecology, climate, population development are common examples; others are social, like the process of higher education, or the Justice system, or a concentration camp; still others, such as the attitude control of a rocket or satellite, or the humble thermostat on the wall of an apartment, are built by man.

If the UFO phenomenon represents a control system, can we test it to determine if it is natural or artificial, open or closed? This is one of the interesting questions about the phenomenon, a question that has never been answered.

The publication of such ideas in *The Invisible College*, a book I wrote in 1975, strongly polarized researchers, because the issue of the psychic nature of the phenomenon naturally had to be raised in the same breath. This was anathema to many people for whom flying saucers could only be nuts-and-bolts spacecraft, an idea we had already left behind.

Coming a few years after *Passport to Magonia*, the publication of *Invisible College* deepened the cleavage between my research and the believers' party line. It has continued to widen to this day and has grown into a chasm.

On one side my colleagues in science "know," or believe they know, that the field is utter nonsense and that the witnesses are either hoaxers or poor observers tricked by hallucinations. As for my friends in ufology, on the other side, they "know" with equal force that these objects are extraterrestrial. I cannot join either camp on the basis of the data I have accumulated. Sometimes I get the awful feeling that I am the only human being who doesn't know what UFOs are.

The UFO occupants described by witnesses in close encounter cases are variously designated in the literature as Aliens, Visitors, Humanoids or Operators. They have continued to behave like the absurd denizens of bad Hollywood movies, giving no sign that their purpose on our planet was related to any sort of rational process. Worse, among the thousand or so abduction cases that have reportedly been analyzed, no pattern has emerged that could be positively correlated with a visitation by extraterrestrial entities. Their technology is a simulacrum—and a very bad one at that—of obsolete human biological and engineering notions. The real mechanism of their elusiveness and their absurdity clearly escapes us. Perhaps this should be taken as a sign that our theories are wrong, that our basic assumptions are flawed?

Before we can proceed we must have a more precise definition of what most ufologists mean by "extraterrestrial." Today the dominant interpretation of the term is still understood at the most obvious level: UFOs are thought to be spacecraft from a civilization that has evolved on another planet. Their pilots are supposed to be humanoids, generally "Short Grays" with large dark eyes who first came here about the time of the Kenneth Arnold sighting of 1947. We are told they are surveying the earth in search of mineral or biological material, that they abduct humans to interbreed with them. It almost sounds rational.

Not all ufologists follow precisely this explanation, of course. There are many variants. Yet the above is a fair summary of the "extraterrestrial" hypothesis prevalent in the current American literature of the field. Its imagery has been reinforced by major motion pictures like *Close Encounters of the Third Kind* (1977) for which Dr. Hynek was an adviser, and *E.T. the Extraterrestrial* (1982). It is also found in countless science-fiction movies and television documentaries.

The fact that many witnesses actually describe something entirely different, which does not trace its beginning to 1947 or even to this century, and only occasionally resembles a spaceship, has been neglected. Even when they describe alien creatures, those do not necessarily follow the standard type of a dwarfish humanoid with grayish skin. The shapes and behaviors offer a bewildering variety.

My own speculation is that UFOs operate in a multi-dimensional reality of which spacetime is a subset. In that sense *I do not completely reject the idea of an extraterrestrial origin*: but I believe that the form of intelligence the phenomenon represents could coexist with us on earth just as easily as it could originate on another planet in our universe, or in a parallel universe.

Scientific training is a heavy burden. I was taught by my mentors that science began with the ability to challenge all theories, including my own. But any questioning of the interplanetary origin of these objects is perceived as a betrayal by those who need extraterrestrial contact as a part of their personal certainty. These people may pretend that they are looking for scientific truth, but in fact they are simply erecting a new dogma.

The stubborn refusal on my part to follow any party line has created some regrettable confusion over the years. Inevitably, various absurd theories have been attributed to me, and statements I never made have been placed in my mouth. For example, my assertion that UFO phenomena were partly psychic in nature was often taken to mean that witnesses were the victims of mere illusions, and that the objects were not physical or material, something I never said, wrote or believed. Later, my observation that a few of the cases were blatantly manipulated by human cults, often inspired by intelligence organizations, was misquoted as a statement that I had renounced my earlier writings and that in my view all "flying saucers" were human secret weapons or instruments of deception. I have made no such statement. To put the matter to rest it is appropriate, then, to restate my position regarding the phenomenon, a position which is consistent with everything I have written before:

The UFO Phenomenon exists. It has been with us throughout history. It is physical in nature and it remains unexplained in terms of contemporary science. It represents a level of consciousness that we have not yet recognized, and which is able to manipulate dimensions beyond time and space as we understand them. It affects our own consciousness in ways that we do not grasp fully, and it generally behaves as a control system.

Because it can manipulate our consciousness in unknown ways, the phenomenon also produces effects that we can only interpret as paranormal in nature. I trust, as Allen Hynek did, that the human science of some future century will account for these effects. Aimé Michel continues to disagree with this assertion, and I understand his objection: no dog in future centuries will ever understand Einstein's relativity, because a dog's brain lacks the adequate structure to do it. Are we in the position of the dog? This is another important issue we have not resolved.

The UFO phenomenon plays a role in many mythological traditions. It has affected our religions and our modern view of the universe. It may well be deceptive in the images it presents to us, masquerading in various guises under different cultures: god-like to the ancient Hebrews or Mesopotamians, elf-like to medieval chroniclers, devilish to Christian inquisitors. It may have manifested in the form of ghosts or rapping spirits for the benefit of our grandfathers at the end of the nineteenth century, or in the form of the Blessed Virgin before devout Catholics. Today we live in the technological civilization of the late twentieth century, and we observe a phenomenon which emulates astronauts in shining space suits.

+ + +

The ***second issue*** *this epilogue must address is that of scientific reaction to the phenomenon. Here again, not much has changed since the Journals were written.*

The only reason the U.S. Air Force was able to get away with its ludicrous treatment of the problem is found in the appalling lack of information, indeed the lack of interest that exists among the academic community in the U.S. and abroad. To most academic thinkers the field of ufology is an aberration. And how could they come to a different conclusion? The genuine data have never been exploited. *The scientific work has never been done*, and the reader of my Diary can plainly see why: a few individual scientists like Hynek and myself and perhaps a dozen others spent their own resources and their spare time documenting tantalizing anecdotes, but the full machinery of science was never brought to bear on the phenomenon. Our greatest failure has been our inability to build a strong enough case before our colleagues, and to get a real investigation started. We simply cannot speculate about what might have been found.

As the Journals show, my early books aroused some measure of private support among a few scientists, of which Fred Beckman at the University of Chicago, Douglas Price-Williams at UCLA and Peter Sturrock at Stanford, who had come to similar conclusions through their own thinking, were courageous examples. A small "Invisible College" continued to develop in later years, but it was not able to undertake long-term research as a group, even when Allen Hynek came out in 1972 with his own perceptive work, *The UFO Experience.* It was a classic effort to launch what he aptly called "the natural history of the phenomenon" but it too failed to create lasting interest among the scientific world.

As I had once predicted to Allen, any effort we made to document the genuine cases and to place them squarely before the public also created a lucrative market for hucksters who could always grab the bigger headlines simply by manufacturing lurid stories, which were eagerly seized upon by television news and by the tabloids because of their shock value. In the minds of many conservative scientists, a phenomenon that was so disgustingly exploited by the media and by wild-eyed, uncritical zealots was automatically deemed unworthy of their time and attention. Hynek's words and mine were simply lost in the noise.

By the mid-seventies those scientists who had a genuine interest in alien intelligence, like Carl Sagan at Cornell and Frank Drake at the University of California, Barney Oliver at Hewlett-Packard and Ronald Bracewell at Stanford, had a less slippery fish to fry: they had become involved in the Seti project, using radiotelescopes in the search for extraterrestrial intelligence through radio signals.

Most astronomers now agree that life must exist throughout the universe, notably on planetary systems surrounding slowly rotating yellow stars like our Sun. Wouldn't an advanced civilization on such a planet use something better than radio to make its presence known far and wide? Someone said derisively that Seti, like its predecessor Project Ozma, was an effort on the part of the long-time-dead to communicate with the not-yet-born! Yet the scientific press gave it much more attention than it did to UFOs. While fascinating as a technical exercise, Seti ignored the many genuine witnesses of UFO encounters who described a form of non-human intelligence interfering with human destiny right here on earth. It still ignores them, perhaps because the raw material of sighting reports does not look "scientific" enough and does not come to us through the channels of pure science. I regard the continuing lack of attention paid to UFOs by science as one of the great intellectual failures of this century.

If the science establishment was turning away, what of private research? From archaeology to medicine, there are numerous examples of rich mavericks or intelligent patrons spearheading novel areas of research that the Establishment has neglected. The names of wealthy families, from Kettering and Ford to Rockefeller and Carnegie, are associated with research foundations representing some of the finest achievements in the arts and the sciences. Unfortunately these institutions never took an interest in this field, although Hynek and others over the years made serious attempts to raise funds for a small, dedicated effort.

Many wealthy patrons have an axe to grind and tend to finance researchers who espouse their theories, investigate their ailments or flatter their eccentricities. Tycoons dying of cancer have occasionally endowed chairs in oncology, and rich families whose children have died young have funded parapsychology projects designed to communicate with departed souls. But the millionaires who truly sponsor speculative, frontier research for its own merits are very rare indeed.

Allen Hynek found this out the hard way. He was repeatedly promised vast sums of money which evaporated as soon as he made it clear he would never compromise his scientific standards. In the summer of 1984, bitterly disappointed with CUFOS (the "Center for UFO Studies" he had established in Evanston), he loaded his personal files into a truck and left for Scottsdale with a promise of support, this time from a wealthy Englishman. Unfortunately things did not go much better in Arizona. Allen asked me to join him there but we soon came to the conclusion that the wealthy patron was only interested in keeping a few scientists in his entourage to promote his personal theories about the phenomenon. Another wild hope bit the dust.

The research situation is very much the same today. Affluent individuals have occasionally provided modest amounts of money, but only to fund those projects that matched their own concept of the extraterrestrial origin of UFOs, to the exclusion of alternative theories. Not only is this an unacceptable bias in any field, but it almost guarantees that even those valid results that might derive from the whole exercise will be rejected as flawed by any scientific committee appointed to assess them. This is akin to someone agreeing to finance a new planetary observatory, but only on the condition that its astronomers be committed to the view that the earth is forever fixed at the center of the universe.

What little research is being done on the subject today is unfortunately exposed to all the vagaries of sectarian thinking. The valuable work of a few private groups continues to be disfigured by intense bickering among various factions. Independent scientists venture into the cross-fire of their vituperative arguments at their own risk, like a tourist caught in a western frontier town bar brawl. Good field research by dedicated investigators (fortunately, they are still numerous and active) rarely sees the light of publication. Many important sightings are buried forever.

By 1976 my own career had evolved. I had become a computer entrepreneur in Silicon Valley, much to Hynek's chagrin: "You should be a professor somewhere," he kept insisting. Well, I had seen first-hand what happened to University professors when they showed too much independence, and I had no inclination to lose the freedom that expertise in high technology afforded me in California.

I saw Hynek often during those years, at my house in Belmont or wherever our travel schedules happened to coincide. Anyone listening in on our discussions would have been surprised to observe that we spent relatively little time on UFOs. To be sure, we did have sharp differences in that regard. He refused steadfastly to question the inconsistencies in the Air Force's policy. As my research led to investigations of ufological sects and cults, I became increasingly aware of the way belief systems were manipulated by outside groups with hidden agendas, although it would be some time before the full horrors of the classified mind-control experiments of that era would be exposed. Hynek and I disagreed on the urgency of pushing for a frank investigation of a cover-up on the part of the agencies I have named in this book. Yet he gradually came to agree with my statement that UFOs were probably not extraterrestrial spacecraft. In October 1976, he courageously told a journalist: "I have come to support less and less the idea that UFOs are 'nuts-and-bolts' spacecraft from other worlds (...) There are just too many things going against this theory. To me, it seems ridiculous that super intelligence would travel great distances to do relatively stupid things like stop cars, collect soil samples, and frighten people. I think we must begin to re-examine the evidence. We must begin to look closer to home."

Invariably our talks veered towards deeper waters: the latest developments in parapsychology, the psychic nature of man and the failure of science to come to grips with higher levels of consciousness. We debated the phenomena of mysticism and the meaning of initiation.

The man I spoke to on such occasions was the true Allen Hynek, and it is a great pity that his colleagues in science and his followers in ufology never heard or acknowledged what he could have offered them in that regard. He knew far more about parapsychology than he publicly revealed. After his death in 1986 his wife Mimi told me that he had wanted me to keep the books he had accumulated on this subject. Today they represent a very special and treasured section of my own library.

In my present professional work in the financing of technology, I often reflect with some bitterness on the lessons of those formative years—

the display of blatant bias and obvious dishonesty given by the various academic projects that dealt with the paranormal, the debacle of Project Blue Book and the spectacle of pettiness and fear given by the pundits of science who prejudged every case without serious consideration, or simply destroyed the data, as my superiors had done when I was a young astronomer on the payroll of the French space committee.

+ + +

The intellectual scandal of the Pentacle document constitutes the ***third issue*** *I must address here. It is hard to excuse the betrayal of science that took place when the Intelligence community decided to bar the Robertson Panel from direct access to the knowledge Pentacle and his group had gained.*

The discovery of the Pentacle document had a major impact on me. It gave me an uncomfortable insight into the practices of government agencies and the high-powered consultants who serve them. If I had remained silent on this issue, as I could have by editing the relevant entries out of the Journal, some of my past actions would have remained incomprehensible. It was the main reason for my return to Europe in 1967. It made obvious some unsavory aspects of scientific policy at the highest level. It provided quite an education for an idealistic young astronomer.

Having said this, I still do not quite know how far one should go in suspecting a sinister design behind this ominous document. A group of CIA officials had convened a panel of the five most eminent physicists in America for what purported to be an objective review of a series of cases of potential importance to science and to national security. The working conclusions reached by a prestigious research institution funded with taxpayers' money and clamoring to be heard were withheld from them, although "Project Stork support at Battelle" was briefly described at the meeting, according to the (now declassified) report on the January 1953 meeting written by F.C. Durant, and addressed to the Assistant Director for Scientific Intelligence of the CIA.

The fact that no member of the Battelle team was asked to elaborate on the findings alluded to in the Pentacle memo is an amazing fact. The reader will recall that the panel was no ordinary group of consultants: Professor Luis Alvarez was awarded the Nobel prize in physics; Lloyd Berkner was a leading space scientist; Sam Goudsmit was one of the acknowledged leaders of American nuclear physics at Brookhaven National Laboratory; Thornton Page was one of the most respected astronomers in the land. The panel chairman, H.P. Robertson, was a world-renowned physicist at the California Institute of Technology.

There will be those who will say that Hynek should have stormed up the steps of the National Academy of Sciences with this document in his hand as soon as I had dredged it up from his archives. They may be right. But Hynek was a quiet man, who disliked confrontation and scandal, feared authority and was in awe of secrecy. He had once told me plainly that *he would not look under the bed even if he knew for certain that something was hidden there.* The document I had found in his files remained poetically ensconced in the frame where I had inserted it, under a color reproduction of a panel from the *Lady and the Unicorn* tapestry. The frame hung for a long time in his office at Corralitos Observatory in the barren mountains of New Mexico, where it was safe from the prying eyes of curious reporters or nosy ufologists. It is only after much soul-searching that I have decided to reveal its existence.

To those who believe in conspiracies, the Pentacle document may come as further indication of a cover-up dating at least from 1953. In a novel in French entitled *Alintel* (published in 1986) I developed a scenario showing how Pentagon-sponsored UFO research could have been taken underground after the Robertson Panel. In that novel I showed how Project Blue Book could have continued as a deliberate ploy designed to deflect away the attention of the technical community and the public, while a very small group of experts went on examining the data. Yet more conservative observers of the UFO scene can justifiably argue that the ominous memorandum is only proof that some important findings were withheld from Alvarez, Robertson, Page and their colleagues, not that an elaborate plot was being hatched. If so, why were the Battelle conclusions not made available? Could it be that Pentacle's clever, detailed recommendations to set up deliberate artificial UFO flaps and simulated cases in selected areas was actually implemented? Is that the explanation for some of the bizarre sightings we were to observe in later years? When I called attention to the blatant manipulation of belief systems that lurked behind such hoaxes, many researchers rejected the whole idea. I found it hard to defend myself, since neither Hynek nor I were prepared to talk publicly about Pentacle.

Now it can no longer be denied that as early as the mid-fifties the Intelligence community was seriously considering precisely this kind of deception, and that it conceived its designs on a grand scale. Later public confessions by independent researcher William Moore concerning some of the covert actions engineered by the Air Force's Office of Special Investigations (OSI) support this scenario. I have reviewed these plots in a recent book called *Revelations* (1991) and it would be useless to belabor the point. I must leave it to future historians of the field to decide objectively whether or not there exists a secret project along the lines of *Alintel*. But the Pentacle document represents the kind of negative factor in the inner workings of science that sociologists would do well to investigate instead of spending their time heaping ridicule on the witnesses of the UFO phenomenon who are only trying to offer their testimony as a gift to modern research.

Today it seems likely to me that the executive branch of the U.S. Government and other major governments do know of the existence, physical reality and awesome implications of the UFO problem. It seems obvious that an agreement is in force among them to keep the data quiet and to discourage independent research. The repeated experiences we had in the sixties when our discreet requests at a very high level within the French government hit a brick wall of secrecy and denial are a strong indication in this direction. Allen Hynek's experiences with Washington were similar. Such withholding of data without Congressional authorization is illegal, of course. The military has no right to deliberately lie to the citizenry or to mislead scientific researchers on such a fundamental issue. But when we look for a deeper, more sinister conspiracy we may simply be underestimating the depth of bureaucratic stupidity. More light should be thrown on the whole problem.

Unfortunately my hands are tied: As a private citizen, I have no authority to reveal the name of the person who signed the memorandum, or his associates. Dr. Hynek never did, although he occasionally hinted that the whole story of Battelle's involvement had not been told. So far, my efforts to establish whether or not the memorandum is still secret have gone nowhere: nobody even seems to have a copy. The Air Force claims it has kept no relevant files from that era. Wright-Patterson Air Force Base says it has no record of Pentacle, and the National Archives respond to my attorney's inquiries with standard brochures about Project Blue Book that are intended for fourth-graders. My hope is that a patient investigator will eventually go through their collections, recover the actual text and obtain its release so it can be debated in the open.

+ + +

*The **fourth issue** that demands to be raised here concerns the sectarian temptation that is undoubtedly present among UFO believers. Anyone who calls himself a serious researcher in this field, and has the courage to confront the skeptics, must also have the courage to expose the dangerous paranoia that is rampant among many zealots of the extraterrestrial Cause.*

When he left the field in April 1991, author Whitley Strieber characterized ufologists as "the cruelest, nastiest and craziest people I have ever encountered." This judgment is too harsh because there are plenty of hard-working, open-minded and generous researchers who have made remarkable contributions to our knowledge of the UFO problem. Unfortunately they are over-shadowed by vocal believers who react to criticism with all the sting and venom of zealots defending a religious dogma.

Any researcher who has not tried to engage the advocates of UFO nuts-and-bolts reality in rational debate can have only a faint idea of what the term "vitriolic" means. These advocates claim that they want to see scientists become involved in the study of the phenomenon, but it does not take long to realize that they only want scientists who agree with their preconceived ideas of its origin and nature.

Some of the more extreme, paranoid views are making an impact upon a wider segment of the public today because the very sensitive, emotionally charged matter of abduction research has become a central obsession for contemporary UFO groups. Various writers who have only limited experience with clinical psychology are allowed to interrogate witnesses under hypnosis, often leading them to fantasize in the direction of their own preconceptions. They spread them to a wider circle through books, films and lectures. Under a warm, comforting and sympathetic, even paternalistic appearance, these writers may be augmenting, rather than healing, the trauma felt by the witnesses; they create a dangerous sense of imminent crisis that heightens the anxiety of their followers.

A summary of the abduction debate may be useful at this point.

It should first be noted that during the period covered by the Journal, abductions were already recognized as one of the most interesting facets of the phenomenon. The case of Vilas Boas in Brazil had been researched by Dr. Olavo Fontes and published in English by Gordon Creighton. The reader will recall the many conversations Hynek had with these men and also with witnesses Betty and Barney Hill, with Dr. Simon and with John Fuller, the gifted writer who must be credited with first calling our attention to the curious fact of "missing time."

There were already a dozen abduction cases in our files by 1970. Some veteran researchers, like Coral and Jim Lorenzen, had accumulated many more. It was clear that abductions had been a part of the mystery since the earliest period. It seemed that the problem we were trying to tackle was a much more formidable one than the arrival on earth of space visitors, impressive as that possibility might be. The phenomenon challenged not only our definitions of physical objects but our concepts of consciousness and reality. At the same time it brought into question the entire history of human belief, the very genesis of religion, the age-old myth of interaction between humans and self-styled superior beings who claimed they came from the sky, and the boundaries we place on research, science and religion.

The abduction experience, in my opinion, is real, traumatic and very complex.

It is unfortunate that the small group of researchers who studied such cases did not stop to carefully develop an appropriate methodology. Instead, abduction research tended to drift into disputes between those who thought the Aliens were here to help us and those who saw them as evil. Betty Hill herself summed up her own disappointment when she retired from the field in September 1991, citing "flaky ideas, fantasies and imagination" in the treatment of abduction stories.

When I search for a reason for this state of affairs I find two major factors. First and foremost is the passing from the scene of those elder statesmen who could have cautioned against hasty conclusions. In particular, the late Dr. Hynek and the Lorenzens had accumulated enough experience with abductions and their investigation under hypnosis to recognize both the implications and the limitations inherent in such cases. But the new self-styled "experts" had no such hesitations.

The second factor that made alien abductions so visible in the late eighties was the sudden rise of tabloid television in the United States. Sensational new interview shows replaced the more sedate afternoon programs of previous years. Television stations concerned with their weekly ratings loved abduction stories because of their high visual and emotional drama.

As a result, when I hear skeptics like Carl Sagan or Paul Kurtz thunder about the growing dangers of the irrational in our society I find it hard to argue against them. Yet the perils created by the hard-core believers are only proportional to the neglect of the problem by those leading scientists who deny the reality of a phenomenon they have never attempted to investigate.

The UFO phenomenon is one of the major scientific and social mysteries the twenty-first century is about to inherit from our own. To investigate it responsibly we need the guidance of open-minded specialists ready to set some standards and to develop new methodology in the treatment of the more traumatic cases. This can only be done in the calm setting of the laboratory, not against a background of televised debates or screaming tabloid headlines.

+ + +

*The **fifth point** I must touch on before bringing this book to an end concerns possible future avenues of research, and my own plans within it.*

I believe that there is a larger issue behind the UFO phenomenon. I continue to be optimistic about science's ability to come to grips with unforeseen observations, paranormal phenomena and radical discontinuities in knowledge.

Witnesses are generous in bringing us some remarkable observations that are begging for an explanation. If the data challenge our view of reality, that is not the witnesses' fault. The burden is squarely on the shoulders of the scientists, to patiently sift through the occasional mistakes in perception that are undoubtedly reported, to eliminate the hoaxes and to document the gems of truly unexplained phenomena. We must do this responsibly, with respect and care for those who offer us their testimony, and with a realistic self-awareness of the limitations of our science. At a time when concepts of the physical universe are undergoing a major revolution, the UFO phenomenon is of unique value for theoretical development. It provides no solution, of course. Even if we were to recover bits of hardware and samples of alien flesh we might not be able to make sense out of them for centuries, but this should not surprise us.

The history of science is filled with anecdotes about phenomena that were clearly recognized, but not applied for a long time. The ancient Egyptians, for example, knew of the magnetic properties of certain metals and their jewelry shows evidence of electroplating, yet they never developed an understanding of simple circuits. The eighteenth-century astronomer Messier observed and named the major nebulae in the northern sky, but it is only in our own century that they were finally recognized as galaxies outside our own. The same even goes for modern technology: the principle of radar had been known for fifty years before it was put into practical application at the end of the Second World War. The list is a very long one. For an anomalous observation to be incorporated into a new theory, many concepts have to mature until a match can be realized. We have not yet reached this point in the study of UFOs. This does not authorize us to throw the data away, or to disregard the phenomenon. On the contrary, carefully guided by the physical parameters from the best cases, research on alternative topologies for our own reality can already proceed.

In the seventies, French author Jacques Bergier, a keen observer of technological and cultural trends, once told me that we must revise the old notion of a single "universe." Perhaps the main lesson to be drawn from UFO sightings, he said, was that we are living instead in what he called a "Multiverse" with many more dimensions than we had suspected. He urged me to think about the numerous ways in which a conscious control system could operate in such a manifold.

Gifted science-fiction author Philip Dick explored a similar notion in a series of stunning stories. He called that superior entity VALIS, for "Vast Alive Living Intelligence System." It is at the level of multiple universes and control systems of consciousness that the UFO phenomenon becomes scientifically interesting, not at the simplistic level of a search for the "propulsion system" of unidentified flying objects. The technology we are witnessing may not be based on what we understand today as propulsion.

Cosmology now recognizes the possibility, indeed the inevitability, of multiple universes with more than four dimensions. Communication and travel within our own universe are no longer thought to be absolutely constrained by the speed of light and a constant arrow of time. Even travel into the past may be considered without necessarily creating insurmountable paradoxes. This is a tremendously exciting development. It opens up vast new realms for theoretical and experimental endeavor.

If we look at the world from an informational point of view, and if we consider the many complex ways in which time and space may be structured, the old idea of space travel and interplanetary craft to which most technologists are still clinging appears not only obsolete but ludicrous. Indeed modern physics has already bypassed it, offering a very different interpretation of what an "extraterrestrial" system might look like.

As I look forward, my goal is to explore some hypotheses about the control system and the forms of communication it may favor. My plan is to quietly reconsider the accumulated data, to take a hard look at my own tentative conclusions and to challenge prevalent theories once again. The time has come to draw the hard lessons from our failure to elucidate the basic nature of the phenomenon. This means seeking advice from a wider circle of experts, reorganizing the work, eliminating a morass of obsolete data.

For some time various knowledgeable friends have urged me to take my research behind the scenes again. I intend to follow their advice. I cannot justify remaining associated with the field of ufology as it presents itself to the public today. Furthermore I suspect that the phenomenon displays a very different structure once you leave behind the parochial disputes that disfigure the debate, confusing the researchable issues that interest me. The truly important scientific questions are elsewhere.

+ + +

This extraordinary adventure has carried its share of sadness, because Janine and I have seen valued friends and colleagues pass away, some of them in tragedy.

One of the few prominent scientists who risked their reputation by studying ufology, Dr. James McDonald has been mentioned at length in these pages. Jim attempted suicide, only succeeding in blinding himself. But he eventually obtained another gun, shot himself again and died in 1971. His suicide was prompted by personal reasons. The UFO phenomenon played only an incidental role in his despair, but there is no question that the rejection of his earnest efforts by the scientific community had contributed to depressing him. Reflecting on our relationship from today's perspective I find it doubtful that I could have worked more closely with him, or that anyone could have influenced his thinking. While he had the highest integrity, his approach to science left no room for compromise and joint research. It made teamwork almost impossible.

Dr. Hynek himself died on 27 April 1986 at his home in Scottsdale, of a brain tumor for which he had been operated on a few months before in San Francisco. We had been very close until the end, and I still miss him every day.

Coral Lorenzen died at sixty-three of respiratory failure, on 12 April 1988. Her husband Jim had fallen victim to cancer two years before, in August 1986. They had perhaps come closest of all of us, in the sixties and seventies, to assembling a complete documentation about the mystery. Their influence on the field is still felt. The organization they founded, the Aerial Phenomena Research Organization (APRO), had a sterling international reputation and worked successfully with foreign specialists.

Donald Keyhoe, the founder of NICAP, died in November 1988, the fourth major pioneer of the field to disappear in two years. He was over ninety at the time. His death brings up another painful point for me, because one of the mistakes I undoubtedly made in the sixties was my failure to seek a meeting with him. When I read Keyhoe's books today I find a ring of truth in them which I had missed in the sixties. I had allowed myself to be repelled too easily by the NICAP officials around him, who fancied complicated titles and seemed to specialize in creating bottlenecks. But Keyhoe himself appears to have known much about the phenomenon, and his insider's understanding of the military was a real asset.

In retrospect there were many other things I could have done during those years that never occurred to me. Primary among them would have been a thorough documentation of the Ruppelt years. I had relied on Allen Hynek's recollection of the early phase of Blue Book, but he himself had told me that Ruppelt had kept many things from him, that he "played his cards close to the vest." The Air Force officer had not trusted his consulting astronomer with all the data. This was an important gap I should have tried to fill, since the Ruppelt archives were accessible. The fact that no one else seems to have thought of it does not justify my omission.

Another important figure, author John Fuller, who popularized the concept of "missing time," died of lung cancer in November 1990, at age seventy-six.

Gérard de Vaucouleurs, who gave me the great opportunity to come to the United States, now serves on the National Academy of Sciences. He has kept his open mind on the subject but his skepticism was still in evidence when he wrote to me after the publication of *Dimensions*, pointing out that the observations I reported merely showed that the human mind was susceptible to extraordinary distortions. In saying this he was summing up the current scientific consensus on the paranormal. Indeed men like Philip Klass and Carl Sagan continue to state publicly that studying UFOs is a waste of time, and that no resources should be channeled to that endeavor.

Fortunately many of those who had a decisive influence on my research in its formative years continue to favor me with their advice and friendship. Aimé Michel and Fred Beckman have remained valuable friends. I still meet with Pierre Guérin whenever I travel to Paris. We sit in a café near the Sorbonne and we argue bitterly about everything from the Big Bang to government cover-ups. After a distinguished career in planetary astronomy, Guérin recently retired from the *Institut d'Astrophysique*, but he has lost none of his argumentative passion. Gordon Creighton, another towering figure with a lovably cantankerous character, has replaced Charles Bowen as editor of the *Flying Saucer Review*, surely one of the most colorful and informative magazines ever published. We correspond regularly, and our meetings are always challenging and fruitful.

Other close friends of that era have simply left the field. Don Hanlon, to whom I am indebted for many important facts and ideas, vanished without a trace in the mid-seventies. Others have gone on with their lives and careers. Bill Powers successfully published his seminal book *Behavior: the Control of Perception* and has been recognized for his work in psychology and in the theory of systems. Sam Randlett teaches music with great talent. Other members of our old Invisible College occasionally come to one another's notice through a published technical paper, a book review, or a lecture.

As for the flying saucers of old, they are still with us in their various forms and guises. Not a day goes by without some notice of a sighting somewhere. The details are seldom catalogued or documented. One of the most profound and puzzling phenomena in the history of man is allowed to exist around us without interference, without even a flicker of acknowledgement or an attempt at intelligent response.

Of that silence of mankind, of that refusal to recognize the unknown, I am still ashamed today. I can only hope that my testimony here may be a challenge to others and that eventually, collectively, we will find the strength to respond.

+ + +

It is the destiny of man to stand always between the certainty of his scientific achievements and the annoying evidence that they do not account for all there is. Other forces manifest. We are quick to give them convenient names and familiar roles. We call them ghosts, spirits, extraterrestrials. When all else fails we abjectly turn them into gods, the better to worship what we fail to grasp, the better to idolize what we are too lazy to analyze.

I am in search of a different truth.

I returned to Pontoise last year to look again at the hills of my childhood, to stand at my father's grave, to examine the steps I had taken when I began this research and to assess what I had learned. I came away with the certainty that, given the chance, I would take the same actions again today, because the only thing that counts in this life is to question the mystery of it, with all the means at our disposal, with every moment of awareness, with every breath.

NOTES AND REFERENCES

Part One: Sub-Space

1. This communication was published in *L'Astronomie*, the Bulletin of the French Astronomical Society, January 1958, pp. 8-9.

2. Michel, Aimé, *Mystérieux Objets Célestes* (Paris: Arthaud, 1958). The text was translated into English by Alex Mebane as *Flying Saucers and the Straight-Line Mystery* (New York: Criterion, 1958). This book introduced the term *orthoteny* (Greek for "drawn in a straight line") to designate the fact that sightings appeared to fall along rectilinear patterns. It was a pioneering work because for the first time it proposed an analytical approach to UFO sightings rather than a value judgment about the witnesses.

3. The French title was *Le Sub Espace* (Paris: Hachette, 1961). It appeared in the book series entitled *"Le Rayon Fantastique"* under my pen name of Jérome Sériel. It received the Jules Verne Prize for that year. A mass paperback edition was later issued by Editions des Champs-Elysées. *Le Réseau Praxitèle* was never published.

4. The *Bourbaki* school of mathematics was born in the nineteen-thirties when a group of young French graduate students decided to challenge the traditional approach to that discipline. They ridiculed their elders by hiring a destitute actor, dressing him as the bearded, solemn Russian visiting "expert" Professor Nicolas Bourbaki, and having him give a long, absurd lecture on advanced mathematics which was favorably received by the assembled Faculty at Ecole Normale Supérieure. When the students revealed their hoax the ensuing scandal was so great that the perpetrators had to leave France in order to pursue their careers elsewhere. From then on their articles, which revolutionized modern mathematics, were always published collectively in international journals under the single name Nicolas Bourbaki.

5. 1939, which saw the start of World War II, was a watershed year in many respects: it marked the founding of Silicon Valley with the first

Hewlett-Packard lab, the beginning of television broadcasting, of modern science fiction and the first color movie, *Gone with the Wind.* It was in 1939 that the first digital computer was built by Professor John V. Atanasoff of Ames, Iowa. The *Batman* comic script and the film *The Wizard of Oz* were also born that year. More significantly, it was on October 11, 1939, that a letter from Albert Einstein and Leo Szilard was delivered to President Roosevelt, announcing that atomic energy could be used to make bombs. At the time of my birth Sigmund Freud was dying in London.

6. There are several organizations calling themselves "Rosicrucian." The one mentioned here is the French Branch of AMORC, the *Antiquus Mysticusque Ordo Rosae Crucis* (Ancient and Mystical Order of the Rosy Cross) founded by Spencer Lewis, with international headquarters in San Jose, California.

7. *Dark Satellite (Le Satellite Sombre)* was published in November 1962 by Editions Denoël in their collection *Présence du Futur,* also under the pen name of Jérome Sériel. A mass paperback edition in Portuguese was later issued in Brazil.

8. Zazzo, René: *Le Devenir de l'Intelligence* (Paris: Presses Universitaires de France, 1945).

9. The *Ecole Polytechnique* was created in Paris by Emperor Napoléon to supply technical graduates in support of his military ambitions. It quickly became a power base for the elite of the French administrative, military and industrial world, nurturing a parochial view of most national problems. The existence of this antiquated mode of education was never challenged, even in the depth of the most "revolutionary" movements in France.

10. Camille Flammarion (1842-1925), founder of the French Astronomical Society, was an astronomer and an author eagerly read by the French public of the "Belle Epoque" for his clear descriptions of planetary and atmospheric phenomena. A keen observer of Mars, he discussed the possibility of life on other planets. He investigated paranormal research, publishing several books about near-death phenomena and ghosts.

11. We are talking in terms of "old" francs here, naturally. A sum of 100,000 francs in 1961 represented about $200, or about one month of an engineer's salary. It later turned out that this "Prize" was in reality an advance on the royalties from the book!

12. Eleven million francs represented approximately $20,000 in 1988 currency.

13. The National Investigations Committee on Aerial Phenomena (NICAP) was formed in Washington in October 1956 by former Navy scientist T. Townsend Brown. Major Donald Keyhoe, a retired Marines officer who wrote several very popular books about flying saucers in the fifties, became the Director in January 1957. Major Keyhoe died on 29 November 1988.

14. Project Blue Book was an official study of unidentified flying objects started in 1953 by the U.S. Air Force. It was based at Wright-Patterson Air Force Base, Dayton, Ohio. It functioned more as a public relations office than as a research project. It was to give American citizens a place to report sightings. It answered periodic Congressional inquiries about Air Force vigilance on the subject. It was generally headed up by a Major and had very limited resources. The project was disbanded in December 1969 following the Condon Report. Its files were transferred to Maxwell Air Force Base in Alabama.

15. To my knowledge, the policemen who had committed the atrocities at the Charonne subway station on Thursday 8 February 1962 were never brought before a Court to answer for their crimes. The dead included a 35-year-old mother of three children, Suzanne Matorell, killed by blows to the head given with a rifle butt. Many of the victims had been crushed to death, according to forensic pathologists. To squelch the media uproar about the killings, Interior Minister Roger Frey seized leftist newspapers the next day and issued a statement whitewashing the police.

Part Two: Blue Book

1. The five short science-fiction stories I wrote in French during this period were all published in the monthly magazine *Fiction.* They were: *Les Calmars d'Andromède* (#94, Sep.1961), *L'Oeil du Sgal* (#107, Oct.1962), *Les Planètes d'Aval* (#110, Jan.1963), *Le Satellite Artificiel* (Special Anthology of French Science Fiction #4, 1963) and *Le Fabricant d'Evènements Inéluctables* (#145, Dec.1965).

2. About the Mars Map see the article "Charting the Martian Surface" by G. de Vaucouleurs in *Sky and Telescope,* Vol.XXX, No.4 (October 1965), notably p.197.

3. Josef Allen Hynek was a first-generation American, the son of Czechoslovakian parents. Born in Chicago in 1910, he became fascinated with astronomy when his mother, a schoolteacher, gave him a book on the subject. He graduated in physics and astronomy at the University of Chicago in 1931, and completed his Ph.D. at Yerkes Observatory. When *Nova Herculis* flared up, Perkins Observatory in Ohio borrowed a spectrograph from Yerkes to study it, and Hynek went along to assist in the project. His work impressed the director of Perkins and he was offered a faculty position at Ohio State University in 1936. He specialized in the study of stellar spectra and in the identification of spectroscopic binaries. He met and married Mimi, his second wife, in Ohio. During their honeymoon they travelled to Washington and he visited a friend who was recruiting scientists for the war effort. He found himself signed to work on a classified project for the development of the radio proximity fuse.

After the war Hynek returned to McMillin Observatory in Ohio. He was contacted by the U.S. Air Force to act as scientific consultant on their investigation of unidentified flying objects. He retained his consulting position with them when he moved to Massachusetts to join Donald Menzel and Fred Whipple at Harvard's Smithsonian Observatory. In 1956 Hynek was in charge of the project to track the future American artificial satellite planned for the International Geophysical Year. When the Russian *Sputnik* was launched on October 4, 1957, he became one of the nation's most "visible" scientists. He moved to Northwestern as chairman of the Astronomy Department in 1960.

4. *Special Report #14* was one of the most important documents in the series issued by the Air Force on the topic of unidentified flying objects. Based on work done in 1953 by one of its contractors, a prestigious Columbus, Ohio, "Think Tank" called the Battelle Memorial Institute, it contained the first serious statistical analysis of UFO sightings. The document was released by Secretary of the Air Force Donald Quarles on 25 October 1955.

5. The Air Technical Intelligence Center (ATIC), formerly the Intelligence Division of the Air Materiel Command, was the Air Force branch in charge of assessing threats to the United States deriving from new, potentially hostile technologies. It focused on Soviet rocketry and aircraft development and later artificial satellite recovery and analysis. It was also charged with screening UFO reports, beginning in 1947. Based at Wright Field in Dayton, Ohio, it supervised the work of the various projects, like Sign (formed in January 1948), Grudge (1948) and Blue Book (1953).

6. Hutin, Serge, *Voyages vers Ailleurs* (Paris: Arthème Fayard, 1962).

7. The "Robertson Panel" convened by the CIA from the 14th to the 17th of January 1953 in Washington under Air Force sponsorship was supposed to assess the relevance of the UFO problem to national security. Chaired by Dr. H. P. Robertson of the California Institute of Technology, a relativity expert and CIA consultant, its members were Dr. Samuel A. Goudsmit of Brookhaven, the discoverer of the electron spin; Dr. Luis Alvarez of the University of California at Berkeley, who was to receive the Nobel Prize in 1968; Dr. Thornton Page of Johns Hopkins, an operations research expert; and Dr. Lloyd Berkner, a director of the Brookhaven National Institute. In addition Dr. Hynek and Frederic C. Durant of Arthur D. Little attended selected portions of the meeting. Also present were Captain Ruppelt, Dewey Fournet of the Ethyl Corporation, General W. M. Garland who was chief of ATIC, Navy photo analysts R.S. Neasham and Harry Woo, and CIA personnel Dr. H. Marshall Chadwell (Assistant Director for Scientific Intelligence), his deputy Ralph L. Clark, and Philip G. Strong, Chief of Operations staff. Also from CIA were Lt. Colonel Frederick Oder (Physics and Electronics Division) and David Stevenson, Weapons and Equipment Division.

8. *Levelland:* On 2 November 1957 near Levelland, Texas, five groups of witnesses independently saw luminous cigar-shaped objects hovering near the highways, interfering with car engines and headlights. *Loch Raven:* On 26 October 1958 two men saw a glowing oval object over a bridge at Loch Raven Dam, Maryland. The Air Force concluded they were not lying, found four groups of independent witnesses and carried the report as unidentified.

9. Several articles appeared under my name in the *Flying Saucer Review* in this period: "Towards a generalization of orthoteny" in March 1962; "Mars and the Flying Saucers" (with Janine) in September 1962; "How to codify and classify saucer sightings" in September 1963; "Recent developments in Orthotenic Research" in November 1963; "A descriptive study of the entities associated with Type-1 sightings" in January and May 1964; "The Menzel-Michel Controversy" in July 1964; "How to

Select Significant UFO Reports" in September 1965; and "UFO Research in the USA" in November 1965 and January 1966.

10. Among General De Gaulle's schemes to enhance France's independence was the *Force de Frappe*, an ambitious plan to develop an independent nuclear capability using the South Pacific as a testing base. He also encouraged the development of nuclear power plants and the Concorde aircraft.

11. *Project Sign* was the first Air Force project designed to deal with the reports of UFOs by the American public following the wave of sightings recorded in 1947. Formed in January 1948, it carried a 2A restricted classification and issued a report in February 1949. By then the name of the project had been changed to *Grudge*. The files remained closed to the public and to scientists.

12. The Aerial Phenomena Research Organization (APRO) was the oldest continuing UFO research group in the United States. It was formed in Wisconsin by Coral Lorenzen in 1952 and continued through 1986. Coral Lorenzen died on 12 April 1988 in Tucson, Arizona, where the organization had moved in the sixties.

13. *Le Parisien Libéré*, 22 May 1952.

14. Dr. Carl G. Jung, the well-known Swiss psychiatrist, wrote a book on the UFO question entitled *Flying Saucers—A Modern Myth of Things Seen In The Skies.* Wilbert Smith was a Canadian government engineer. Captain Edward J. Ruppelt was the head of UFO investigations for the Air Force from 1951 to 1953. He died in 1960.

15. The participants were: Major Robert J. Friend, ATIC; Arthur C. Lundahl, CIA, Chairman; Commander Julius M. Larsen, ONI; Lt-Commander D. W. Luiber, CIA; Lt-Commander R. S. Neosham, CIA; Mr. C. F. Camp, CIA; Mr. H. F. Schemfele, CIA; Mr. J. W. Cain, CIA; Mr. W. S. Stahlings, CIA. Colonel Friend retired as head of Blue Book in 1963.

16. On 17 May 1965 Lt-Col. Spaulding wrote to Lt-Col. Carl C. Arnold, director of information, 3AF, to inquire about British UFO activity and to ask "if they had a program comparable to Blue Book" and if so, "do they have a scientific consultant," adding that Dr. Hynek "would like to correspond with him on a personal basis."

17. These observations were cut from the NBC documentary but both cases were described in detail in *Challenge to Science.* The sighting at

Vins was the basis for a celebrated scene in the movie *Close Encounters of the Third Kind*, where metallic objects vibrate wildly while a saucer hovers nearby.

18. Lt. Colonel John F. Spaulding had his office in the Pentagon and was responsible for overseeing Project Blue Book. His title was Chief, Civil Branch, Community Relations Division, Office of Information.

19. Ivan T. Sanderson was a gifted biologist and naturalist, widely travelled, who wrote several books on the fauna of Africa and the Caribbean (notably *Caribbean Treasure* [New York: Viking, 1939]). He later became fascinated with UFOs.

20. Other members were Jesse Orlansky, Launor Carter, Richard Porter and Willis Ware.

21. The paper with Hynek appeared in the August 1966 issue of the *Publications of the Astronomical Society of the Pacific* (PASP) under the title "An Automatic Question-Answering System for Stellar Astronomy," while the paper with Krulee appeared later in the *Information Storage and Retrieval Journal* (No.4, p.13, 1968) under the title "Retrieval Formulae for Inquiry Systems."

22. The article with John Welch was entitled "Respiratory Mechanics following Major Surgery." It was printed in the *Proceedings of the Sixth International Conference on Medical Electronics*, p. 497 (1965).

Part Three: Pentacle

1. Realizing at last that computers were now vital to a country's national security and status in the world, the General placed a high priority on internal French computer development when he drafted the Plan Calcul. His ambitious plan floundered under the greed of the industrialists at semi-public companies Bull, CGE and Thomson, who rushed to obtain subsidies under the plan and squandered millions of dollars on worthless projects, destroying in the process most of the small independent computer firms that were beginning to appear in France. Bull emerged with the biggest piece of the pie, and proceeded to impose its awkward "French computers" based on manufacturing licenses from the U.S. Honeywell corporation. No genuine French computer approximating the performance of the CDC 6600 was ever developed.

2. Bill Powers eventually published his theory as a book entitled *Behavior—the Control of Perception* (Chicago: Aldine, 1973).

3. The budget for the project was eventually expanded to $513,000.

4. Max Heindel, Manly Hall and Rudolf Steiner represent three major currents in twentieth-century esoteric scholarship. Heindel created a Rosicrucian movement that was heavily oriented towards spiritism and astrology. His influence throughout Latin America has remained considerable. Hynek owned nearly all of his books. Manly Hall is the founder and director of the Philosophical Research Society in Los Angeles. Among his many works is an amazing compilation entitled *Secret Teachings of All Ages*. Rudolf Steiner, whose influence on Hynek was the deepest, thought and spoke as a scientist. Born near Vienna, Austria, in 1861, he first joined the theosophical movement of Helena Blavatsky, then went on to start his own movement called *Anthroposophy*. The Nazis set fire to his headquarters building, the beautiful Goetheanum, on 31 December 1922. Steiner died on 30 March 1925, leaving many books and an active network of disciples all over the world.

5. *Le Comte de Gabalis*, by the Marquis Montfaucon de Vilars, subtitled *Entretiens sur les Sciences Secrètes* (Paris: Claude Barbin, 1652), has long been a classic among occult books because it describes various orders of paranormal beings and man's relationship to them.

6. Jean Baudot is also the author of *La Machine à Ecrire* (Montreal: Editions du Jour, 1964), the first volume of poems written by a computer. The machine composed some pearls of wisdom that would have enchanted Jean Cocteau, such as "The plaintive coffins no longer furnish interesting tears," or "Wealth and a pleasure filled with joy waste away sadness together with its roots."

7. This article, entitled "Théorie des Systèmes Autocodeurs," appeared in *Revue d'Informatique et de Recherche Opérationnelle* (RIRO) No. 3, pp. 63-70, 1967.

8. This was written at a time when practically all computer work took place in batch mode, meaning that programs punched in the form of decks of cards were read into the computer and run by an operator, who later separated the output listings for each user to pick up. The programmer had nothing to do until his output came back. Typical turnaround time was 24 hours, although we were often able to get two runs a day at Northwestern. There was no such thing as an interactive program, and time-sharing only existed as an experiment in a few computer laboratories.

9. Allen Newell and Herbert Simon were among the founders of Artificial Intelligence in the United States. Their work, which was mainly conducted at Carnegie-Mellon University in Pittsburgh, is summarized (among other places) in *The Sciences of the Artificial*, by H. Simon (Cambridge: MIT Press, 1969).

10. The Tunguska explosion took place in Central Siberia shortly after 7 a.m. on 30 June 1908. This thirty-megaton blast, which devastated an area of hundreds of square miles of pine forest, is still unexplained. Various theories have been proposed, including the idea that the exploding object might have been a giant meteor or the nucleus of a comet, but they fail to explain the data since no fragments were recovered and no one had observed an approaching comet, which presumably would have been a spectacular sight for weeks before the collision. Alexander Kazantsev speculated that the object may have been a spaceship propelled by antimatter.

11. M.K. Jessup, trained in astronomy, later served as a photographer with a Department of Agriculture expedition to the rubber plantations of the Amazon. Deeply fascinated by the UFO mystery, he wrote four books on the subject before committing suicide in April 1959.

12. The prospector who was the witness in the Falcon Lake case was named Steven Michalak. A 51-year-old mechanic, he was looking for minerals when he suddenly observed two glowing red objects, one of which landed. When he approched and touched it he received severe burns and a scorching pain on his chest while the disk took off.

13. Wilbert Brockhouse Smith was a Canadian radio engineer, born in Lethbridge, Alberta, in 1910. He held a Master's degree from the University of British Columbia. In 1939 he joined the Department of Transport. He engineered Canada's wartime monitoring service. At the end of the war he established a network of ionospheric measuring stations. In 1952 he became a member of a special government group to investigate UFO Phenomena, known as Project Second Storey. He lost credibility with his colleagues when he began claiming that he was in communication with UFO occupants. Smith died in 1962.

14. Journalist John Fuller published many non-fiction books including *Gentlemen Conspirators* and *The Money Changers*. From 1957 to 1966 he wrote the well-known *Trade Winds* column for the *Saturday Review*. Before researching the Barney and Betty Hill story he had published

Incident at Exeter, a study of UFO sightings in New Hampshire.

15. In spite of repeated requests on the part of my attorneys under the Freedom of Information Act during 1989, the Air Force has proven unable to locate this document or to tell me whether or not it has been automatically declassified. The chief of the Records Section at Wright-Patterson Air Force Base wrote back on 19 July 1989 that "As of the current date most records with a date of 1953 and earlier would probably be destroyed or retired to the appropriate Federal Records Center." We were not able to find out what the appropriate Center was for Project Stork.

On 22 August George Conner of the Records Management group at Wright-Patterson forwarded our request to the Foreign Technology Division (FTD/IMD) for processing. There the chief of the Freedom of Information Branch, Sergeant Tammy K. McDonough, indicated to us that their files included no information on Project Grudge or Sign, and that the codename used for the research on unidentified flying objects was Blue Book: "In 1970 all Project Blue Book material was forwarded to the Albert F. Simpson Research Center, Maxwell AFB, Alabama, for storage. In 1976, because of the historical value, all UFO material was assessioned (sic) by the National Archives. We possess no further information on Project Blue Book or Unidentified Flying Objects." Sergeant McDonough provided an address at the National Archives.

On 31 August 1989 we redirected our request to the National Archives, which has not located the relevant papers at this point. Accordingly I have decided to err on the side of safety by refraining from publishing the names of the authors and by deleting some passages. Perhaps other researchers will be able to obtain the release of the information in full.

16. Frank Edwards (1908-1967) was a popular radio commentator and author who often attacked on the air what he saw as the official debunking of the UFO question.

17. Hynek conducted a covert study of astronomical opinion while attending the joint meeting of the American Astronomical Society and the Astronomical Society of the Pacific in Victoria, British Columbia, approaching 44 astronomers between June 23 and July 7, 1952. The attitude ranged from "I would not say anything about it if I did see one" to a definite, sympathetic interest in the problem. The astronomers interviewed were not aware that anything more than a personal private talk between scientists was involved. Five out of forty-four had seen a UFO. Among these were Otto Struve and Adel, who had taken five bearings on a passing object they saw in 1950 in Flagstaff, Arizona. La Paz described seeing his famous green fireballs, Clyde Tombaugh said that he and his wife had once seen an unusual light they could not identify, and Desbins said he had seen two lights that were "too slow for a meteor and too fast for an aircraft."

18. American journalist John Keel published *Jadoo*, an interesting book about the Near East and his experiences among the Yezidis before becoming interested in UFOs. He wrote several books (*The Eighth Tower, Operation Trojan Horse*) and many articles about UFOs.

19. Hynek's letter was finally published in *Science* magazine, October 21, 1966, p. 329.

20. See Eisenbud, Jule: *The World of Ted Serios* (New York: William Morrow, 1967).

21. Dr. Bill Olle was one of the first computer scientists to propose the notion of "non-procedural languages." This concept was especially important in the retrieval of files and the manipulation of large data-bases, since it saved the long and tedious process of writing and debugging a special program for every interrogation of the files. Infol was the first of many systems which are now called DBMS, or "database management systems."

22. "What is Flying in our Skies?" appeared first in *Tekhnika Molodezhi* (*Young Technology* magazine) No. 8, Moscow, August 1967. It was reprinted in *Trud* on 24 August 1967.

Part Four: Magonia

1. This article by William Markowitz appeared in *Science*, Vol. 157, pp. 1274-1279, 15 September 1967. It was entitled *The Physics and Metaphysics of Unidentified Flying Objects*, with this absurd banner, "Reported UFOs cannot be under extraterrestrial control if the laws of physics are valid."

2. Institut de Recherche en Informatique et Automatique (IRIA) has since grown into a "National" Institute and is now known as INRIA. One of its notable members is Professor Ichbiah, who developed the Ada language.

3. In fact I published a number of articles in *Flying Saucer Review* during this period. Following "A ten-point research proposal" which appeared in September 1966, "The Pattern behind UFO Landings" in October 1966, "Airships over Texas" (with Don Hanlon) in January 1967, and "A Survey of French UFO Groups" in September 1967, I wrote several pieces while in France, notably: "Analysis of 8,260 sightings" which appeared in May 1968 and "A Catalogue of 923 landing reports" in July 1969. In addition I wrote an article entitled "The UFO Phenomenon: A Scientific Problem" for publication in *UFOs Around the World* in September 1966.

4. Hynek's article in *Playboy* appeared in the December 1967 issue under the title: *The UFO Gap* (p. 143).

5. Andrew Tomas was working at the time on the manuscript of *The Time Barrier*, which was published as *La barrière du Temps* (Paris: Julliard, 1969). He also published *Sur le Rivage des Mondes Infinis* (Paris: Albin Michel and London: Souvenir Press, 1974, under the title *On the Shores of Endless Worlds*).

6. *Planète* #2. Paris: Editions Retz, Dec. 1961. "La Sociologie," p. 133.

7. The "Service de Documentation Extérieure et de Contre-Espionnage" (SDECE) was the major French secret service. It has been reorganized several times since 1968 and is now called "Direction Générale de la Sécurité Extérieure" (DGSE).

8. General Ailleret's four-engine DC-6 military aircraft operated by the GLAM (the French Cabinet's air transportation group) crashed as it was leaving the island of La Réunion, in the Indian Ocean. Also killed in the crash were his wife, daughter and fifteen staff members. There was one survivor, Michèle Renard, a nurse. The airplane turned inland and hit a mountain five km away from Saint-Denis airfield instead of banking towards the sea after taking off. Rumors variously attributed the accident to pilot error, to bad weather or to sabotage. General Ailleret had served as commander of "Special Weapons" from 1956 to 1960. In that capacity he supervised the work leading to the first two French A-bomb tests, working with Professor Rocard.

9. Michel and Françoise Gauquelin were French psychologists trained in statistics. As an experiment they once "disproved" the alleged correlation between the Zodiac and human destiny. However, Michel Gauquelin also noticed an unexpected effect linking the position of certain planets above the horizon at the time of birth with the careers pursued by remarkable individuals. For instance, there is a significant tendency for champion athletes to have been born at the time of either the rise or the upper culmination of Mars. Among the Gauquelins' books published in the U.S. are *The Cosmic Clocks* (Chicago: Henry Regnery, 1967) and *Cosmic Influences on Human Behavior* (NY: Stein & Day, 1973) with a Foreword by Dr. Hynek. More recently Michel Gauquelin (who divorced Françoise) published an article entitled "Is there a Mars effect?" in the *Journal of Scientific Exploration,* Vol. 2, No.1, pp. 29-51, 1988. He died on 20 May 1991.

10. A member of the French Committee for National Defense, Professor Yves Rocard was director of the physics laboratory at Ecole Normale Supérieure from 1945 to 1973. He wrote several textbooks on physics and was regarded as an authority on mechanical vibrations and instability, working notably on the Tancarville bridge near Le Havre and on the suspension system for Citroën cars. From 1944 to 1951 he was head of the French Navy research services, and he remained a scientific consultant for them throughout his career.

11. The sighting by Father Gill in Papua New Guinea on 26 June 1959 has remained a classic of UFO literature. Father Gill was the head of the Anglican mission at Boianai. Together with dozens of witnesses, he saw an orange object hovering over the sea at 6:45 pm. Four occupants were visible on its "deck" and a beam of blue light was emitted upward from it. When the witnesses waved, one of the occupants was seen to wave back.

12. This think tank swore that the CDC6600 computer would not be used to compute an H-bomb, a pledge that the Americans presumably took with a grain of salt.

13. Marcel Granger, *Adieu a Machonville,* published in *Minute,* 26 September 1968.

14. The *Barbouzes,* in French slang, are "the bearded ones," the agents of the multiple, shadowy intelligence organizations that operate under the umbrella of the French government, often in competition with each other.

15. Jacques Bergier, French nuclear chemist and prolific writer, is best-known as co-author of *The Morning of the Magicians* with Louis Pauwels. He also wrote several books on modern espionage and was a co-founder of *Planète,* a monthly magazine similar in contents and orientation to the later *Omni.* A man of seemingly universal interest, he avidly read scientific and science-fiction magazines in several languages, notably Russian which he spoke fluently. He had access to De Gaulle in matters of French national security, especially when they involved high technology.

16. Stanislas de Guaita assumed the leadership of the French Rosicrucian movement in 1887, at the age of twenty-four, when he founded the *Ordre Kabbalistique de la Rose+Croix.* The descendant of a line of Florentine aristocrats, the Marquis de Guaita had gained a reputation as a minor poet when he first arrived in Paris. In 1884, when he read Peladan's *Vice Suprème,* he began his own esoteric research. In his luxurious apartment of 20 Avenue Trudaine he accumulated a magnificent library with the help of high-ranking Mason Oswald Wirth. His major work is his three-volume *Essais de Sciences Maudites.* Maurice Barrès, who was a childhood friend of Guaita, wrote that he occasionally fought against "larvae" by firing his revolver at them, and spent weeks in his library making heavy use of morphine and hashish while performing his magical experiments. Legend has it that he was strangled by a "flying spirit" in 1897.

17. American film-maker Kenneth Anger directed *Scorpio Rising, Invocation of My Demon Brother, Lucifer Rising* and *The Inauguration of the Pleasure Dome,* among other works inspired by the magical theories of Aleister Crowley. He also authored a two-volume exposé of the steamy underside of the movie business, *Hollywood Babylon.*

18. The "assembly code" of a particular computer is a language in which one statement generally corresponds to one machine instruction. Writing in that language gives systems programmers greatest power because it makes available every feature of the computer. At the same time, however, it is long, cumbersome and error-prone, hence very expensive. In contrast the so-called "higher-level languages" like Fortran and Cobol are more general languages, closer to English and easier to use. They mask the true power of the machine and cannot be used to develop the basic tools like compilers and the operating systems that manage the flow of "jobs" through the computer. The "Implementation Language-1" of RCA was a brilliant attempt to give programmers a language that was both elegant and powerful, but it was never released for marketing reasons. A year or so after I left the company RCA decided to get out of the computer business altogether.

19. A few years later the term "software" was translated into French as "*logiciel*," a word that has become standard.

20. Jung, Carl Gustav: *Alchemical Studies.* Princeton University Press, Bollingen Series XX, 1967, pp. 15-16.

21. The two photographs in question were published by Allen Hynek in his book *The UFO Experience*, Chicago: Henry Regnery, 1972, facing page 53.

INDEX